Lecture Notes in Computer Science 13427

More information about this series at https://link.springer.com/bookseries/558

Christine Strauss · Alfredo Cuzzocrea ·
Gabriele Kotsis · A Min Tjoa ·
Ismail Khalil (Eds.)

Database and Expert Systems Applications

33rd International Conference, DEXA 2022
Vienna, Austria, August 22–24, 2022
Proceedings, Part II

Editors
Christine Strauss
University of Vienna
Vienna, Austria

Alfredo Cuzzocrea
University of Calabria
Rende, Italy

Gabriele Kotsis
Johannes Kepler University of Linz
Linz, Austria

A Min Tjoa
Vienna University of Technology
Vienna, Austria

Ismail Khalil
Johannes Kepler University of Linz
Linz, Austria

ISSN 0302-9743 ISSN 1611-3349 (electronic)
Lecture Notes in Computer Science
ISBN 978-3-031-12425-9 ISBN 978-3-031-12426-6 (eBook)
https://doi.org/10.1007/978-3-031-12426-6

This Springer imprint is published by the registered company Springer Nature Switzerland AG
The registered company address is: Gewerbestrasse 11, 6330 Cham, Switzerland

Preface

Welcome to the two-volume edition of the proceedings of the 33rd International Conference on Database and Expert Systems Applications (DEXA 2022). After a break of two years due to the COVID-19 pandemic situation, which forced us to use online formats, we were happy that we could finally meet in person in Vienna, Austria, during August 22–24, 2022. The wide variety of the topics, as well as the depth of the presented research, revealed, that sound research in the field of database and expert systems applications was not at all shut down by the pandemic. The papers accepted and presented at DEXA 2022, which are collated in these two volumes of proceedings, are an impressive collection of the research and development performed during the challenging recent times.

This year, the DEXA Program Committee accepted 43 full papers and 20 short papers, leading to an acceptance rate of 35%. The total number of submissions was comparable with recent DEXA editions, and we are proud to see again that the DEXA community is global as we received contributions from all around the world (Europe, America, Asia, Africa, Oceania). Our Program Committee performed more than 500 reviews, which not only serve the purpose of quality control for the conference but also contained valuable feedback and insights for the authors. We would like to sincerely thank our Program Committee members for their rigorous and critical, and at the same time motivating, reviews of DEXA 2022 submissions.

As is the tradition of DEXA conference series, all accepted papers were published in Lecture Notes in Computer Science (LNCS) and made available by Springer. Authors of selected papers presented at the conference will be invited to submit substantially extended versions of their conference papers for publication in special issues of two international journals: Knowledge and Information Systems (KAIS) and Transactions of Large Scale Data and Knowledge Centered Systems (TLDKS), both published by Springer. The submitted extended versions will undergo a further review process.

DEXA 2022 covered a wide range of relevant topics: (i) big data management and analytics, (ii) consistency, integrity, and quality of data, (iii) constraint modeling and processing, (iv) database federation and integration, interoperability, and multi-databases, (v) data and information semantics, (vi) data integration, metadata management, and interoperability, (vii) data structures and data management algorithms, (viii) graph databases, (ix) incomplete and uncertain data, (x) information retrieval, (xi) statistical and scientific databases, (xii) temporal, spatial, and high dimensional databases, (xiii) query processing and transaction management, (xiv) visual data analytics, data mining, and knowledge discovery, (xv) WWW and databases, as well as web services.

We would like to express our gratitude to the distinguished keynote speakers for their presented leading edge topics:

- Ricardo Baeza-Yates, Institute for Experiential AI, Northeastern University, USA
- Sabrina Kirrane, Institute for Information Systems and New Media, Vienna University of Economics and Business, Austria
- Philippe Cudré-Mauroux, University of Fribourg, Switzerland

DEXA 2022 also featured six international workshops that focused the attention on a variety of specific topics:

- The 2nd International Workshop on AI System Engineering: Math, Modelling and Software (AISys 2022);
- The 1st International Workshop on Applied Research, Technology Transfer and Knowledge Exchange in Software and Data Science (ARTE 2022);
- The 1st International Workshop on Distributed Ledgers and Related Technologies (DLRT 2022);
- The 6th International Workshop on Cyber-Security and Functional Safety in Cyber-Physical Systems (IWCFS 2022);
- The 4th International Workshop on Machine Learning and Knowledge Graphs (MLKgraphs 2022);
- The 2nd International Workshop on Time Ordered Data (ProTime 2022).

Like the success of every conference, DEXA's success is also built on the continuous and generous support of its participants and contributors and their perpetual and sustained efforts. Our sincere thanks go to the loyal and dedicated authors, distinguished Program Committee members, session chairs, organizing and steering committee members, and student volunteers who worked hard to ensure the continuity and the high quality of DEXA 2022.

We would also like to express our thanks to all institutions actively supporting this event, namely

- Software Competence Center Hagenberg (SCCH), Austria;
- Institute of Telecooperation, Johannes Kepler University Linz (JKU), Austria;
- Web Applications Society (@WAS);
- Austria Society for Artificial Intelligence (ASAI), Austria;
- Vienna University of Economics and Business (WU), Austria; and
- Austrian Blockchain Center (ABC Research), Austria.

We hope you enjoyed the DEXA 2022 conference: not only as an opportunity to present your own work to the DEXA community but also as an opportunity to meet new peers and foster and enlarge your network. We are looking forward to seeing you again next year!

August 2022

Christine Strauss
Alfredo Cuzzocrea

Organization

Program Committee Chairs

Christine Strauss	University of Vienna, Austria
Alfredo Cuzzocrea	University of Calabria, Italy

Steering Committee

Gabriele Kotsis	Johannes Kepler University Linz, Austria
A Min Tjoa	Vienna University of Technology, Austria
Robert Wille	Software Competence Center Hagenberg, Austria
Bernhard Moser	Software Competence Center Hagenberg, Austria
Alfred Taudes	Vienna University of Economics and Business and Austrian Blockchain Center, Austria
Ismail Khalil	Johannes Kepler University Linz, Austria

Program Committee

Sonali Agarwal	IIIT Allahabad, India
Riccardo Albertoni	IMATI-CNR, Italy
Toshiyuki Amagasa	University of Tsukuba, Japan
Idir Amine Amarouche	USTHB, Algeria
Rachid Anane	Coventry University, UK
Mustafa Atay	Winston-Salem State University, USA
Ladjel Bellatreche	LIAS, ISAE-ENSMA, France
Nadia Bennani	LIRIS, INSA de Lyon, France
Karim Benouaret	Université Claude Bernard Lyon 1, France
Djamal Benslimane	Université Claude Bernard Lyon 1 , France
Vasudha Bhatnagar	University of Delhi, India
Andreas Both	DATEV eG, Germany
Athman Bouguettaya	University of Sydney, Australia
Omar Boussaid	ERIC, Université Lumiere Lyon 2, France
Stephane Bressan	National University of Singapore, Singapore
Pablo Garcia Bringas	University of Deusto, Spain
Barbara Catania	Università degli Studi di Genova, Italy
Ruzanna Chitchyan	University of Bristol, UK
Soon Chun	City University of New York, USA
Deborah Dahl	Conversational Technologies, USA

Jérôme Darmont	Université Lumiere Lyon 2, France
Soumyava Das	Imply Data, USA
Vincenzo Deufemia	University of Salerno, Italy
Dejing Dou	University of Oregon, USA
Cedric Du Mouza	Cnam, France
Johann Eder	University of Klagenfurt, Austria
Andreas Ekelhart	Secure Business Austria, Austria
Markus Endres	University of Passau, Germany
Noura Faci	Université Claude Bernard Lyon 1, France
Bettina Fazzinga	University of Calabria, Italy
Flavio Ferrarotti	Software Competence Centre Hagenberg, Austria
Flavius Frasincar	Erasmus University Rotterdam, The Netherlands
Bernhard Freudenthaler	Software Competence Centre Hagenberg, Austria
Steven Furnell	University of Nottingham, UK
Manolis Gergatsoulis	Ionian University, Greece
Vikram Goyal	IIIT-Delhi, India
Sven Groppe	University of Lübeck, Germany
Wilfried Grossmann	University Vienna, Austria
Francesco Guerra	Università di Modena e Reggio Emilia, Italy
Giovanna Guerrini	University of Genoa, Italy
Allel Hadjali	LIAS, ISAE-ENSMA, France
Abdelkader Hameurlain	IRIT, Paul Sabatier University, France
Sven Hartmann	Clausthal University of Technology, Germany
Manfred Hauswirth	Technical University Berlin, Germany
Ionut Iacob	Georgia Southern University, USA
Hamidah Ibrahim	Universiti Putra Malaysia, Malaysia
Sergio Ilarri	University of Zaragoza, Spain
Abdessamad Imine	Loria, France
Ivan Izonin	Lviv Polytechnic National University, Ukraine
Stephane Jean	LIAS, ISAE-ENSMA and University of Poitiers, France
Peiquan Jin	University of Science and Technology of China, China
Anne Kayem	Hasso Plattner Institute, University of Potsdam, Germany
Elmar Kiesling	Vienna University of Economics and Business, Austria
Uday Kiran	University of Tokyo, Japan
Carsten Kleiner	Hannover University of Applied Science and Arts, Germany
Henning Koehler	Massey University, New Zealand
Michal Kratky	VSB-Technical University of Ostrava, Czech Republic

Petr Kremen	Biomax Informatics AG and Czech Technical University in Prague, Czech Republic
Josef Küng	Johannes Kepler Universitaet Linz, Austria
Lenka Lhotska	Czech Technical University in Prague, Czech Republic
Chuan-Ming Liu	National Taipei University of Technology, Taiwan
Jorge Lloret	University of Zaragoza, Spain
Hui Ma	Victoria University of Wellington, New Zealand
Qiang Ma	Kyoto University, Japan
Elio Masciari	University of Naples Federico II, Italy
Jun Miyazaki	Tokyo Institute of Technology, Japan
Lars Moench	University of Hagen, Germany
Riad Mokadem	Pryamide, Paul Sabatier University, France
Anirban Mondal	University of Tokyo, Japan
Yang-Sae Moon	Kangwon National University. South Korea
Franck Morvan	IRIT, Paul Sabatier University, France
Philippe Mulhem	LIG-CNRS, France
Enzo Mumolo	University of Trieste, Italy
Francesco D. Muñoz-Escoí	Universitat Politècnica de València, Spain
Ismael Navas-Delgado	University of Málaga, Spain
Javier Nieves	Azterlan, Spain
Brahim Ouhbi	Université Moulay Ismail, Morocco
Marcin Paprzycki	Systems Research Institute, Polish Academy of Sciences, Poland
Louise Parkin	LIAS, ISAE-ENSMA, France
Iker Pastor-Lovez	University of Deusto, Spain
Dhaval Patel	IBM, USA
Clara Pizzuti	ICAR-CNR, Italy
Elaheh Pourabbas	IASI-CNR, Italy
Simone Raponi	Hamad Bin Khalifa University, Qatar
Claudia Roncancio	Grenoble Alpes University, France
Massimo Ruffolo	ICAR-CNR, Italy
Marinette Savonnet	LE2I, University of Burgundy, France
Florence Sedes	IRIT, Paul Sabatier University, France
Hossain Shahriar	Kennesaw State University, USA
Michael Sheng	Macquarie University, Australia
Patrick Siarry	Université Paris-Est Créteil, France
Tarique Siddiqui	Microsoft Research, USA
Gheorghe Cosmin Silaghi	Babes-Bolyai University, Romania
Srinath Srinivasa	International Institute of Information Technology, Bangalore, India
Bala Srinivasan	Monash University, Australia

Christian Stummer	Bielefeld University, Germany
Olivier Teste	IRIT, Université Toulouse Jean Jaurès, France
Jean-Marc Thevenin	Université Toulouse 1 Capitole, France
A Min Tjoa	Vienna University of Technology, Austria
Vicenc Torra	University of Skövde, Sweden
Traian Marius Truta	Northern Kentucky University, USA
Krishnamurthy Vidyasankar	Memorial University, Canada
Piotr Wisniewski	Nicolaus Copernicus University, Poland
Ming Hour Yang	Chung Yuan Chritian University, Taiwan
Haruo Yokota	Tokyo Institute of Technology, Japan
Yan Zhu	Southwest Jiaotong University, China
Qiang Zhu	University of Michigan - Dearborn, USA
Ester Zumpano	University of Calabria, Italy

External Reviewers

Amani Abusafia, Australia
Ahoud Alhazmi, Australia
Abdulwahab Aljubairy, Australia
Balsam Alkouz, Australia
Radim Baca, Czech Republic
Saidi Boumediene, France
Bernardo Breve, Italy
Taotao Cai, Australia
Luciano Caroprese, Italy
Dipankar Chaki, Australia
Peter Chovanec, Czech Republic
Gaetano Cimino, Italy
Stefano Cirillo, Italy
Matthew Damigos, Greece
Kaushik Das Sharma, India
Yves Denneulin, France
Kong Diison, China
Nabil El Malki, France
Marco Franceschetti, Austria
Myeong-Seon Gil, South Korea
Ramon Hermoso, Spain
Akm Tauhidul Islam, USA
Eleftherios Kalogeros, Greece
Sharanjit Kaur, Malaysia
Julius Koepke, Austria
Abdallah Lakhdari, Australia
Hieu Hanh Le, Japan
Ji Liu, China
Qiuhao Lu, USA
Josef Lubas, Austria
Petr Lukas, Czech Republic
Jorge Martinez-Gil, Austria
Amin Mesmoudi, France
Gabriele Oligeri, Qatar
Shaowen Peng, Japan
Gang Qian, USA
Savio Sciancalepore, The Netherlands
Babar Shahzaad, Australia
Zheng Song, USA
Piotr Sowinski, Poland
Junjie Sun, Japan
Raquel Trillo-Lado, Spain
Zheni Utic, USA
Eugenio Vocaturo, Italy
Alexander Voelz, Austria
Kai Wang, China
Yi-Hung Wu, Taiwan
Chengyang Ye, Japan
Chih-Chang Yu, Taiwan
Feng George Yu, USA
Xiao Zhang, China

Organizers

scch { }
software competence
center hagenberg

JKU
Institute for
Telecooperation

Web Applications Society
@WAS

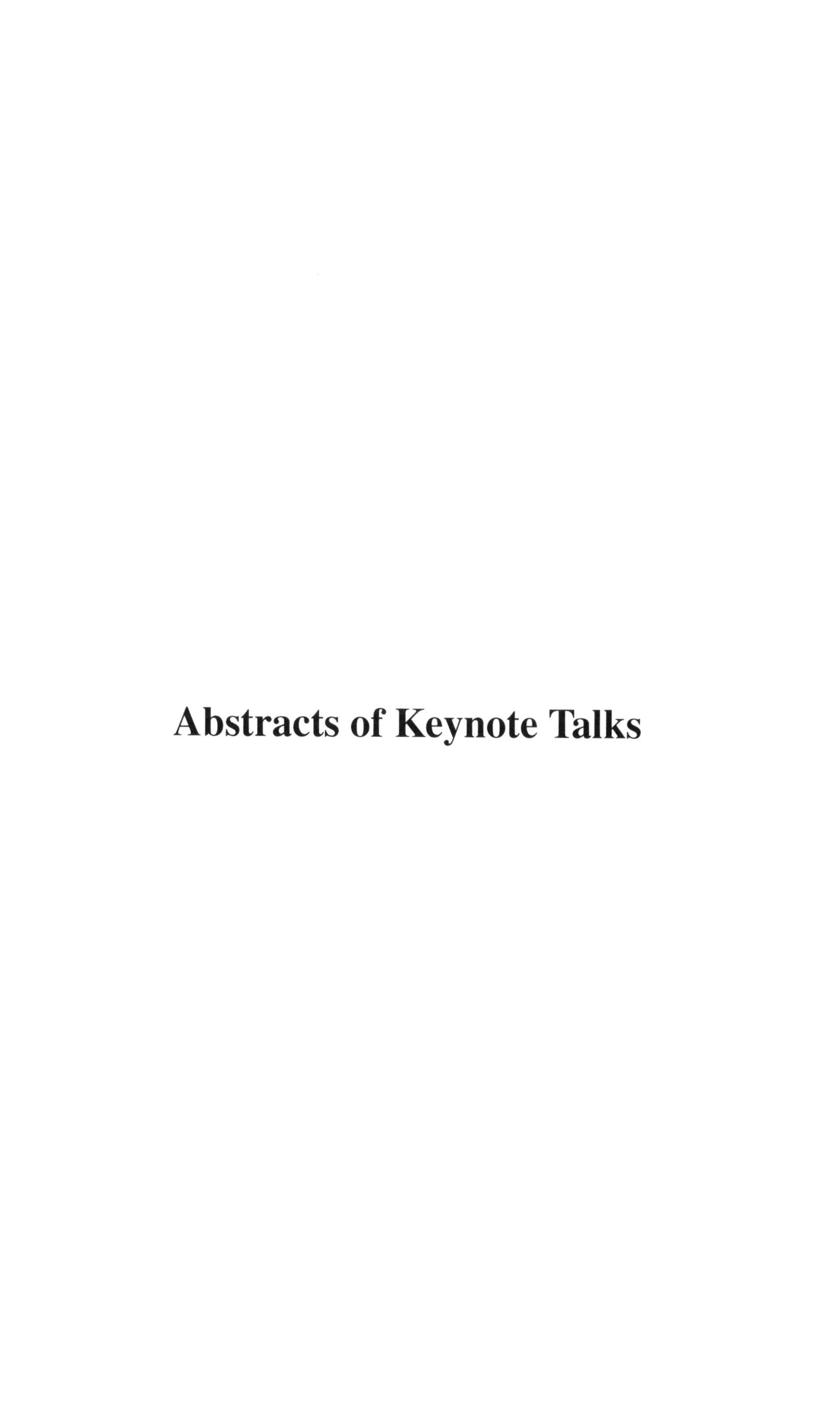

Abstracts of Keynote Talks

Responsible AI

Ricardo Baeza-Yates

Institute for Experiential AI @ Northeastern University

Abstract. In the first part we cover five current specific problems that motivate the needs of responsible AI: (1) discrimination (e.g., facial recognition, justice, sharing economy, language models); (2) phrenology (e.g., biometric based predictions); (3) unfair digital commerce (e.g., exposure and popularity bias); (4) stupid models (e.g., minimal adversarial AI) and (5) indiscriminate use of computing resources (e.g., large language models). These examples do have a personal bias but set the context for the second part where we address four challenges: (1) too many principles (e.g., principles vs. techniques), (2) cultural differences; (3) regulation and (4) our cognitive biases. We finish discussing what we can do to address these challenges in the near future to be able to develop responsible AI.

Following the Rules: From Policies to Norms

Sabrina Kirrane

Institute for Information Systems and New Media @ Vienna University of Economics and Business

Abstract. Since its inception, the world wide web has evolved from a medium for information dissemination, to a general information and communication technology that supports economic and societal interaction and collaboration across the globe. Existing web-based applications range from e-commerce and e-government services, to various media and social networking platforms, many of whom incorporate software agents, such as bots and digital assistants. However, the original semantic web vision, whereby machine-readable web data could be automatically actioned upon by intelligent software web agents, has yet to be realized. In this talk, we will show how rules, in the form of policies and norms, can be used to specify a variety of data usage constraints (access policies, licenses, privacy preferences, regulatory constraints), in a manner that supports automated enforcement or compliance checking. Additionally, we discuss how, when taken together, policies, preferences, and norms can be used to afford humans more control and transparency with respect to individual and collaborating agents. Finally, we will highlight several open challenges and opportunities.

Contents – Part II

Neural Networks

Efficient Data Processing Techniques

Contents – Part I

Deep Learning

Smart Cities and Human Computing

Advanced Machine Learning

Warehousing Methodologies

Time Series, Streams and Event Data

Clustering-Based Cross-Sectional Regime Identification for Financial Market Forecasting

Rongbo Chen[1], Mingxuan Sun[1], Kunpeng Xu[1], Jean-Marc Patenaude[2], and Shengrui Wang[1](✉)

[1] University of Sherbrooke, Sherbrooke, QC, Canada
{rongbo.chen,mingxun.sun,kunpeng.Xu,shengrui.wang}@usherbrooke.ca
[2] Laplace Insights, Sherbrooke, QC, Canada
jeanmarc@laplaceinsights.com

Abstract. Regime switching analysis is extensively advocated in many fields to capture complex behaviors underlying an ecosystem, such as the economic or financial system. A regime can be defined as a specific group of complex patterns that share common characteristics in a specific time interval. Regime switch, caused by external and/or internal drivers, refers to the changing behaviors exhibited by a system at a specific time point. The existing regime detection methods suffer from two drawbacks: they lack the capability to identify new regimes dynamically or they ignore the cross-sectional dependencies exhibited by time series data for the forecasting. This promoted us to devise a cluster-based regime identification model that can identify cross-sectional regimes dynamically with a time-varying transition probability, and capture cross-sectional dependencies underlying financial time series for market forecasting. Our approach makes use of a nonlinear model to account for the cross-sectional regime dependencies, neglected by most existing studies, that can improve the performance of a forecasting model significantly. Experimental results on both synthetic and real-world dataset demonstrate that our model outperforms state-of-the-art forecasting models.

Keywords: Regime switch analysis · Cross-sectional regime identification · Financial market forecasting

1 Introduction

Financial markets may dynamically exhibit abrupt behavior changes. While some of these may be transitory, often the time-evolving behavior of market prices may persist over some specific time intervals [3,9]. For example, the mean, volatility, and correlation patterns in stock returns may change dramatically in a fluctuating stock market [6,20]. These sudden changes or structural break can be viewed as regime switches, some of which may be recurring and some of which may be permanent. Such switches are prevalent in the dynamic financial market. Regime switching analysis [12,13] is extensively advocated for its ability to

C. Strauss et al. (Eds.): DEXA 2022, LNCS 13427, pp. 3–16, 2022.
https://doi.org/10.1007/978-3-031-12426-6_1

capture these sudden changes or structural breaks in market behaviors hidden in financial data, making it a promising approach in financial analysis and market studies. Understanding the phenomenon of how new dynamics of prices and fundamentals persist for a certain length of time after a change helps explore financial behaviors for market forecasting.

Due to their ability to parsimoniously capture stylized behaviors of many financial series, including fat tails, persistently occurring periods of fluctuation followed by periods of low volatility, skewness, and time-varying correlations, regime switching models continue to gain in popularity [2,3,12]. In finance, a regime can be defined as a specific group of complex patterns that share the same characteristics in a specific time interval. Regime switching refers to the changing behaviors exhibited by time series transiting from one regime to another at a specific time points; such changes can be caused by external and/or internal drivers. Existing regime models are designed to predict the likelihood of a structural break resulting in a regime switch that is driven by a combination of driving variables, corresponding to pressures from within or outside the market [4,5,8]. Most, however, handle only single time series, and are not capable of dealing with multiple time series, which is more complex due to the statistically cross-sectional correlations underlying the multiple time series. Such is often the case with financial data, for example, market prices often have positive and negative correlations to one another, and stocks as broad asset classes have exhibited prolonged periods of negative correlation [11].

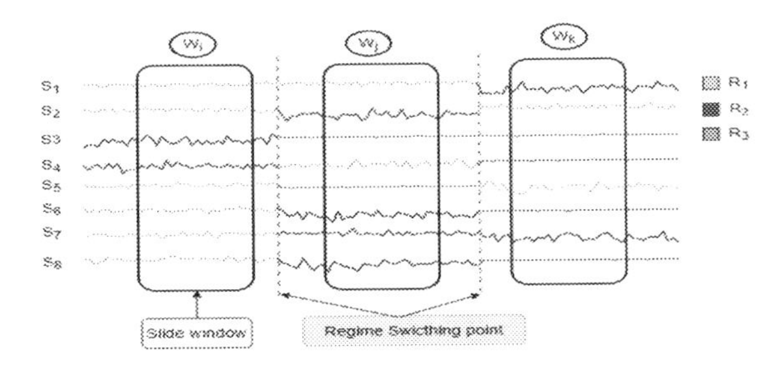

Fig. 1. An example of a multiple time series set consisting of 3 distinct regimes.

Though some regime models [14,23] have been proposed for modeling multiple time series, most of them suffer from issues such as having to specify the number of regimes and lacking the capability to identify regimes dynamically. This prevents them from achieving good performance, since regimes themselves are often thought of as approximations to underlying states that are unobserved and may arrive at any future time in the time-varying financial market, such as financial crisis. Moreover, time series may not synchronize with regime switching. For example, in Fig. 1, the two sub-sequences $\{S_3, S_4\}$ belong to regime R_2 and the others belong to regime R_1 in window W_i. However, most existing regime models [14,23] would identify the whole set of sub-sequences in window W_i (resp. W_j and W_k) as regime R_1 (resp. R_2 and R_3), and miss out on the regimes of sub-sequences $\{S_3, S_4\} \in R_2$ in window W_i, $\{S_1, S_4\} \in R_1$ and $\{S_3, S_5\} \in R_3$ in window W_j, and $\{S_1, S_7\} \in R_2$ and $\{S_2, S_5\} \in R_1$ in window W_k. Consequently, these models can not achieve satisfactory forecasting performance.

To address the issues above, we propose a financial market forecasting model utilizes clustering-based cross-sectional regime identification model. We proposed clustering-based regime identification model can identify cross-sectional regimes dynamically along with a forecasting process based on a time-varying transition probability matrix, to address the problem of specifying a fixed number of regimes and switching among in a fixed set of regimes with a static transition probability matrix. We then devise a non-linear model to capture cross-sectional regime dependencies (as presented in Fig. 1) on multiple time series for financial market forecasting. The significance of this work can be summarized as follows:

- We propose a clustering-based cross-sectional regime identification model on multiple time series, which allows the identification of multiple cross-sectional regimes dynamically, with a time-varying transition probability matrix, along with the forecasting process in the time evolving financial market, bypassing the need to specify a fixed number of regimes that switch within a fixed set of regimes with a static transition probability matrix.
- We propose a non-linear financial market forecasting model relying on a clustering-based regimes identification model, which can capture the cross-sectional dependencies among financial time series to generate forecast for the time-evolving financial market.
- We validate our model by implementing it on synthetic and real-world datasets, comprehensive experimental results, compared with state-of-the-art forecasting algorithms, demonstrate the suitability and promising performance of the proposed model.

The remainder of this paper is organized as follows. In Sect. 2, we discuss related work on financial market forecasting. Section 3 presents the proposed forecasting model in detail. Section 4 provides comprehensive experimental results on synthetic and real-world data and compares the results with other baselines. Finally, conclusions are given in Sect. 5.

2 Related Work

Financial time series forecasting is undoubtedly a hot topic for finance researchers in both academia and the finance industry due to its potential financial gain. Recent literature reports a number of methods applied to financial time series forecasting, including statistical models such as VARMA [20], TRMF [29] and the GARCH family models [2], However, most of these methods are based on linear equations, which are incapable of modeling financial data governed by complex non-linear dynamic patterns. Methods based on deep learning and graph neural networks [9,17,19,27,31] have also been proposed for financial time series forecasting, due to their capability of exploiting long-term and/or short-term dependencies and non-linear dynamic patterns underlying complex data.

Though these approaches can achieve relatively good performance, at the cost of high time complexity due to the overwhelming number of parameters, they are lacking in terms of model interpretability.

On the other hand, among the existing models employed in computational economics and econometric time series analysis, regime-switching models have proved the most preferable, due to their ability to capture non-linear patterns in the market, coupled with heightened model interpretability [12]. Boudt et al. [7] proposed a two-regime model with two state process for funding and market liquidity and TED spread. Alan et al. [1] proposed a multi-regime model to forecast the impact on volatility in global equity markets during the COVID-19 pandemic. Mahmoudi et al. [21] proposed a Markov regime-switching model for detection of structural regimes to analyze the impact of the crude oil market on the Canadian stock market. These methods require to specify the number of regimes manually. Sanquer et al. [25] proposed a hierarchical Bayesian model for automatically identifying hidden regimes. Note that all of these methods are focused on single time series with a static transition probability matrix, and can not be easily applied on multiple time series.

To model multiple time series, Hochstein et al. [14] proposed a regime switching vector autoregressive model that can deal with the changing dependency structures of multivariate time series. Matsubara [23] proposed a regime shift forecasting model on co-evolving data streams. Though it can identify regimes dynamically, it can not capture multiple regimes in one slide window as shown in Fig. 1. Tajeuna et al. [28] proposed a regime shift model for multiple time series forecasting, but it focused on regime identification on discontinuous windows and ignore continuity of time series. Overall, many authors have contributed to advances in handling cross-sectional regime identification on multiple time series, but it remains a challenging task, as many issues have not yet been addressed.

3 The Proposed Model

In this section, we give an overview of the proposed model, followed by detailed description of the cross-sectional regime identification, model description and estimation, and financial market forecasting.

3.1 Overview of the Proposed Model

This subsection gives an overview of the proposed financial market forecasting model relying on clustering-based cross-sectional regime identification, using the scenario shown in Fig. 2. We start by identifying regimes via clustering methods from the first slide window, where the optimal number of clusters in each window is determined by a silhouette score [24], and we then build a non-linear model on each of the identified regimes and obtain the regime parameters and transition probability. Finally, we make a forecast based on a non-linear regime model. Forecasting is performed on the window at the next timestamp. At this iteration step, we need to evaluate whether or not the regimes identified in the window exist in the regime database (RB) by comparing with their cluster centers. If they exist, we need to add the data to the corresponding regime and update the regime parameters; if not, we will add them to the regime database and estimate the regime parameters. The scenario of the proposed model is shown in Fig. 2.

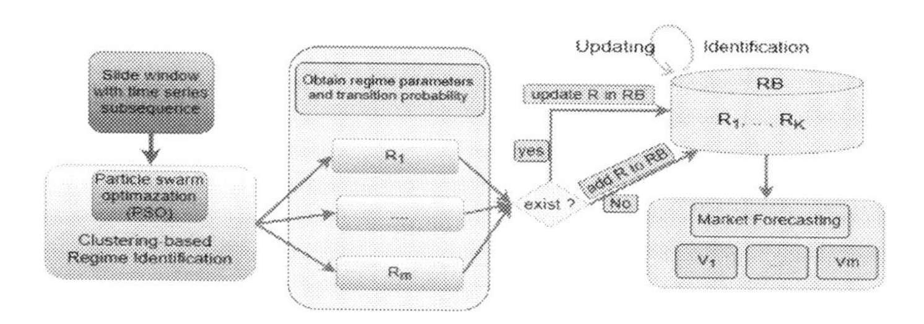

Fig. 2. Overview of the proposed model.

3.2 Cluster-Based Regime Identification

To identify the regimes in each slide window, we use a fuzzy C-Means to cluster time series to reveal the available structures within the data. Each of these structures is defined as a regime that shares some common patterns hidden in the data. To overcome the variable bias in time series data, we here used an extended FCM [18] of squared Euclidean distance to control the impact of each variable in evaluating the similarity between time series, the distance between a time series x_i with length L and a cluster center c_k is defined as follows:

$$d^2(x_i, c_k) = \sum_{j=1}^{L} \lambda_j ||x_{ij} - c_{kj}||^2 \lambda_j \geq 0, \sum_{j=1}^{L} \lambda_j = 1 \tag{1}$$

where λ_i is the importance of the ith variable, and the larger the value of λ_i, the greater the importance of the ith variable in the clustering process. This approach, to some extent, balances the noise underlying the data, achieving better clustering results. The coefficient $\lambda_j, (1 \leq j \leq L)$ can be estimated by Particle Swarm Optimization (PSO) algorithm, which is a tool for searching for optimal values by using a flock of particles, further details can be found in [15,30]. The objective function for the Sum of Error (SE) is defined as follows:

$$SCE = \sum_{k=1}^{K} \sum_{i=1}^{N} u_{ki}^m d^2(x_i, c_k) \tag{2}$$

where K is the number of clusters, $m (m > 1)$ is the fuzzification coefficient, and N is the number of time series. U and c_i are the partition matrix and center of the ith cluster, respectively. By optimizing the objective function Eq. 2, the partition matrix and cluster centers can be calculated as follows:

$$v_k = \frac{\sum_{i=1}^{N} u_{ik}^m w_i}{\sum_{i=1}^{N} u_{ik}^m} \qquad U_{ki} = \frac{1}{\sum_{m=1}^{K} (\frac{||c_k - x_i||}{||c_k - x_i||})^{2/(m-1)}}, m > 1 \tag{3}$$

Thus, we can identify the regimes using the method presented above, and then build non-linear regime models based on these identified regimes, optimize the parameters of the regime models and estimate the transition probability. Finally, we make market forecast based on the regime models.

3.3 Regime Modeling and Transition Probability Estimation

The clustering approach described above only allows use to identifying regimes; that is; groups of time series which share similar patterns. For a better forecasting, we need to build an effective non-linear regime model on these clusters. Based on the work presented in [23,26], we can build a single non-linear regime model on each cluster in order to make a forecast. Thus, the non-linear regime model is defined as follows:

$$\frac{ds(t)}{dt} = \mu + \mathcal{G}g(s(t)) + \mathcal{F}f(s(t)) \tag{4}$$

$$v(t) = \epsilon_k + \mathcal{E}s(t) \tag{5}$$

where $s(t)$ is a hidden vector that evolves over time and describes the potential behaviors in the corresponding regime, and $v(t)$ is the actual observed value. $ds(t)/dt$ denotes the derivative with respect to time t. $g(\cdot)$ is a linear function, while $f(\cdot)$ is non-linear. Here, μ, $\mathcal{G}$ and $\mathcal{F}$ describe the potential activities $s(t)$, capturing linear and non-linear dynamic patterns of the regime. For parameters optimization, readers are referred to [23].

Regime transition probability describes the likelihood that the current regime stays the same or switch to another. In fact, we need to investigate whether the regimes in one window will change in the subsequent window or not. Rather than calculating static transition probabilities as elaborated in existing model, we can track the regime transition trajectory of each time series; the regime transition from regime R_i to regime R_j can be estimated as follows [28]:

$$\mathbb{Q}_1(i,j) = \begin{cases} 0 & if \sum_{i=1}^{K}\sum_{j=1}^{K} \aleph(i,j)\mathbb{N}(i,j) = 0 \\ \frac{\sum_{k=1}^{K} \aleph(i,k)\mathbb{N}(k,j)}{\sum_{i=1}^{K}\sum_{j=1}^{K} \aleph(i,j)\mathbb{N}(i,j)} & else \end{cases} \tag{6}$$

where $\aleph(i,j) = \frac{|\mathcal{N}(R_i,R_j)|}{\mathbf{N}}$ is the risk of suddenly switching from regime R_i to R_j, while $\mathbb{N}(i,j) = \frac{|\mathbf{N}_i \cap \mathbf{N}_j|}{|\mathbf{N}_i \cup \mathbf{N}_j|}$ is the probability of switching from R_i to R_j. $\mathcal{N}(R_i, R_j)$ is the number of time series appearing in the trajectory from regime R_i to R_j for the two windows. $\mathbf{N}_i$ and $\mathbf{N}_j$ are the numbers of time series present in regimes R_i and R_j, and $\mathbf{N}$ is the total number of time series. To further improve the above estimate, we also consider the difference between two cluster centers c_i and c_j in clustering-based regime identification: $\mathbb{Q}_2(i,j) = \frac{1}{|c_i - c_j|}$ for $i \neq j$ otherwise $\mathbb{Q}_2(i,j) = \frac{1}{|c_i|}$ ensuring the probability of staying at the same regime, and the effect of $s_i(t)$ and $s_j(t)$ that describes the potential behaviors in regime R_i and R_j: $\mathbb{Q}_3(i,j) = \frac{1}{|s_i - s_j|}$ for $i \neq j$ otherwise $\mathbb{Q}_3(i,j) = \frac{1}{|s_i|}$. All of these are the underlying drivers that may result in a regime switch. Thus, the transition probability $\mathbb{Q}$ of switching from regime R_i to regime R_j can be defined as follows:

$$\mathbb{Q}(i,j) = \mathbb{Q}_1(i,j)\frac{\mathbb{Q}_2(i,j)}{\sum_{i=1}^{k}\mathbb{Q}_2(i,j)}\frac{\mathbb{Q}_3(i,j)}{\sum_{i=1}^{k}\mathbb{Q}_3(i,j)}(1 \leq i,j \leq K) \tag{7}$$

Note that we can get the regime transition probability of each time series occurring within each slide window.

3.4 Financial Market Forecasting

Based on the previous section, once we build the non-linear model of each clustering-based regime, we can learn all the parameters. We make use of the fourth-order Runge-Kutta method [16] to generate l/γ ($\gamma = 3$) potential activity sets $\mathcal{S} = [s(t+\gamma), s(t+2\gamma)\cdots, s(t+l)]$ to estimate the $[\{s(t+1),\cdots, s(t+l)]$, as presented in [23], for the forecasting step. The process is defined as follows:

$$s(t+\gamma) = s(t) + \frac{1}{6}(K_1 + 2K_2 + 2K_3 + K_4) + \mathcal{O}^5 \tag{8}$$

where $ds(t)/dt = F(s(t))$, $K_1 = \gamma F(s(t))$, $K_2 = \gamma F(s(t)+\frac{1}{2}K_1)$, $K_3 = \gamma F(s(t)+\frac{1}{2}K_2)$, $K_4 = \gamma F(s(t) + K_3)$. Thus, once we obtain the potential activity set $\mathcal{S}$, the regime model Eq. 5 can be used to make l-steps-ahead-forecast $\mathcal{V}$, as follows:

$$\mathcal{V} = \epsilon_k + \mathcal{E}\mathcal{S} \tag{9}$$

The detailed framework of our forecasting model utilizing clustering-based regime identification is shown in Algorithm 1.

4 Experiments

4.1 Dataset Description

To test the performance of our model, we used one synthetic dataset and three real-world financial datasets. The synthetic dataset (SyD) consists of 400 time series of length 1350, governed by 6 distinct regime. They are generated by the regime functions RF as $RF(t) = \sum_{k=1}^{6}\alpha_k\varpi_k(t)fcn_k(t)$, where $\alpha \in \{1,0\}$ ($\sum_{k=1}^{6}\alpha_k = 1$) allows having one regime to be exhibited in a time interval. $\varpi_k(t) \in [0,1]$ is to exhibit regimes with constraint $\sum_{k=1}^{6}\varpi_k(t)$=1. The 6 regime functions are defined as follows:

$$\begin{aligned}
fcn_1(t) &= \cos(\frac{2\pi t}{5}) + cos(\pi(t-3)) \qquad fcn_2(t) = \cos(\frac{2\pi t}{5}) - cos(2\pi(t-3)) \\
fcn_3(t) &= \sin(\frac{2\pi t}{5} - 3) - \sin(\frac{\pi t}{6}) \qquad fcn_6(t) = \cos(\frac{3\pi t}{5}) + \sin(\frac{2\pi t}{5} - t) \\
fcn_4(t) &= \tan(\frac{\pi t}{2} - 3) - \frac{1}{2}\cos(\frac{\pi(t-3)}{6}) + \cos(\pi(t-13)) \\
fcn_5(t) &= \tan(\frac{\pi t}{2} - 3) * \cos(\frac{\pi(t-3)}{6}) + \cos(\pi(t-13))
\end{aligned} \tag{10}$$

Algorithm 1: Framework of the Proposed Model

Input: Financial time series: X, Slide window length: w, Threshold: η
Forecasting window: l, maximal number of regimes in a window: m
Output: Forecasting time series: F, Transition probability: τP

```
begin
    /* Initialization                                                    */
      - tc = w, current time point;
      - RB = [], regime database;
      - F = [], forecasting time series;
      - τP = [], transition probability;

    repeat
        /* Get slide window data X_w from X                              */
        X_w = X[tc-w:tc];
        /* Identify regimes on X_w by clustering-based method            */
        Obtain a regime set RS;
        if len(RB) ==0 then
            Add all the regimes in RS to RB;
            for R in RB do
                /* Regime estimated value on R                           */
                Obtain parameters on regime R by Eq. 4 and 5;
                Generate S by Eq. 8 set on regime R;
                Obtain forecast value on regime R by Eq. 9;
                Obtain Q transition probability by Eq. 7;
            F.append(v^l);
            τP.append(Q);
        for R_RS in RS do
            Err =[];
            Obtain C_RS center of regime R_RS;
            for R_RB in RB do
                Obtain C_RB center of regime R_RB;
                Err.append(d^2(C_RS − C_RB));
            if min(Err) > η then
                /* Identified a new regime R_RS                          */
            else
                /* Regime R_RS already existed                           */
                Find the best R_Rm of R_RS in RB and add R_Rm data into R_RS;
            Obtain parameters on regime R_RS by Eq. 4 and 5;
            Generate S by Eq. 8 set on R_RS;
            Obtain forecast value on regime R_RS by Eq. 9;
            Obtain Q transition probability by Eq. 7;
            Replace R_Rm by R_RS;
        F.append(v^l);
        τP.append(Q);
    until iterate for next window;
```

Three real-world datasets were selected from financial markets. The first one (Stocks) consists of 200 stocks selected from top 500 companies including AAPL, IBM, BAC, MSFT, WMT and so on. The second (Sectors) is comprised of 9 financial sector SPDR Funds: XLB, XLE, XLF, XLI, XLK, XLP, XLU, XLV, XLY. The last (ETFs) contains 18 ETF funds: EWA, EWC, EWD, EWG, EWH, EWJ, EWS, EWW, EWP, EWQ, EWM, EWL, EWI, EWN, EWO, EWK, EWU, SPY. These three datasets consist of daily frequencies (for business days only), comprising over 22 years' data, available on the yahoo finance website[1]. It is worth noting that the experimental financial data were converted into volatility based on the log-return of the close price, as described in [10].

4.2 Performance Metrics

There are many metrics for evaluating the performance of a forecasting model. Here, due to space limition, we report the results obtained on the most popular metric for evaluating forecasting performance: the root mean square error (RMSE) [22]. The performance metrics is defined as follows:

$$RMSE = \sqrt{\frac{1}{T}\sum_{t=1}^{T}(\sigma_i - \hat{\sigma_i})^2} \tag{11}$$

where T is the length of the forecasting window, σ_i and $\hat{\sigma_i}$ are the ground truth and predicted value at time t respectively. The smaller value, the better the performance of a forecasting model, meaning that the predicted value is closer to the ground truth.

4.3 Experimental Results and Discussion

In this section, we present the results of our model, and evaluate its performance against some comparable state-of-the-art methods on synthetic and real-world datasets. We tested our model using a slide window of half a year (126 business days) to forecast one month ahead (21 business days) for the real-world data, and a slide window of length 75 to forecast 15 steps ahead on the synthetic data.

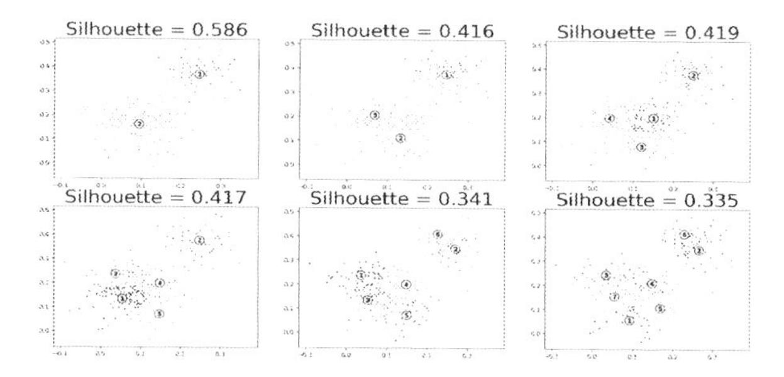

Fig. 3. The clustering results of the first window in stock dataset

[1] https://ca.finance.yahoo.com/.

Regime Identification Analysis. An accurate, effective regime identification model is the key to achieving good performance of our forecasting model, since the latter is built on the regime identification results. We therefore start by analyzing the regimes identified by our clustering-based regime identification model. First, we need to identify the regimes in the first slide window and then iterate to the next slide window. Note that the synthetic dataset SyD was generated with $K = 6$ known regimes by Eq. 10, but for our real-world datasets, we do not have ground truth information such as the number of regimes. To validate the performance of our regime identification model, we therefore test it for numbers of regimes varying from 2 to the maximal value 7 on the synthetic and real-world datasets, and make use of the silhouette score to find the optimal number of regimes in each slide window. For example, the results for the first slide window of the Stock dataset is shown Fig. 3. It is clear that all 200 time series in the window are clustered into 2 groups with the largest silhouette score (0.586), which means that there are two regimes identified in the first slide window. Look at the regimes identified in the synthetic and real-world datasets as shown in Fig. 4, it can be seen that 6 distinct regimes are identified in the synthetic data, which is corresponds to the true number of regimes generated. There are 6 distinct regimes identified both in Stock and Sector dataset. While 8 different regimes are identified in the ETFs dataset as in Fig. 4. In summary, our clustering-based regime identification model can find the regime groups accurately and effectively.

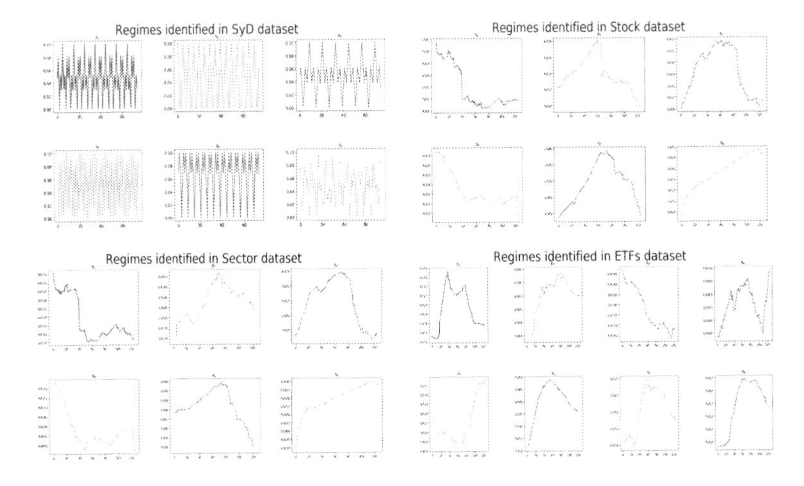

Fig. 4. The regimes identified in four datasets respectively.

Market Forecasting Analysis. To validate the performance of our model, we use a slide window of half a year with 126 business days to forecast one month with 21 days ahead for the real-world dataset, while a window of 75 steps to forecast 15 steps ahead for the synthetic dataset. For page limitation, we randomly selected two time series from the Sector and ETFs datasets, respectively, as examples to show the performance of regime identification and value forecasting of our proposed model. The results are as shown in Fig. 5.

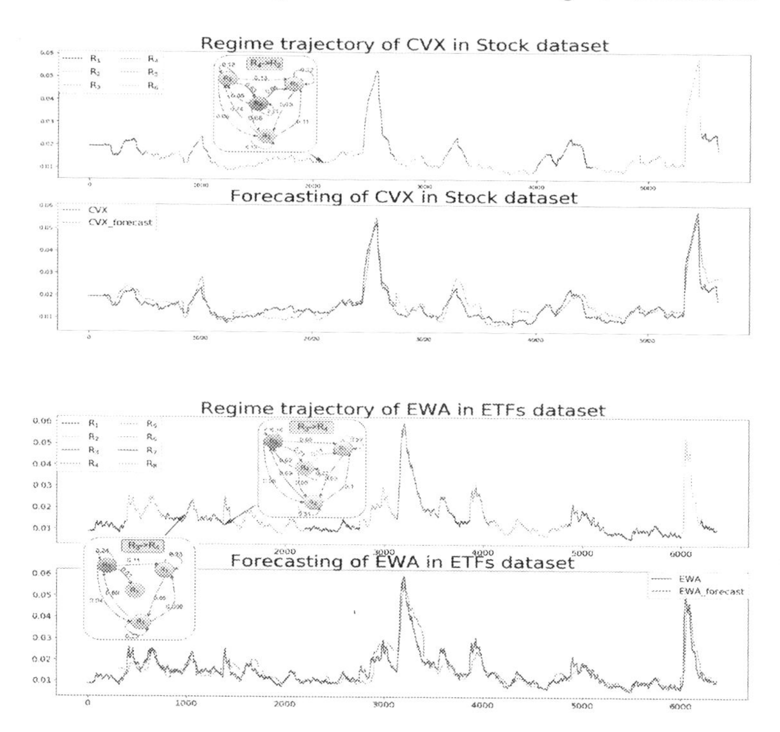

Fig. 5. Examples of regime trajectories and market forecasting, and also including three regime transition processes with the time-varying transition probabilities in trajectory.

Look at the regime trajectories in Fig. 5, it can be seen that there are six and eight regimes identified in stock and ETFs dataset respectively, and the identified regimes, marked with corresponding colors, are roughly synchronized to the regime exhibited in the real market. Moreover, we present regime switching process on trajectories of stock CVX and etf EWA. When comparing with two switches from R_3 to R_4 in trajectory of etf EWA as in Fig. 5, the probabilities of R_4 ($p_{R_3->R_4} = 0.61$) in the first switching are completely different from the second one ($p_{R_3->R_4} = 0.71$) due to our unique time-varying learning mechanism that is lacked by existing methods. Furthermore, this kind of time-varying transition probabilities can be used to explicitly explain the regime switching mechanism from the model interpretability. What the most important is that the ground truth is obviously well matched by our forecasting on stock CVX and ETF EWA as shown in Fig. 5. In summary, our model can identify regime accurately and demonstrates promising performance for the market forecasting.

Performance Analysis. To validate the forecasting quality of our model, we compared our experimental results with that generated by the baselines on the four test datasets. The baselines for the result comparison are as follows: VARMA [20] is a classical statistical model for analyzing and forecasting time series data.

TRMF [29] is a regularized matrix factorization based auto-regressive prediction model. VAR-MLP [31] is a hybrid model that combines auto-regressive and multi-layer model for time series forecasting. MAGNN [9] is a graph neural network based methods for financial time series prediction. DSTP-RNN [19] is an attention-based recurrent neural network method for multivariate time series forecasting. RegimeCast [23] is a regime shifts based forecasting model for co-evolving real-time data streams. The results are shown in Table 1; the best results are highlighted in bold and the second best are underlined.

Table 1. Comparison of RMSE on test datasets.

Models	SyD	Stocks	Sectors	ETFs
VARMA	.35303	.00349	.00316	.00328
TRMF	.26986	.00307	.00275	.00262
VAR-MLP	.23388	.00226	.00232	.00245
MAGNN	.18652	<u>.00161</u>	.00121	<u>.00135</u>
DSTP-RNN	.20421	.00172	<u>.00086</u>	.00237
RegimeCast	<u>.12516</u>	.00205	.00143	.00169
Ours	**.08072**	**.00103**	**.00041**	**.00082**

We can see that our model outperforms the baselines, earning the smallest (best) values on the performance metric, and shows a significant improvement over the second best baselines (underlined) in Table 1. The deep learning methods outperform the traditional methods (VARMA, TRMF), as the latter cannot harness the non-stationary and non-linear dependencies for the prediction. However, the deep learning based VAR-MLP can not explicitly model the cross-sectional dependencies for the prediction, putting it at disadvantage compared to the graph neural network based methods (DSTP-RNN, MAGNN). RegimeCast can capture the non-stationary and non-linear dependencies for the value prediction, but it ignores the cross-sectional regime dependencies, which is a degenerated version of our model that operates without considering multiple regimes in slide windows. In sum, our model shows promising performance on financial market forecasting.

5 Conclusion

In this paper, we have proposed a financial market forecasting model utilizing clustering-based cross-sectional regime identification. Our proposed model not only captures cross-sectional dependencies in multiple time series, but also identifies cross-sectional regimes dynamically along with the time-evolving financial markets, using a time-varying transition probability matrix. In addition, we have built a non-linear forecasting model based on a clustering-based cross-sectional regime model for financial market forecasting. Experimental results on synthetic and real-world datasets demonstrate the promising performance of our model. However, we will improve the performance of our model and test it on hourly and minutes financial time series or sensor data. Moreover, we also may apply our model into other domains, such as the energy consumption and mechanical fault diagnosis. In short, we see the significant challenges for our future work, but we are confident that the proposed method has great potential in real applications.

Acknowledgements. This work was supported by NSERC CRD - Quebec Prompt - Laplace Insights - EAM - joint program under the grants CRDPJ 537461-18 and 114-IA-Wang-DRC 2019 to S. Wang and Chinese Scholarship Council scholarship to R. Chen.

References

1. Alan, N.S., Engle, R.F., Karagozoglu, A.K.: Multi-regime forecasting model for the impact of COVID-19 pandemic on volatility in global equity markets. NYU Stern School of Business (2020)
2. Ali, G., et al.: EGARCH, GJR-GARCH, TGARCH, AVGARCH, NGARCH, IGARCH and APARCH models for pathogens at marine recreational sites. J. Stat. Econ. Methods **2**(3), 57–73 (2013)
3. Ang, A., Timmermann, A.: Regime changes and financial markets. Annu. Rev. Financ. Econ. **4**(1), 313–337 (2012)
4. Baillie, R.T., Morana, C.: Modelling long memory and structural breaks in conditional variances: an adaptive FIGARCH approach. J. Econ. Dyn. Control **33**(8), 1577–1592 (2009)
5. Banerjee, A., Urga, G.: Modelling structural breaks, long memory and stock market volatility: an overview. J. Econom. **129**(1–2), 1–34 (2005)
6. Bollerslev, T., Engle, R.F., Nelson, D.B.: Arch models. Handb. Econom. **4**, 2959–3038 (1994)
7. Boudt, K., Paulus, E.C., Rosenthal, D.W.: Funding liquidity, market liquidity and ted spread: a two-regime model. J. Empir. Financ. **43**, 143–158 (2017)
8. Charfeddine, L., Khediri, K.B.: Financial development and environmental quality in UAE: cointegration with structural breaks. Renew. Sustain. Energy Rev. **55**, 1322–1335 (2016)
9. Cheng, D., Yang, F., Xiang, S., Liu, J.: Financial time series forecasting with multi-modality graph neural network. Pattern Recogn. **121**, 108218 (2022)
10. Christensen, B.J., Prabhala, N.R.: The relation between implied and realized volatility. J. Financ. Econ. **50**(2), 125–150 (1998)
11. Faniband, M., Faniband, T.: Government bonds and stock market: volatility spillover effect. Indian J. Res. Capital Markets **8**(1–2), 61–71 (2021)
12. Hamilton, J.D.: A new approach to the economic analysis of nonstationary time series and the business cycle. Econometrica J. Econom. Soc. 357–384 (1989)
13. Hamilton, J.D.: Regime switching models. In: Durlauf, S.N., Blume, L.E. (eds.) Macroeconometrics and Time Series Analysis. TNPEC, pp. 202–209. Palgrave Macmillan UK, London (2010). https://doi.org/10.1057/9780230280830_23
14. Hochstein, A., Ahn, H.I., Leung, Y.T., Denesuk, M.: Switching vector autoregressive models with higher-order regime dynamics application to prognostics and health management. In: 2014 International Conference on Prognostics and Health Management, pp. 1–10. IEEE (2014)

15. Hu, M., Wu, T., Weir, J.D.: An adaptive particle swarm optimization with multiple adaptive methods. IEEE Trans. Evol. Comput. **17**(5), 705–720 (2012)
16. Jackson, E.A.: Perspectives of Nonlinear Dynamics: Volume 1, vol. 1. CUP Archive (1989)
17. Lai, G., Chang, W.C., Yang, Y., Liu, H.: Modeling long-and short-term temporal patterns with deep neural networks. In: The 41st International ACM SIGIR Conference on Research and Development in Information Retrieval, pp. 95–104 (2018)
18. Li, J., Izakian, H., Pedrycz, W., Jamal, I.: Clustering-based anomaly detection in multivariate time series data. Appl. Soft Comput. **100**, 106919 (2021)
19. Liu, Y., Gong, C., Yang, L., Chen, Y.: DSTP-RNN: a dual-stage two-phase attention-based recurrent neural network for long-term and multivariate time series prediction. Expert Syst. Appl. **143**, 113082 (2020)
20. Lütkepohl, H.: Forecasting with VARMA models. Handb. Econ. Forecast. **1**, 287–325 (2006)
21. Mahmoudi, M., Ghaneei, H.: Detection of structural regimes and analyzing the impact of crude oil market on Canadian stock market: Markov regime-switching approach. Studies in Economics and Finance (2022)
22. Makridakis, S.: Accuracy measures: theoretical and practical concerns. Int. J. Forecast. **9**(4), 527–529 (1993)
23. Matsubara, Y., Sakurai, Y.: Regime shifts in streams: real-time forecasting of co-evolving time sequences. In: Proceedings of the 22nd ACM SIGKDD International Conference on Knowledge Discovery and Data Mining, pp. 1045–1054. ACM (2016)
24. Rousseeuw, P.J.: Silhouettes: a graphical aid to the interpretation and validation of cluster analysis. J. Comput. Appl. Math. **20**, 53–65 (1987)
25. Sanquer, M., Chatelain, F., El-Guedri, M., Martin, N.: A smooth transition model for multiple-regime time series. IEEE Trans. Signal Process. **61**(7), 1835–1847 (2012)
26. Scheffer, M., Carpenter, S., Foley, J.A., Folke, C., Walker, B.: Catastrophic shifts in ecosystems. Nature **413**(6856), 591 (2001)
27. Shih, S.Y., Sun, F.K., Lee, H.Y.: Temporal pattern attention for multivariate time series forecasting. Mach. Learn. **108**(8), 1421–1441 (2019)
28. Tajeuna, E.G., Bouguessa, M., Wang, S.: Modeling regime shifts in multiple time series. arXiv preprint arXiv:2109.09692 (2021)
29. Yu, H.F., Rao, N., Dhillon, I.S.: Temporal regularized matrix factorization for high-dimensional time series prediction. In: Advances in Neural Information Processing Systems 29 (2016)
30. Zhan, Z.H., Zhang, J., Li, Y., Chung, H.S.H.: Adaptive particle swarm optimization. IEEE Trans. Syst. Man Cybern. Part B (Cybern.) **39**(6), 1362–1381 (2009)
31. Zhang, G.P.: Time series forecasting using a hybrid ARIMA and neural network model. Neurocomputing **50**, 159–175 (2003)

Alps: An Adaptive Load Partitioning Scaling Solution for Stream Processing System on Skewed Stream

Beiji Zou[1,3], Tao Zhang[1,3], Chengzhang Zhu[1,2,3](✉), Ling Xiao[1,3], Meng Zeng[1,3], and Zhi Chen[1,3]

[1] School of Computer Science and Engineering, Central South University, Changsha 410083, China
{bjzou,194711055,194701006,zengmeng,chen.zhi}@csu.edu.cn, anandawork@126.com

[2] The College of Literature and Journalism, Central South University, Changsha 410083, China

[3] Mobile Health Ministry of Education-China Mobile Joint Laboratory, Changsha 410008, China

Abstract. The distributed stream processing system suffers from the rate variation and skewed distribution of input stream. The scaling policy is used to reduce the impact of rate variation, but cannot maintain high performance with a low overhead when input stream is skewed. To solve this issue, we propose Alps, an Adaptive Load Partitioning Scaling system. Alps exploits adaptive partitioning scaling algorithm based on the willingness function to determine whether to use a partitioning policy. To our knowledge, this is the first approach integrates scaling policy and partitioning policy in an adaptive manner. In addition, Alps achieves the outstanding performance of distributed stream processing system with the least overhead. Compared with state-of-the-art scaling approach DS2, Alps reduces the end-to-end latency by 2 orders of magnitude on high-speed skewed stream and avoids the waste of resources on low-speed or balanced stream.

Keywords: Data streams · Stream processing system · Adaptive scaling policy

1 Introduction

Streaming data is generated in a consecutive way and has the demand of processing in time. The traditional batch processing system such as MapReduce [1] and Spark [2,3] based on "process after store" prototype [4] cannot meet the demand of low latency. Hence, distributed stream processing system (DSP) such as Storm [5] and Flink [6] is proposed to solve the high latency problem. DSP also has the scalable and fault-tolerant properties to process stream 24-hours-a-day.

The rate variation and data distribution of stream is a great challenge for DSP [7], since DSP suffers the high latency and low throughput due to the

C. Strauss et al. (Eds.): DEXA 2022, LNCS 13427, pp. 17–31, 2022.
https://doi.org/10.1007/978-3-031-12426-6_2

congestion and backpressure on high-speed and skewed stream. Furthermore, it is crucial to maintain the characteristics of low latency and high throughput of DSP [4,8], and consequently the scaling policy is indispensable. However, the manual tuning of scaling parameters is time-consuming and error-prone [9]. Therefore, DSP abandons the manual tuning and adopts the advanced automatic scaling policy.

Recently, several approaches are proposed to tackle the above performance degradation issues. Rule-based approaches like StreamCloud [10], which are simple and effective, use some predefined rules to generate the parallelism configuration. However, the accurate prior knowledge is required by rule-based approaches and is hard to acquire in special circumstances. Compared with rule-based approaches, model-based approaches are more flexible and robust. The state-of-the-art model-based approaches DS2 [11] tunes the parallelism configuration of DSP by the linear performance model, and consequently it maintains the high throughput of DSP on balanced stream in a simple manner. However, DS2 is hard to handle the skewed stream with a few resources, because of the limitation of the linear performance model. When using DS2 on skewed stream, backpressure appears more frequently and causes the performance degradation of DSP. In a word, maintaining an outstanding performance of DSP with least overhead on skewed stream is a meaningful problem. However, the advanced scaling policy does not deal with the performance degradation on skewed stream.

We propose Alps, an Adaptive Load Partitioning Scaling system, to tackle the above issues by means of integrating scaling policy and partitioning policy adaptively. The usage of partitioning policy reduces the impact of skewness, and consequently the scaling policy maintains the high throughput and low latency on the skewed stream. Due to the adaptive algorithm, Alps also avoids the unnecessary waste of resources. Therefore, Alps has two key features. First, to avoid the unnecessary waste of resources, Alps compares the waiting time caused by imbalance with the estimated partitioning duration to determine whether to use the partitioning algorithm. Second, for the purpose of high throughput and low latency, Alps derives parallelism configuration from the instrumentation-driven linear performance model.

We implement and evaluate Alps on Flink. Compared with state-of-the-art scaling approach DS2, Alps reduces the end-to-end latency by 2 orders of magnitude on high-speed skewed stream and avoids the waste of resources on low-speed stream. The main contributions are stated as follows:

- We explore the impact of rate variation and data distribution on DSP. Based on these findings, we propose and implement an adaptive partitioning scaling algorithm. It compares the waiting time caused by imbalance with partitioning duration and determines whether to use a partitioning policy, rather than using expensive partitioning policy all the time. Therefore, it maintains an outstanding performance of DSP and avoids the waste of resources.
- We integrate the scaling policy and partitioning policy into Alps to achieve the high throughput and low latency of DSP on skewed stream. Besides, Alps has the capability to tune the parallelism configuration automatically.

- We evaluate Alps on synthesized Zipf datasets and Reddit datasets. And Alps achieves good results on both synthesized and realistic datasets.

The rest of the paper is organized as follows. The related work is discussed in Sect. 2. We present the key algorithm of Alps exhaustively in Sect. 3. Alps is evaluated and analyzed in Sect. 4 by abundant experiments. We conclude this paper in Sect. 5.

2 Related Work

Elastic scaling is a crucial technology in DSP, because of dynamic characteristics of stream. The elastic scaling approaches are categorized into 2 types: rule-based and model-based. And model-based approaches mainly contain queue-theory, control-theory and performance-model approaches.

Rule-based approaches rely on heuristic rules. They need to define several effective rules, and then apply these rules on collected metrics to generate actions used to tune the DSP. Backpressure [9], memory usage [10], CPU usage [10,12], network bandwidth [13], throughput [14] and congestion [14] are widely used as instructive metrics. Besides, Stela [15] designs and implements a special metric ETP as the instructive metrics. Rule-based approaches can tune DSP in a prompt manner, with the cost of generalizability. Furthermore, Rule-based approaches are hard to acquire the effective rules used to instruct the scaling, especially on skewed stream.

Queue-theory approaches view the DSP as an open queue network [16]. Some approaches consider M/M/k queue is adequate to estimate the waiting time compared with GI/G/k queue in DSP. [17] uses Erlang latency formula to estimate the waiting time, then builds and solves the optimization problem about estimated waiting time and parallelism configuration to find the best parallelism. The other approaches [18] view each operator as a GI/G/1 queue, and use Kingman's formula [19] to estimate the waiting time. Parallelism configuration is generated by optimizing the latency metrics related to the waiting time. Queue-theory approaches highly depend on the accuracy of queue model and the rigorous hypotheses. However, the rigorous hypotheses are always not satisfied on skewed stream.

Control-theory approaches employ several control frameworks to optimize the parallelism configuration. Most control-theory approaches regard the DSP as a black-box model, which means there is a static but unknown relationship between the input and output of the system. [20] considers that the system model is a linear function. Then the feedback loop is used to tune the parallelism. [21] adopts GI/G/1 queue model as system model, then uses MPC framework to optimize parallelism. Although control-theory approaches are robust against the dynamics of input stream, overshoot and settling time damages the performance of DSP. Furthermore, skewed stream makes the trade-off between overshoot and settling time harder.

Performance-model approaches thoroughly study the prototype of DSP, and thus propose the accurate performance model used to calculate the optimal parallelism. [11] utilizes the useful time, which excludes the waiting time, and regards the system model as a linear model. [22] adopts Gaussian Process Model as the performance model instead of linear performance model. Meanwhile, Bayesian optimization is used to search a feasible parallelism configuration. Performance-model approaches achieves an outstanding performance of DSP on balanced stream. However, skewed stream leads to the imbalanced utilization of resources, and consequently ruins the effects of exact resources provision.

3 Model and Algorithm Design

3.1 Algorithm Foundation

The main reason of performance degradation in DSP is the congestion and backpressure caused by the rate variation and data distribution. To explore the impact of rate variation and data distribution, Case 1 and 2 in Appendix are constructed. Based on the experiments in these cases, some observations which are the foundation of the algorithm are as follows:

Observation 1. Scaling policy is poor to deal with the skewed stream. And partitioning policy reduces the system performance if not integrated with the scaling policy. Only using partitioning policy and scaling policy simultaneously achieves the best performance on high-speed and skewed stream.

Observation 2. Both input rate and skewness of the stream have a great impact on system performance. And partitioning policy is not always necessary on skewed stream, since the scaling policy does well on low-speed and skewed stream.

Table 1. The symbols used in the paper.

Symbol	Explanation
$N_{proc;i}$	The number of data processed by instance i
$T_{useful;i}$	The processing time of instance i without waiting time
$R_{true-proc;i}$	The true processing rate of instance i
$R_{obs-proc;i}$	The observed processing rate of instance i
$R_{true-input;i}$	The true input rate of instance i
R_{source}	The expected output rate of source operator
$T_{partition}$	The partitioning duration
$T_{obs-wait}, T_{queue}$	The observed waiting time and ideal queuing time
$W^{-}_{skewness}, W_{rate}, W_{partition}$	The willingness of skewness, rate and partition
α	The weight of skewness in willingness function
$\theta, \theta_{upperbound}, \theta_{lowerbound}$	The threshold of rate in willingness function
$Topo$	The topology of job
$S_{partition}$	The signal of partition

Alps utilizes the adaptive partitioning scaling algorithm to improve the performance of DSP. Inspired by Observation 1 and 2, the scaling policy and partitioning policy are integrated into the adaptive partitioning scaling algorithm. Before exploiting the scaling policy to generate the optimal parallelism configuration, the adaptive partitioning scaling algorithm determines whether it is worthy to partition stream in a balanced way. The detail of adaptive partitioning scaling algorithm is presented in Sect. 3.3. For concise presentation, several symbols used in the paper exhibit in Table 1.

3.2 Operator Performance Model

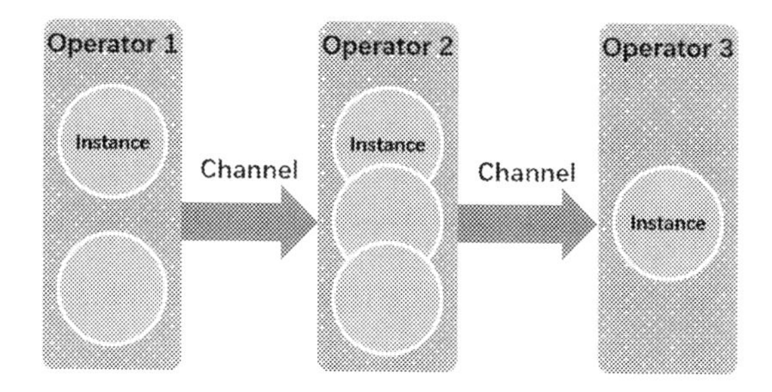

Fig. 1. Logical graph.

A DSP job can be extracted as a logical graph (see Fig. 1). The vertex in logical graph is an operator used to execute a subtask and the edge is the channel of transporting stream between adjacent operators. Each operator utilizes several instances to implement parallel processing.

Alps is based on the assumption that the instances in DSP are homogeneous and adequate. Hence, the performance is proportional to the parallelism of operators. Alps also assumes the only factor influencing performance is CPU, in other words, bandwidth is not the bottleneck of DSP. Based on the above assumptions, Alps regards DSP performance model as linear model. Inspired by [11], Alps uses useful time to represent the duration of processing stream without the waiting time. Then the true processing rate of each instance i is calculated as follows:

$$R_{true-proc;i} = \frac{N_{proc;i}}{T_{useful;i}} \tag{1}$$

The true processing rate is more accurate than observed processing rate since it is not affected by the waiting time resulting from backpressure. According to the logical graph, the parallelism of each operator j is calculated as follows:

$$Parallelism_{optimal;j} = \frac{\sum_{i \in O_j} R_{true-input;i}}{\sum_{i \in O_j} R_{true-proc;i}/|O_j|} \tag{2}$$

where $Parallelism_{optimal;j}$ is the optimal parallelism of operator j, $R_{true-input;i}$ is the true input rate of instance i derived from the true output rate of upstream instances and O_j is all the instances of operator j.

3.3 Adaptive Partitioning Scaling

As stated in Observation 1 and 2, scaling policy is poor to handle the skewed stream. Besides, using partitioning policy rashly leads to a performance degradation. Hence, when to use partitioning policy is a crucial problem. We introduce an adaptive partitioning scaling algorithm (APS) in Alps to solve this problem. The algorithm determines whether to use the partitioning policy, then uses scaling policy to generate the parallelism configuration.

We design the APS algorithm based on a plain idea. Partitioning policy produce a positive effect when the partitioning duration is lower than the waiting time caused by backpressure and congestion. Besides, the partitioning policy is more likely to produce a positive effect on high-speed stream and waste resources on low-speed stream. Therefore, the APS algorithm uses a willingness function to represent the willingness of using the partitioning policy.

The willingness function in APS algorithm consists of the impact of skewness and rate. And APS regards the willingness function as a linear combination of skewness willingness and rate willingness. The willingness function is as follows:

$$W_{partition} = \alpha * W^{-}_{skewness} + (1 - \alpha) * W_{rate} \tag{3}$$

where $W_{partition}$ is the willingness of using partitioning algorithm, $W^{-}_{skewness}$ is the unwillingness caused by skewness, W_{rate} is the willingness caused by rate and α is the weight of skewness.

The unwillingness caused by skewness $W^{-}_{skewness}$ is represented by the ratio of the difference between waiting time and partitioning duration to partitioning duration (see Eq. 4). The waiting time T_{wait} caused by backpressure is the difference between observed waiting time and the ideal queuing duration (see Eq. 5). The observed waiting time $T_{obs-wait}$ is calculated by the maximum difference between observed processing duration (see Eq. 6). And the ideal queuing duration T_{queue} is estimated by Kingman's formula (see Eq. 7). $T_{partition}$ is related to the partitioning algorithm itself, and thus it is a static value. The unwillingness formula of skewness is as follows:

$$W^{-}_{skewness} = \begin{cases} \frac{T_{wait} - T_{partition}}{T_{partition}}, & T_{wait} < T_{partition} \\ 0, & T_{wait} \geq T_{partition} \end{cases} \tag{4}$$

$$T_{wait} = T_{obs-wait} - T_{queue} \tag{5}$$

$$T_{obs-wait} = \max_{i \in O_j} \frac{1}{R_{obs-proc;i}} - \min_{i \in O_j} \frac{1}{R_{obs-proc;i}} \tag{6}$$

$$T_{queue} \approx \left(\frac{\rho}{1-\rho}\right)\left(\frac{c_a^2 + c_s^2}{2}\right)\left(\frac{1}{\sum_{i \in O_j} R_{true-proc;i}/|O_j|}\right) \tag{7}$$

where O_j is the instances of operator j, $\rho = \frac{R_{obs-input}}{R_{true-proc}}$ is the utilization, and c_a, c_s are the coefficients of variation for arrival and service time.

The willingness caused by input rate is represented by the maximum utilization (see Eq. 8). The utilization in APS algorithm is represented by the ratio of observed processing rate to true processing rate (see Eq. 9).

$$W_{rate} = \begin{cases} 1, & U \geq \theta_{upperbound} \\ \frac{U-\theta}{\theta_{upperbound}-\theta}, & \theta_{upperbound} > U \geq \theta \\ \frac{U-\theta}{\theta-\theta_{lowerbound}}, & \theta > U > \theta_{lowerbound} \\ -1, & \theta_{lowerbound} \geq U \end{cases} \tag{8}$$

$$U = \frac{\max_{i \in O_j} R_{obs-proc}}{R_{true-proc}} \tag{9}$$

where θ is the threshold of rate, U is the maximum utilization and O_j is the instances of operator j.

Algorithm 1. Adaptive Partitioning Scaling.

Input: α, $T_{partition}$, $Topo$, R_{source}
Output: $Parallelism$, $S_{partition}$
1: Collect rate files of all the instances
2: **if** All files have been collected **then**
3: Parse $R_{obs-input}$, $R_{obs-proc}$, $R_{true-proc}$, $R_{true-output}$ from rate files
4: Calculate the willingness function of partitioning $W_{partition}$
5: **if** $W_{partition} > 0$ **then**
6: $S_{partition}$ = **true**
7: **else**
8: $S_{partition}$ = **false**
9: **end if**
10: **end if**
11: $Topo.Source.R_{true-output} = R_{source}$
12: **for all** $op \in Topo.Operators$ **do**
13: Calculate the optimal parallelism of op and add to $Parallelism$
14: $op.next.R_{true-input} = op.R_{true-output}$
15: **end for**
16: **return** $S_{partition}, Parallelism$

The APS algorithm is listed in Algorithm 1. Before applying the policy, the APS waits for all the rate files. Then APS calculates the willingness of partitioning when all rate files are collected. If willingness is positive, it means using partitioning algorithm is worthy. Or it means no need to use partitioning algorithm. At last, APS calculates the parallelism of all operators using Eq. 2.

4 Evaluation

4.1 Setup

We evaluate Alps on a cluster composed of 14 Dell OptiPlex 7080 (8-core CPU, 32 GB DDR4 RAM). The 11 hosts are used to deploy a customized Flink 1.12.1 cluster with 20 slots. The other 3 hosts deploy Zookeeper 3.4.9 and Kafka 2.8.1 to simulate the real environment. For the purpose of increasing the throughput, Kafka is configured with 3 partitions. Alps is allocated to the host owning the job-manager in Flink cluster, where can re-configure DSP conveniently. As the metrics are distributed on the task-managers of Flink, the Samba network file system is used to collect metrics for Alps.

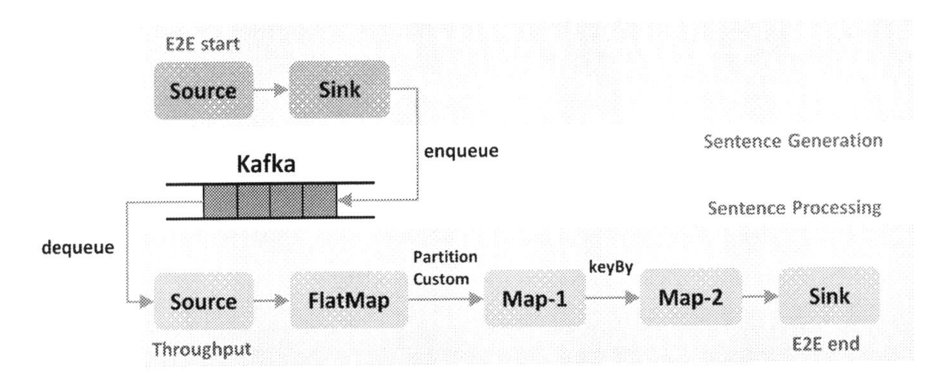

Fig. 2. The topology of Sentence-Processing and Sentence-Generation job.

For comparison, a Sentence-Processing job (see Fig. 2), that consists of Source, FlatMap, Map-1, Map-2 and Sink, is utilized to evaluate the performance of Alps. This job can balance the input stream with the W-C partitioner [23] between the FlatMap operator and Map-1 operator. The input stream of Sentence-Processing job is generated by a Sentence-Generation job. The execution of Sentence-Generation job is after the warming up of Sentence-Processing job to obtain an accurate end-to-end latency. The generation of input stream is not affected by the congestion and backpressure of DSP, since Kafka decouples the production and consumption of stream. Besides, synthesized Zipf datasets and realistic Reddit datasets are used in these experiments (see Table 2).

Table 2. The experiment datasets.

Dataset	Keys	Tuples	Skewness
Zipf	10K	10M	0.4, 0.6, 0.8, 1.0, 1.2, 1.4, 1.6
Reddit	1.2M	21M	Unknown

We compare Alps with DS2, which is the concise and state-of-the-art approach based on performance model. The main criteria of evaluation are end-to-end latency, throughput and overhead. The end-to-end latency is represented by the duration from the time when stream is generated in source operator (E2E start) to the time when stream is consumed in sink operator (E2E end). And the throughput is represented by the observed output rate of source operator in Sentence-Processing job. In order to make a comprehensive evaluation, both Alps and DS2 are evaluated on streams with different rates and different skewness. Based on such experiment configurations, the benefits of Alps are highlighted.

4.2 Adaptive Partitioning Scaling

To have a precise evaluation of Alps on skewed stream, the experiments are constructed on the streams with the skewness ranging from 0.4 to 1.6. Furthermore, Alps is evaluated under the conditions of low- (90K), medium- (180K) and high-speed (240K) skewed streams to indicate the advantages of dealing with the combined effects of rate variation and data distribution. The empirical parameters of adaptive partitioning scaling algorithm in Alps are 0.55 (α), 0.6 (θ), 0.5 ($\theta_{lowerbound}$) and 0.7 ($\theta_{upperbound}$) respectively.

End-to-End Latency. Both average and p99 end-to-end latency are used to indicate the performance of DSP. The result of average and p99 end-to-end latency is shown in Fig. 3. We use magnitude of end-to-end latency metrics to represent the performance instead of end-to-end latency. As shown in Fig. 3, both DS2 and Alps maintains an ideal performance when input rate is 90K sentences/s, since DSP has the adequate computational capacity to tolerate the imbalance. However, it is obvious that DS2 suffers a drastic degradation while Alps still maintains a low end-to-end latency when input rate is 240K sentences/s. DS2 cannot cope with the slight imbalance on high-speed stream, since the huge difference of quantity between each parallel instance causes the congestion. On the contrary, Alps avoids the performance degradation on high-speed and skewed stream, because of the combined effects of partitioning policy and scaling policy. Compared with state-of-the-art DS2, Alps decreases the end-to-end latency by 2 orders of magnitude when input stream is high-speed and skewed.

Throughput. The adaptive partitioning scaling algorithm can also maintain a high throughput, though the input stream is high-speed and skewed. Both Alps and DS2 are experimented on streams with different rates and skewness, and the results represented by throughput are drawn in Fig. 4. DS2 maintains a high throughput when input rate is low-speed or the stream is balanced. However, the degradation appears when skewness is large than 0.6 on medium-speed streams, because of the congestion and backpressure. Furthermore, the throughput on high-speed stream decreases more rapidly than on medium-speed stream for the reason that the impact of imbalance is amplified with the increase of rate. On the contrary, Alps maintains a high throughput among all the streams. As Alps integrates the partitioning policy and scaling policy, Alps protects DSP from

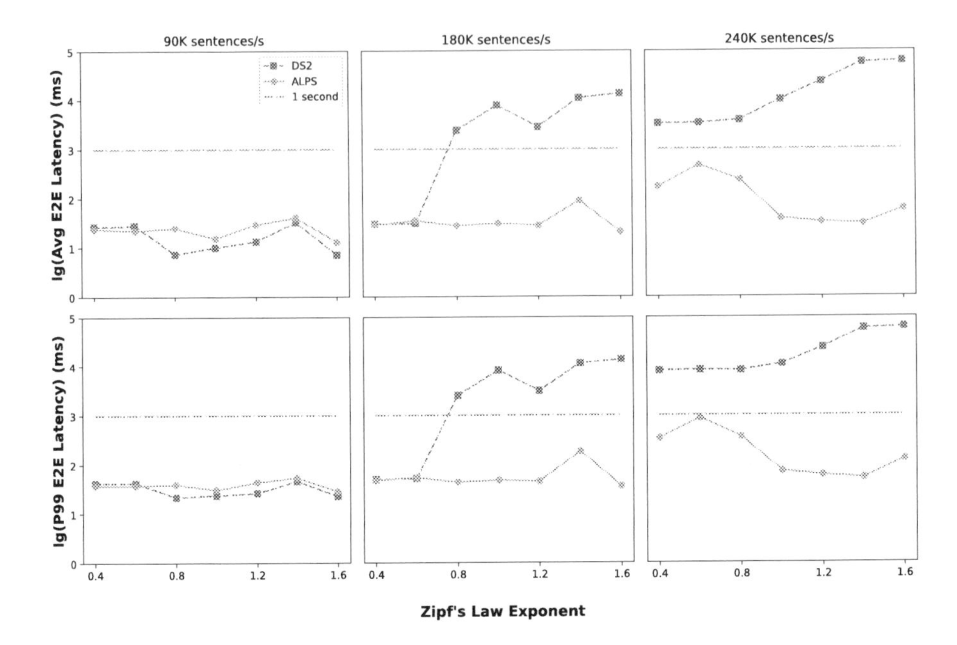

Fig. 3. The end-to-end latency with different input rates and skewness.

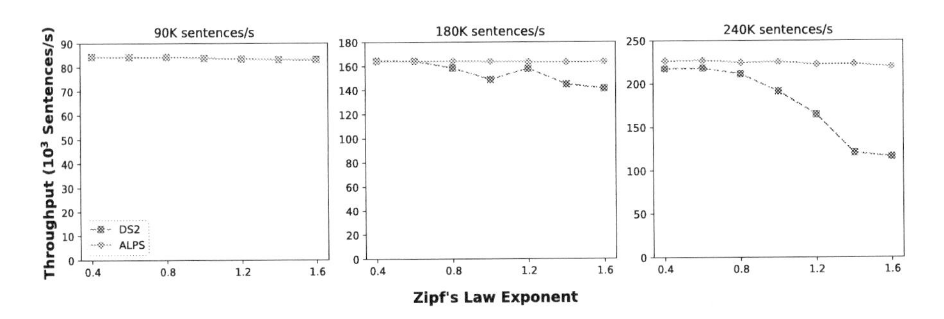

Fig. 4. The throughput with different input rates and skewness.

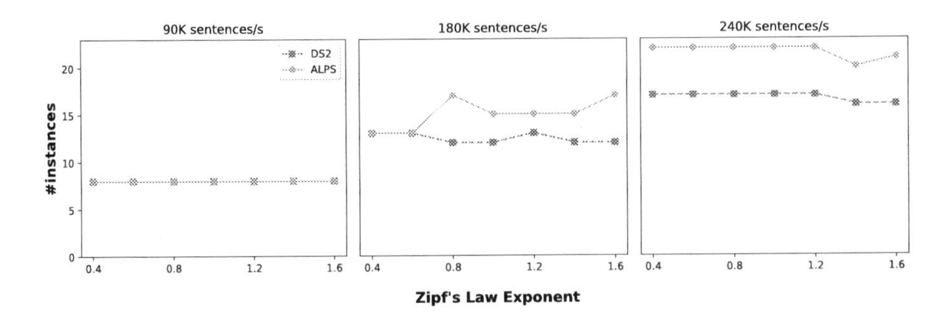

Fig. 5. The overhead with different input rates and skewness.

the impact of congestion and backpressure. Therefore, Alps has the ability to maintain the high throughput on high-speed and skewed stream.

Overhead. As shown in Appendix Case 1, partitioning policy maintains the low end-to-end latency and high throughput with the cost of additional overhead. It is worthy to keep an outstanding performance of DSP with the acceptable overhead. Figure 5 shows the overhead of Alps and DS2 on streams with different rates and skewness. As shown in Fig. 5, DS2 keeps a steady overhead whatever the skewness is due to the linear performance model. Actually, DS2 does not maintain a great performance after invoking the scaling policy on high-speed stream. However, Alps keeps an outstanding performance with the cost no more than 30% additional overhead on high-speed stream. Besides, the overhead does not increase with the growth of skewness when input rate is constant. The adaptability of Alps is exposed on low- and medium-speed streams. It is obvious that using partitioning policy (rate = 180K, skewness ≥ 0.8) causes more overhead than without partitioning policy (rate = 180K, skewness ≤ 0.6). Comparing to the common approach that using partitioning policy all the time, Alps saves the resources by means of ignoring the partitioning policy on low-speed stream. Although the partitioning policy is omitted on skewed and low-speed stream, Alps still maintains a wonderful performance. On medium-speed stream, the partitioning policy is invoked when skewness is larger than 0.6. It is necessary to use the partitioning policy with the cost of additional overhead to prevent the performance degradation of DSP. Therefore, Alps adaptively invokes the partitioning policy to maintain the performance of DSP and avoid the unnecessary overhead.

4.3 Alps on Realistic Datasets

To evaluate the practical effectiveness of Alps, the input stream is generated with different rates from the Reddit datasets. The end-to-end latency and throughput are used to evaluate the performance of Alps and the results are shown in Fig. 6.

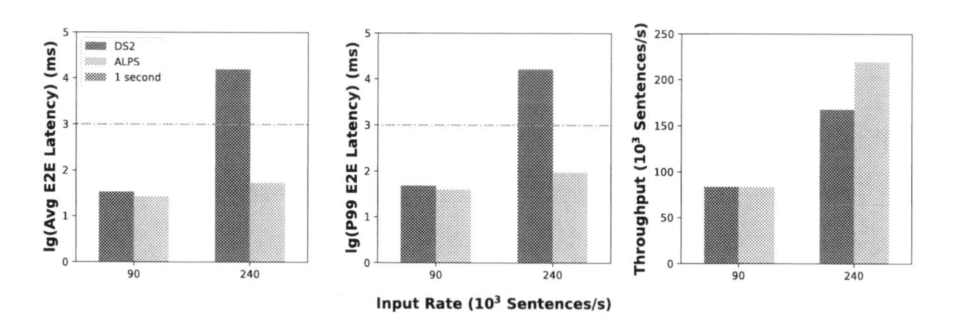

Fig. 6. The performance of Alps on Reddit datasets.

As shown in Fig. 6, both Alps and DS2 keep the low latency and high throughput on low-speed stream (90K sentences/s), since no congestion and backpressure appears. However, DS2 suffers the high latency and low throughput on high-speed stream (240K sentences/s) because of the congestion and backpressure. On the contrary, Alps invokes the partitioning policy on high-speed and skewed stream to reduce the impact of imbalance. Hence, Alps achieves the tens of milliseconds latency and 219K sentences/s throughput on high-speed Reddit stream. Compared with DS2, Alps decreases the end-to-end latency by 2 orders of magnitude and increases 30% throughput. Therefore, Alps can maintain an outstanding performance with the indispensable resources on realistic stream.

5 Conclusion

In this paper, we propose Alps, an adaptive partitioning scaling system to maintain the low latency and high throughput of DSP and reduce the overhead on skewed stream. Alps determines whether to use the partitioning policy by means of the adaptive partitioning scaling algorithm. The partitioning policy is utilized to keep an outstanding performance with the cost of acceptable overhead on high-speed and skewed stream, while it is omitted to avoid the additional overhead on low-speed or balanced stream. To evaluate the effectiveness of Alps, several experiments are conducted to compare with state-of-the-art approach DS2. The results on synthesized and realistic datasets show that Alps decreases the end-to-end latency by 2 orders of magnitude when consuming high-speed and skewed stream and avoids the waste of resources when input stream is low-speed or balanced.

For future work, we plan to investigate a more general and robust model of adaptive partitioning scaling algorithm and apply Alps to other DSP. Predicting the rate and distribution of input stream is also an interesting problem that we plan to explore.

Acknowledgements. This work is supported by the National Key R&D Program of China (2018AAA0102100), the Scientific and Technological Innovation Leading Plan of High-tech Industry of Hunan Province (2020GK2021), the National Natural Science Foundation of China (No. 61902434), International Science and Technology Innovation Joint Base of Machine Vision and Medical Image Processing in Hunan Province (2021CB1013).

Appendix

Case 1. The impact of partitioning policy.

Four experiments are conducted to reveal the scaling and partitioning issues on skewed stream. The scaling policy uses state-of-the-art approaches DS2 and partitioning policy uses state-of-the-art partitioner W-C [23]. The synthesized input stream obeys Zipf's Law and the rate is 240K sentences/s. Exp1,2,4 start with 1 parallelism, except the source whose parallelism is 3. Exp3 using the optimal parallelism configuration of Exp2. The remaining configurations are in Table 3 and the results are shown in Fig. 7.

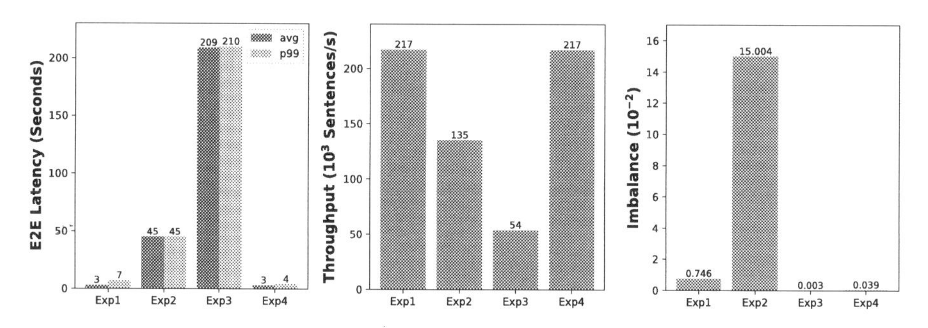

Fig. 7. The results of scaling policy and partitioning policy.

Table 3. The configurations of Sentence-Processing job (see Fig. 2) with scaling policy and partitioning policy.

Experiment	Exponent of Zipf's Law	Scaling policy	Partitioning policy
Exp1	0.0	DS2	Default
Exp2	1.0	DS2	Default
Exp3	1.0	-	W-C
Exp4	1.0	DS2	W-C

Case 2. The impact of different input rates and skewness.

We run DS2 on several input streams with different exponents of Zipf's Law, ranging from 0.4 to 1.8. And the input rate ranges from 30K sentences/s to 240K sentences/s. Therefore, we conduct 64 experiments to find out the impact of input rate and data distribution. The remaining configurations are the same with Case 1. The end-to-end latency is used to evaluate the performance of DS2. The results are shown in Fig. 8.

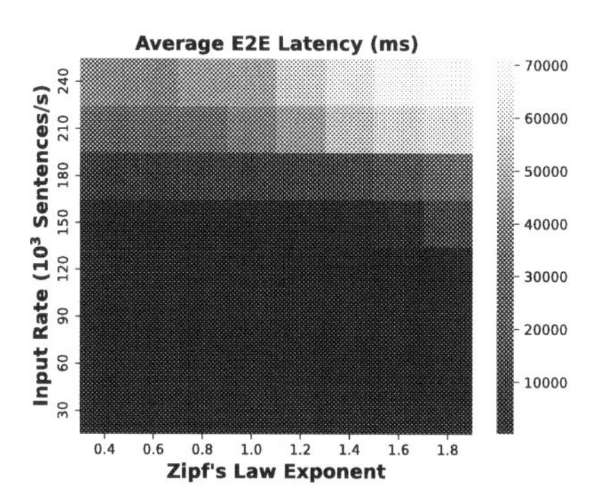

Fig. 8. The end-to-end latency of DS2 with different input rates and skewness.

References

1. Dean, J., Ghemawat, S.: Mapreduce: simplified data processing on large clusters. Commun. ACM **51**(1), 107–113 (2008)
2. Zaharia, M., Chowdhury, M., Franklin, M.J., Shenker, S., Stoica, I.: Spark: cluster computing with working sets. In: Proceedings of the 2nd USENIX Conference on Hot Topics in Cloud Computing, HotCloud 2010, p. 10. USENIX Association, USA (2010)
3. Zaharia, M., et al.: Resilient distributed datasets: a fault-tolerant abstraction for in-memory cluster computing. In: Proceedings of the 9th USENIX Conference on Networked Systems Design and Implementation, NSDI 2012, p. 2. USENIX Association, USA (2012)
4. Stonebraker, M., Çetintemel, U., Zdonik, S.: The 8 requirements of real-time stream processing. SIGMOD Rec. **34**(4), 42–47 (2005)
5. Apache storm homepage. https://storm.apache.org
6. Carbone, P., Katsifodimos, A., Ewen, S., Markl, V., Haridi, S., Tzoumas, K.: Apache flink: stream and batch processing in a single engine. Bull. IEEE Comput. Soc. Tech. Committee Data Eng. **36**(4) (2015)
7. De Matteis, T., Mencagli, G.: Keep calm and react with foresight: strategies for low-latency and energy-efficient elastic data stream processing. In: Proceedings of the 21st ACM SIGPLAN Symposium on Principles and Practice of Parallel Programming, PPoPP 2016. Association for Computing Machinery (2016)
8. Hirzel, M., Soulé, R., Schneider, S., Gedik, B., Grimm, R.: A catalog of stream processing optimizations. ACM Comput. Surv. **46**(4), 1–34 (2014)
9. Floratou, A., Agrawal, A., Graham, B., Rao, S., Ramasamy, K.: Dhalion: self-regulating stream processing in heron. Proc. VLDB Endow. **10**(12), 1825–1836 (2017)
10. Gulisano, V., Jiménez-Peris, R., Patiño-Martínez, M., Soriente, C., Valduriez, P.: Streamcloud: an elastic and scalable data streaming system. IEEE Trans. Parallel Distrib. Syst. **23**(12), 2351–2365 (2012)
11. Kalavri, V., Liagouris, J., Hoffmann, M., Dimitrova, D., Forshaw, M., Roscoe, T.: Three steps is all you need: fast, accurate, automatic scaling decisions for distributed streaming dataflows. In: Proceedings of the 13th USENIX Conference on Operating Systems Design and Implementation, OSDI 2018, pp. 783–798. USENIX Association, USA (2018)
12. Heinze, T., Roediger, L., Meister, A., Ji, Y., Jerzak, Z., Fetzer, C.: Online parameter optimization for elastic data stream processing. In: Proceedings of the Sixth ACM Symposium on Cloud Computing, SoCC 2015, pp. 276–287. Association for Computing Machinery, New York (2015)
13. Heinze, T., Jerzak, Z., Hackenbroich, G., Fetzer, C.: Latency-aware elastic scaling for distributed data stream processing systems. In: Proceedings of the 8th ACM International Conference on Distributed Event-Based Systems, DEBS 2014, pp. 13–22. Association for Computing Machinery, New York (2014)
14. Gedik, B., Schneider, S., Hirzel, M., Wu, K.L.: Elastic scaling for data stream processing. IEEE Trans. Parallel Distrib. Syst. **25**(6), 1447–1463 (2014)
15. Xu, L., Peng, B., Gupta, I.: Stela: enabling stream processing systems to scale-in and scale-out on-demand. In: 2016 IEEE International Conference on Cloud Engineering (IC2E), pp. 22–31 (2016)
16. Fu, T.Z.J., Ding, J., Ma, R.T.B., Winslett, M., Yang, Y., Zhang, Z.: DRS: auto-scaling for real-time stream analytics. IEEE/ACM Trans. Netw. **25**(6), 3338–3352 (2017)

17. Tijms, H.C.: Stochastic Modelling and Analysis: A Computational Approach. Wiley, Hoboken (1986)
18. Lohrmann, B., Janacik, P., Kao, O.: Elastic stream processing with latency guarantees. In: 2015 IEEE 35th International Conference on Distributed Computing Systems, pp. 399–410 (2015)
19. Kingman, J.F.C.: The single server queue in heavy traffic. Math. Proc. Cambridge Philos. Soc. **57**(4), 902–904 (1961)
20. Khoshkbarforoushha, A., Khosravian, A., Ranjan, R.: Elasticity management of streaming data analytics flows on clouds. J. Comput. Syst. Sci. **89**, 24–40 (2017)
21. De Matteis, T., Mencagli, G.: Elastic scaling for distributed latency-sensitive data stream operators. In: 2017 25th Euromicro International Conference on Parallel, Distributed and Network-based Processing (PDP), pp. 61–68 (2017)
22. Zhang, L., Zheng, W., Li, C., Shen, Y., Guo, M.: Autrascale: an automated and transfer learning solution for streaming system auto-scaling. In: 2021 IEEE International Parallel and Distributed Processing Symposium (IPDPS), pp. 912–921 (2021)
23. Anis Uddin Nasir, M., De Francisci Morales, G., Kourtellis, N., Serafini, M.: When two choices are not enough: balancing at scale in distributed stream processing. In: 2016 IEEE 32nd International Conference on Data Engineering (ICDE), pp. 589–600 (2016)

Latent Relational Point Process: Network Reconstruction from Discrete Event Data

Guilherme Augusto Zagatti[1(✉)], See-Kiong Ng[1,2], and Stéphane Bressan[1,2]

[1] Institute of Data Science, National University of Singapore, Singapore, Singapore
gzagatti@u.nus.edu, {seekiong,steph}@nus.edu.sg
[2] School of Computing, National University of Singapore, Singapore, Singapore

Abstract. Digital interactions, such as Wi-Fi access, financial transactions, and social media activities, leave crumbs in their path. These vast quantities of fine-granularity data generated by complex real-world relational systems propound unprecedented alternatives to expensive and laborious surveys and studies for researchers who want to learn and understand the underlying network models. Point processes are well-suited for modelling the discrete data commonly observed from digital interactions of today's complex systems. In this work, we present a latent relational point process framework for recovering the posterior probability of latent relations from discrete event data effectively and efficiently. Our proposed framework comprises a general definition of the latent relational point process, an algorithm for fitting the parameters of an evolutionary version of the model to the data, and goodness-of-fit tests to quantify the suitability of the model to the data. The proposed framework is evaluated for the modelling of a social network from the observations of social interactions.

Keywords: Point process · Networks · Knowledge extraction

1 Introduction

Networks are widely used to model complex real-world relational systems such as technological, biological, and social systems. However, real-world networks are often large, noisy and dynamic. Thus, direct observation can be costly and laborious. Digital interactions leave digital traces that are abundant and easy-to-collect. These discrete event data can be used to infer the underlying connections between the elements in the complex system. The challenge is thus to devise a principled yet effective and efficient framework to learn the network models from this data. We propose a latent relational point process framework for reconstructing networks from discrete event data. The proposed framework specialises on the seminal framework introduced by Mark Newman [14,15] for reconstructing network structure from data with a new data model, point processes, to better represent how the observed data as generated by the underlying network model were manifested.

C. Strauss et al. (Eds.): DEXA 2022, LNCS 13427, pp. 32–46, 2022.
https://doi.org/10.1007/978-3-031-12426-6_3

To develop our proposed framework, we put forward the general definition of the latent relational point process, a new family of doubly stochastic point processes where the distribution over point processes depends on a model for latent relations. Then, we develop an algorithm for fitting the parameters of the evolutionary version of the model to the data. Evolutionary point processes evolve along a single dimension, such as time. Since the likelihood of all such processes can be expressed in terms of their conditional intensity function, we modify Newman's expectation-maximisation algorithm to directly use such log-likelihood. Finally, we devise goodness-of-fit tests into our framework to evaluate the performance of the fit. Since latent networks are not directly observable, these tests provide important evidence to assess the fitted model.

The remainder of this paper proceeds as follows. In Sect. 2, we present the state-of-the-art and related work to position our work and contribution. Then, Sect. 3 introduces the latent relational framework and narrow our discussion to evolutionary point processes. Thereafter, in Sect. 4 we demonstrate the proposed framework and evaluate its performance on a model of encounters. In such a model, individuals who share a latent relation meet more often than those who do not. To evaluate the performance of the model, we conduct simulations and revisit the MIT Reality Mining data set [4]. Finally, Sect. 5 concludes by summarising our contribution and pointing to future research directions.

2 Related Work

Network reconstruction is a problem pervasive to many disciplines. One of the main challenges remains the lack of ground truth data [1]. Therefore, robust reconstruction frameworks and identifying assumptions are required to mitigate these challenges. Previous work on reconstruction frameworks can be broadly classified under three categories [6]: link prediction, correlation analysis, and tomographic inference.

Link prediction takes a more traditional machine learning approach to predict the existence of an edge between two nodes based on the proximity of their characteristics on a latent space. It requires data about nodes and edges for inference and prediction. In contrast, correlation analysis does not require any knowledge of network edges, and it relies on the correlation of node attributes to establish which nodes are connected via statistical testing. In contrast, network tomography takes an experimental approach by probing nodes in the network to develop hypothesis about the network topology.

A key missing piece of the above categorisation of network reconstruction is a methodology that focuses on interaction attributes. Techniques that rely on thresholding do not take into account the sampling properties of the interaction process itself. A common approach in this field is to coarse grain interactions into a sequence of static networks. However, the selection of the optimal window size can be challenging [7,17] and these approaches can produce unreliable estimates [20].

The formalism developed by Mark Newman [14,15] embraces the fact that data about interactions could be a reflection of latent networks. Newman proposes a reconstruction method that requires prior assumptions about the network model "that represents how the network structure is generated" [14] and the data model "that represents how that structure maps onto observed data" [14] to define the data likelihood. Using a Bayesian approach, he proposes an expectation-maximisation algorithm for estimating model parameters and the latent network posterior probability. In subsequent work, Newman and colleagues [19,20] explore automatic discovery via Monte Carlo posterior approximation.

Many of the case studies presented in the studies above involve point processes, however a framework that focuses especifically on these class of data models was not introduced. This paper bridges the research from Newman and colleages with that of temporal networks [5,12]. We propose an alternative data model, point processes, that is well-suited for the discrete data that are commonly observed from digital interactions of today's complex systems. We thus name our framework the latent relational point process framework. Point process characterize discrete data either as a random collection of points or as a counting measure [2]—*i.e.* the random number of events in an interval. Association matrices can describe interdependent point processes. Examples where point process and network models come together include the study of neuronal activity [16], social media activity [5] and epidemics [11]. Our framework also applies to exponential random graph models [18].

3 Framework

3.1 General Framework

The latent relational point process is a doubly stochastic point process which is driven by the latent relations of nodes in the model. A point process is a random bounded finite integer-valued measure over a measurable space. We say that a point process is doubly stochastic, when there is a distribution over possible point processes. This distribution is parameterized by the unobserved relations of nodes in the model.

Let $\mathbb{V}$ be a countable set of nodes and the identity random variable $\mathcal{E}$ be the set of latent relations on the probability space of relations between nodes $(\mathbb{E}, \mathcal{E}, \mu_{\mathcal{E}})$. We denote $E \in \mathbb{E}$ an outcome of $\mathcal{E}$. A relation $e \in E$ indicates the subset of nodes in $\mathbb{V}$. Let the interaction space $\mathcal{K}_{\mathbb{V}}$ be a set of subsets of $\mathbb{V}$ and let $\mathbb{X}$ represent the location space. We assume that both spaces are complete separable metric spaces. Let $\mathbb{Y} = \mathbb{X} \times \mathcal{K}_{\mathbb{V}}$ represent the underlying topological space. We construct a measurable space over $\mathbb{Y}$ with the Borel algebra, $(\mathbb{Y}, \mathcal{B}(\mathbb{Y}))$.

Definition 1 (latent relational point process). *A latent relational point process on* $\mathbb{Y}$ *is the point process* $\mathcal{N}$ *defined on the measurable space of bounded finite integer-valued measures on* $\mathbb{Y}$, $\mathbb{M}(\mathbb{Y})$, *with probability:*

$$\mu_{\mathcal{E}}(B) \equiv E_{\mathcal{E}}[\Pr(\mathcal{N}(\cdot \mid E) \in B)] = \int_{\mathbb{E}} \Pr(\mathcal{N}(\cdot \mid E) \in B) \, d\mu_{\mathcal{E}}(E), \forall B \in \mathcal{B}(\mathbb{M}(\mathbb{Y})) \tag{1}$$

The measurable space $\mathbb{M}(\mathbb{Y})$, is the space of all measures η on $\mathbb{Y}$ such that $\eta(A) \in \mathbb{N}, \forall\, A \subseteq \mathbb{Y}$. We assume that there is a family of point processes $\{\mathcal{N}(\cdot \mid E) : E \in \mathbb{E}\}$ on $\mathbb{Y}$ indexed by the elements $E \in \mathbb{E}$. Following Proposition 6.1.II of [2], we have that this family forms a *measurable family* if for each set $B \in \mathcal{B}(\mathbb{M}(\mathbb{Y}))$ the function $\Pr(\mathcal{N}(\cdot \mid E) \in B)$ is $\mathcal{B}(\mathbb{E})$-measurable. Therefore, the $\mu_{\mathcal{E}}$ defines a probability measure on $\mathcal{B}(\mathbb{M}(\mathbb{Y}))$.

Except when $\mathcal{K}_{\mathbb{V}}$ is not finite, the latent relational point process is a marked point process with ground process on the location space $\mathbb{X}$ (see definitions in [2]). This representation is useful when deriving estimators and goodness-of-fit tests.

The definition of latent relation $\mathbb{E}$ is deliberately general. It allows for all types of graphs and hypergraphs. We only impose the condition that the probability space of relations be well-defined and the set of relations be fixed during the period of observation. However, in subsequent sections we impose further restrictions to operationalise the definition and to derive valid estimators.

3.2 Evolutionary Framework

To operationalise latent relational point processes, we investigate the subclass of evolutionary point processes, which are processes whose location space can be naturally ordered. In other words, the process naturally evolves. We will narrow the focus of this section to the case when the location space represents continuous time and $\mathbb{X} = \mathbb{R}_0^+$. These processes are particularly important in theory because their likelihood can be expressed in terms of the conditional intensity [2].

The hazard function is the probability that the ground process will take place within the next infinitesimal period of time given that it has not occurred until then. Let $H_{t-} = \{(t_n, k_n) \mid 0 \leq t_n < t\}$ denote the internal history of the process up to but not including time t. Note that $t_0 \leq 0 < t_1 < \cdots < t_n < \cdots < t_N < t$ is a natural ordering of the realization of the point process on $[0, t)$ and that $N \equiv \eta(t)$ is the total number of realizations from time 0 up to but not including time t.

The hazard function $h_g(t \mid H_{t-})$ is the probability that the next event takes place in t given no changes to the internal history. We construct the conditional intensity function of the ground process as $\lambda_g^*(t) \equiv \lambda_g(t \mid H_{t-}) = \sum_{n=1}^{N} \mathbb{1}(t_{n-1} < t \leq t_n)\, h_g(t \mid H_{t-})$. Let the conditional mark distribution $f_K^*(k \mid t) \equiv f_K(k \mid t, H_{t-})$. Assume that the evolutionary model is parameterized by a vector of parameters θ. As shown in [2], the likelihood of this marked point process can be expressed as:

$$\mathcal{L}(H_{t^-} \mid \theta) \equiv \Pr(H_{t^-} \mid \theta) = \prod_{n=1}^{N} \lambda_g^*(t_n \mid \theta)\, f_K^*(k_n \mid t_n, \theta)\, \exp\left(-\int_0^t \lambda_g^*(u \mid \theta)\, du\right) \tag{2}$$

Consider the realization of a set of relations E in the latent relational point process, extend the definition of θ to include the parameterization of the relational model, then the likelihood of the latent relational evolutionary point process becomes:

$$\mathcal{L}(H_{t^-} \mid \theta) \equiv \Pr(H_{t^-} \mid \theta) = \int_{\mathbb{E}} \mathcal{L}(H_{t^-} \mid E, \theta)\, d\mu_{\mathcal{E}}(E \mid \theta)$$
$$= \int_{\mathbb{E}} \prod_{n=1}^{N} \lambda_g^*(t_n \mid E, \theta)\, f_K^*(k_n \mid t_n, E, \theta) \times \exp\left(-\int_0^t \lambda_g^*(u \mid E, \theta)\, du\right) d\mu_{\mathcal{E}}(E \mid \theta) \tag{3}$$

3.3 Estimation via Expectation Maximisation

The observation model for the evolutionary latent relational point process assumes that an observer captures every interaction that takes place over a fixed duration of time $[0, T]$, such that observation n consists of the tuple (t_n, k_n). We denote the observed data as H_{T^-} since it coincides with the definition of internal history presented above.

We collect all model parameters in a single vector θ. Our objective is to compute the maximum *a posteriori* estimator $\hat{\theta} = \text{argmax}_\theta \Pr(\theta \mid H_{T^-})$. The derivation proposed in this Subsection is inspired by the derivation in [14,15]. Using Bayes' theorem, we can decompose the posterior probability as following:

$$\Pr(\theta \mid H_{T^-}) = \frac{\left[\int_{\mathbb{E}} \mathcal{L}(H_{T^-} \mid E, \theta)\, d\mu_{\mathcal{E}}(E \mid \theta)\right] \Pr(\theta)}{\Pr(H_{T^-})} \tag{4}$$

In order to evaluate the posterior we need to be able to marginalize the likelihood of the data over the distribution of all possible relations $\mu_{\mathcal{E}}(E \mid \theta)$. This could be a complex distribution which depends on the values of the model parameters θ. To simplify the problem we can approximate $\mu_{\mathcal{E}}(E \mid \theta)$ with a simpler probability measure that is independent of model parameters. Using the duality formula introduced in [9], we can find a lower bound for the posterior in Eq. 4—see Appendix A.1 for the full derivation:

$$\begin{aligned}\log \Pr(\theta \mid H_{T^-}) \geq & \int_{\mathbb{E}} \left[\ell(H_{T^-} \mid E, \theta) + \log(f_{\mathcal{E}}(E \mid \theta))\right] \hat{f}_{\mathcal{E}}(E)\, d\iota(\epsilon) \\ & - \int_{\mathbb{E}} \log(\hat{f}_{\mathcal{E}}(E)) \hat{f}_{\mathcal{E}}(E)\, d\iota(\epsilon) + \log \Pr(\theta) - \log \Pr(H_{T^-})\end{aligned} \tag{5}$$

where $\ell(\cdot) = \log \mathcal{L}(\cdot)$, $\hat{f}_{\mathcal{E}}(E)$ is a surrogate distribution that approximates the true distribution $f_{\mathcal{E}}(E \mid \theta)$ and ι is a dominating measure. As discussed in [9], the Lebesgue or counting measure are commonly used as the dominating measure ι. When ι is the Lebesgue measure, all the duality formula including the KL divergence can be expressed in terms of the probability density function. This is the assumption mainly explored in [9]. In our case, since the space of relations $\mathbb{E}$

is discrete, it is more appropriate to use the counting measure as the dominating measure ι which allow us to replace most integrations involving ι with a sum involving probability masses f.

To maximize the right-hand side of the inequality in Eq. 5 we take the derivative of this expression with respect to θ and set it equal to 0:

$$\int_{\mathbb{E}} \left[\nabla_\theta \ell(H_{T^-}, E \mid \theta) + \nabla_\theta \log f_{\mathcal{E}}(E \mid \theta)\right] \hat{f}_{\mathcal{E}}(E)\, d\iota(\epsilon) + \nabla_\theta \log \Pr(\theta) = 0 \quad (6)$$

The duality formula [9] also states that the right-hand side of the above inequality is maximized whenever:

$$\hat{f}_{\mathcal{E}}(E) = \frac{\mathcal{L}(H_{T^-} \mid E, \theta)\, f_{\mathcal{E}}(E \mid \theta)}{\int_{\mathbb{E}} \mathcal{L}(H_{T^-} \mid E, \theta)\, f_{\mathcal{E}}(E \mid \theta)\, d\iota(\epsilon)} \quad (7)$$

Applying Bayes' theorem we find that the above expression is nothing more than the posterior distribution over the possible networks $\Pr(E \mid H_{T^-}, \theta)$. See Appendix A.2 for the derivation.

We are now in a position to develop the expectation-maximization algorithm to compute the *a posteriori* estimator $\hat{\theta}$. Equation 5 defines a lower-bound for the posterior probability with respect to any choice of approximating probability measure $\hat{\mu}_{\mathcal{E}}$. At the same time, Eq. 7 defines the condition that $\hat{\mu}_{\mathcal{E}}$ should meet such that the lower-bound is maximal given a choice of parameters θ. To compute $\hat{\theta}$ we can proceed iteratively by computing a candidate for $\hat{\theta}$ with Eq. 6 and then choosing the distribution over $\mathcal{E}$ according to Eq. 7. We stop the iteration once the candidate for $\hat{\theta}$ changes less than a fixed threshold ϵ between iterations.

The expectation-maximization algorithm returns the approximate distribution of relations between nodes $\hat{\mu}_{\mathcal{E}}$ under the maximum *a posteriori* parameter vector estimate $\hat{\theta}$. This distribution determines the probability that a set of nodes are connected given the observed data which allow us to reconstruct the latent social network. For instance, prior to observing the data we might assign the same probability of a relationship between i and j, and i and l. However, if after observing the data we see that i meets more often with j than with l, the posterior will assign a higher probability to there being a relation between i and j than between i and l.

3.4 Goodness-of-Fit Test

Evolutionary point processes lend themselves to an elegant testing framework [2]. An important result from point process theory is Theorem 7.4.1 from [2] which states that a simple point process $\mathcal{N}$ can be transformed to a Poisson process $\tilde{\mathcal{N}}$ with unit rate via the transformation $\tilde{\mathcal{N}}(t) \equiv \mathcal{N}(\Lambda^{*-1}(t))$ where $t \mapsto \Lambda^*(t) \equiv \int_0^t \lambda^*(s)du$, Λ^* is known as the compensator. Applying Proposition 7.4.VI (b) of [2], we define the following compensator:

$$\hat{\Lambda}^*(t) = \int_{\mathcal{K}_\mathbb{V}} \int_0^t \int_{\mathbb{E}} \lambda_g^*(u \mid E, \hat{\theta})\, \hat{f}_K^*(k \mid u, E, \hat{\theta}) d\hat{\mu}_{\mathcal{E}}(E)\, du\, d\tilde{\mu}_K \quad (8)$$

where $\tilde{\mu}_K$ is a stationary mark distribution that does not depend on time t and network configuration E. Then, we have that the evolutionary latent relations point process can be transformed to a compound point process with unit intensity rate via the compensator defined in Eq. 8 and stationary mark distribution $\tilde{\mu}_K$.

If $\hat{f}^*_K(k \mid t)$ does not depend on time, then the compensator of the ground process can be used to transform the ground process to a Poisson process with unit intensity rate. Moreover, the distribution $\hat{f}_K(k)$ can be used to mark this process. Under this assumption, we can develop a goodness-of-fit test to determine whether the model estimated from the observed data indeed follows an evolutionary latent relational point process.

We adapt Algorithm 7.4 V of [2]. First, compute the transformed time-sequence ($\tau_i = \Lambda^*_g(t_i)$) using the estimates $\hat{\theta}$, $\hat{\mu}(E)$. The cumulative step-function $(\tau_i / \tau_T, i / N)$ lies in the unit square. Since it can be approximated by the Gaussian process $y = x + \epsilon$, $\epsilon \sim Normal(0, T^{-1})$, [2] proposes an approximate goodness-of-fit test. The null hypothesis that the transformed process follows a unit-rate Poisson process implies that with a probability equal to $100(1-\alpha)\%$ all points of the transformed time-sequence should fall inside the $100(1-\alpha)\%$ confidence band drawn around $y = x$. Second, we test that the empirical distribution of interactions follow the estimated distribution $\hat{f}_K$. Assuming that the interaction space $\mathcal{K}_{\mathbb{V}}$ is finite, this amounts to a multinomial test which can be approximated by a likelihood ratio test with test statistic distributed according to a χ^2 distribution:

$$-2 \sum_{k \in |\mathcal{K}_{\mathbb{V}}| \text{s.t.} N_k > 0} N_k \left[\log \hat{f}_K(k) - \log N_k / N\right] \sim \chi^2 \left(|\mathcal{K}_{\mathbb{V}}| - 1 - |\theta|\right) \quad (9)$$

In practice and since $\hat{f}_K$ is independent of each other, we quantize the observations to a fixed number of groups of approximate equal size—in our experiments we use ten groups—to reduce spurious effects from low interaction volumes. The LR-test can be very sensitive to expected probabilities very close to 0.

4 Empirical Evaluation

4.1 Simple Pair Model

We first consider a simple evolutionary model—the *simple pair model*—that assumes that those who share a relationship meet at a different rate than those who do not. We assume that $\mathcal{E}$ is distributed according to the Gilbert random graph model which assumes that for each pair of nodes $v_i, v_j \in \mathbb{V}$ there is a probability p that they form an edge $e_{i,j} = \{v_i, v_j \mid i < j\}$. Let E be a realization of the random graph and let A be the associated $|\mathbb{V}| \times |\mathbb{V}|$ adjacency matrix such that $a_{ij} = 1$ if $e_{i,j} \in E$ and 0 otherwise. There is a one-to-one mapping between the set E and A, thus the log-probability mass function is equal to:

$$\log f_{\mathcal{E}} (E \mid \theta) = \sum_{i<j} a_{i,j} \log p + \sum_{i<j} (1 - a_{i,j}) \log(1-p) \quad (10)$$

The interaction space is the set of all 2-permutations of $\mathbb{V}$, thus $|\mathcal{K}_{\mathbb{V}}| = \binom{|\mathbb{V}|}{2}$. Let $\mathcal{N}_{i,j}(t)$ denote the random counting measure that represent the number of encounters between i and j during the period $[0,t)$ and $\lambda_1 \neq \lambda_2$, we assume that $\mathcal{N}_{i,j}(t) \sim \text{Poisson}(\lambda_1 t)$ if $e_{i,j} \in E$ and $\mathcal{N}_{i,j}(t) \sim \text{Poisson}(\lambda_2 t)$ otherwise.

The simple pair model is a superposition of $|\mathcal{K}_{\mathbb{V}}|$ independent homogeneous Poisson processes. We assume that all constituting processes are simple, *i.e.* $\Pr(\mathcal{N}_{i,j}\{t\} \geq 2) = 0$, which implies that the ground process itself is also simple whenever $|\mathcal{K}_{\mathbb{V}}|$ is finite[1]. The rate of the ground process is the weighted average of the different encounter rates:

$$\lambda_g^*(t \mid E) = \lambda_g(t \mid E) = |E|\,\lambda_1 + (|\mathcal{K}_{\mathbb{V}}| - |E|)\,\lambda_2 = \lambda_1 \sum_{i<j} a_{i,j} + \lambda_2 \sum_{i<j} (1 - a_{i,j}) \tag{11}$$

Encounter sampling follows a two-step procedure. First, we determine whether to sample an encounter from the set of pairs of nodes connected by a latent edge or from the remaining pairs. Second, we sample from the selected set uniformly at random. Therefore, the probability distribution over the marks follows a categorical distribution:

$$\begin{aligned} f_K^*(k_{i,j} \mid E) &= f_K(k_{i,j} \mid t, E) \\ &= \mathbb{1}(e_{i,j} \in E)\frac{|E|\,\lambda_1}{\lambda_g |E|} + \mathbb{1}(e_{i,j} \notin E)\frac{(|\mathcal{K}_{\mathbb{V}}| - |E|)\,\lambda_2}{\lambda_g(|\mathcal{K}_{\mathbb{V}}| - |E|)} = a_{i,j}\frac{\lambda_1}{\lambda_g} + (1 - a_{i,j})\frac{\lambda_2}{\lambda_g} \end{aligned} \tag{12}$$

To fit the *simple pair model* to observed data, we again follow the procedure detailed in Subsect. 3.3. First, note that $\theta = (\lambda_1, \lambda_2, p)^\top$. As detailed in Appendix A.3, we obtain the following expressions for updating θ during the maximisation step:

$$\hat{\lambda}_1 = \frac{\sum_{i<j} \hat{f}_{\mathcal{E}}(a_{i,j}) N_{i,j}}{T \sum_{i<j} \hat{f}_{\mathcal{E}}(a_{i,j})}, \; \hat{\lambda}_2 = \frac{\sum_{i<j} (1 - \hat{f}_{\mathcal{E}}(a_{i,j})) N_{i,j}}{T \sum_{i<j} (1 - \hat{f}_{\mathcal{E}}(a_{i,j}))}, \; \hat{p} = \frac{\sum_{i<j} \hat{f}_{\mathcal{E}}(a_{i,j})}{\binom{|\mathbb{V}|}{2}} \tag{13}$$

The expressions for iterating $\hat{\lambda}_1$ and $\hat{\lambda}_2$ above resemble in large part the maximum likelihood estimator for the homogeneous Poisson model. It pools all the encounters together and estimate the intensity as the ratio between the total accumulated time and the total number of encounters. The key difference is that the present estimator weights each encounter according to the probability that each pair is connected. Similarly, we find that the expression for iterating $\hat{p}$ resembles the estimator for the Gilbert random graph model.

In the expectation step, we substitute our estimates into Eq. 7 to update the value of $\hat{f}_{\mathcal{E}}$—see derivation details in Appendix A.3:

$$\hat{f}_{\mathcal{E}}(A) = \prod_{i<j} \left[\hat{f}_{\mathcal{E}}(a_{i,j})\right]^{a_{i,j}} \left[1 - \hat{f}_{\mathcal{E}}(a_{i,j})\right]^{(1-a_{i,j})} \tag{14}$$

[1] A Poisson process will always be simple if its intensity measure λ is diffuse, *i.e.* $\forall t \in \mathbb{X}, \lambda\{t\} = 0$. See proposition 6.9 in [8]. When $|\mathcal{K}_{\mathbb{V}}|$ is finite, the ground intensity is also diffuse, since $\lambda_g\{t\} = |E|\,\lambda_1\{t\} + (|\mathcal{K}_{\mathbb{V}}| - |E|)\,\lambda_2\{t\} = |E|\,0 + (|\mathcal{K}_{\mathbb{V}}| - |E|)\,0 = 0$.

where:

$$\hat{f}_{\mathcal{E}}\left(a_{i,j}\right) = \frac{\left(\hat{\lambda}_1^{N_{i,j}} \exp(-T\hat{\lambda}_1)\,\hat{p}\right)}{\left(\hat{\lambda}_1^{N_{i,j}} \exp(-T\hat{\lambda}_1)\,\hat{p}\right) + \left(\hat{\lambda}_2^{N_{i,j}} \exp(-T\hat{\lambda}_2)\,(1-\hat{p})\right)} \tag{15}$$

The expressions for iterating $\hat{\theta}$ and $\hat{f}_{\mathcal{E}}$ resemble those obtained in [15]. If the encounters indeed follow a homogeneous Poisson process, the characterization proposed in [15] is enough to determine the whole process. In essence, the author models the *avoidance function* $\Pr(\mathcal{N}(A) = 0)$ on a suitably rich class of Borel sets. As proven in the Rényi-Mönch Theorem 9.2.XII in [3] this is enough to determine the distribution of the whole process. The advantage of using the *avoidance function* is its simplicity.

4.2 Synthetic Data

We simulate the simple pair model. First, we simulate a Gilbert random graph with $p = 0.2$ and $|\mathbb{V}| = 100$. The network degree histogram is depicted in Fig. 1 (i). With only 100 nodes, the size of the encounter space is equal to $|\mathcal{K}_{\mathbb{V}}| = 4,950$ pairs which illustrates the rapid growth of the encounter space relative to the number of nodes. In our experiments the encounter rate of pairs that share a latent connection is held fixed, $\lambda_1 = 0.25$, meaning that an encounter is expected every 4 units of time. We experiment with three different rates for those not connected which entail a high ($\lambda_2 = 0.01$), medium ($\lambda_2 = 0.1$) and low ($\lambda_2 = 0.15$) degree of separation between the total number of encounters for those connected and disconnected. Figure 1(a)–(c) depicts the histogram of the total number of encounters at the end of the simulation for each experiment. We run the simulation for 100 units of time ($T = 100$).

We estimate the parameters of each experiment using the expectation-maximisation algorithm described in Sect. 4.1. The estimated values are available in the first 3 rows of Table 1. Column MSE refers to the mean-squared error between true link $a_{i,j}$ and the posterior $f_{\hat{\mathcal{E}}}(a_{ij})$ (lower is better), Λ_g p-value refers to the goodness-of-fit test of the compressed ground process (lower is better), LR-test quantized refers to the LR-test for the compressed mark distribution (lower is better). In all the cases, the parameters λ_1, λ_2 and p are recovered with high accuracy. Our main interest is in recovering the latent network. The mean-squared error by construction ranges between 0 and 1. A random guess such that all entries are equal to 0.5 always yields a mean-squared error equal to 0.25. When the rate of encounter for those who do not share a latent relation is very low ($\lambda_2 = 0.01$), the mean-squared error is very close to 0, the mean-squared error rises to 0.017 and 0.065 as the degree of mixing increases. Figure 1(d) depicts the posterior probability of latent relation given the data $\hat{f}_{\mathcal{E}}(a_{i,j})$ for different λ_2. There is a sharp phase transition on the posterior probability between those connected and disconnected when the degree of separation is high ($\lambda_2 = 0.01$). As the degree of mixing increases, the model increasingly struggle to separate the data. The posterior probability becomes less sharp when $\lambda_2 = 0.15$ with a larger number of observations falling on the transition between 0 and 1.

We conduct an ablation test to understand the impact of the prior p on model estimates. Results are shown in Fig. 1(e)–(f) which depict, respectively, the mean-squared error (lower is better) between the true link $a_{i,j}$ and the posterior $f_{\hat{\mathcal{E}}}(a_{i,j})$, and, the LR-ratio quantized test statistic for the compressed mark distribution (lower is better). The mean-squared error is the lowest around the true p. The prior tends to play a more important role when the relative difference in the rates of encounter decreases. We also see a similar effect for the quantized LR-test statistics. In summary, we find that the prior on the network model tends to be more important when the rates of encounter between the groups are closer. To understand, the role of the relative difference between λ_2 and λ_1 we conduct an ablation study in which we vary λ_2 from 0.25 to 0.01. Results are depicted in Fig. 1(g)–(h), which depict, respectively, the mean-squared error, and the LR-quantized test statistic. Subfigure (g) shows that the mean-squared error decreases sharply as the relative difference between the rates of encounter increases. Although the quantized LR-test statistic in Subfigure (h) declines fast, it quickly climbs up again. This is due to the fact that the test statistic tends to magnify errors when the expected probability of encounter is low.

4.3 Revisiting MIT Reality Mining

We evaluate the performance of our model on the MIT Reality Mining data set [4] with parameters $T = 9$, $|\mathbb{V}| = 71$, $|\mathcal{K}_{\mathbb{V}}| = 2{,}485$ and compare the results with those from [14,15]. Estimates are listed in the last two rows of Table 1, symbols "-" and "*" denote, respectively, that the true network is not available in the data (since it is latent), and, that the item is not computable as is not part of the model. The results obtained from our Poisson model are very similar to that of the Bernoulli model. Figure 2(a) depicts the posterior probability of latent relation given the data $\hat{f}_{\mathcal{E}}(a_{ij})$ for the Bernoulli model [15] and the Poisson model. The advantage of our proposed methodology is that it allow us to further test whether the Poisson model is a good candidate for the data. Figure 2(b) shows the scatter plot of the estimated compressed ground process along with its expected value and the 95% confidence interval. Note that all points lie completely inside the confidence interval. Moreover, the quantized LR-test in Table 1 also falls within the same values of our simulation. Thus, the simple pair model is indeed a good candidate for the data.

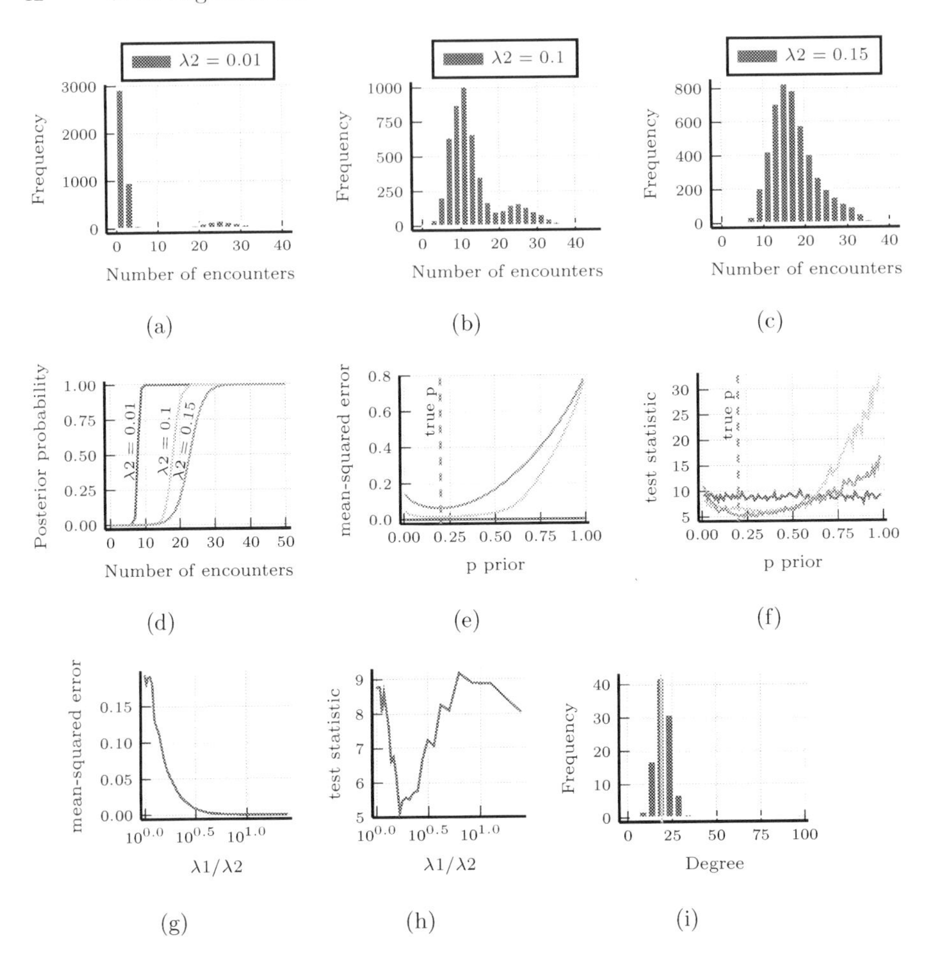

Fig. 1. Simple pair model simulation, description in Sect. 4.2.

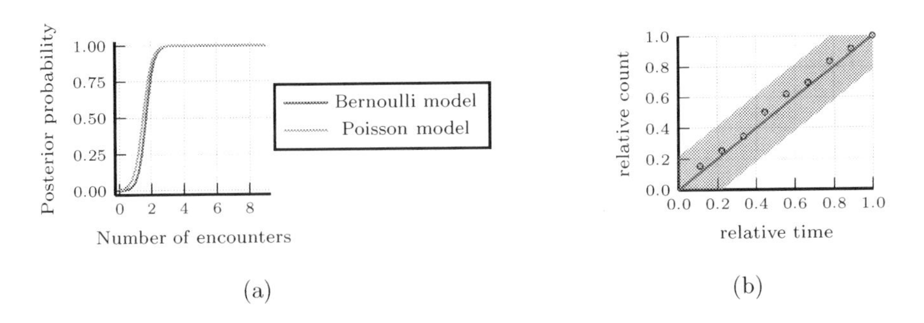

Fig. 2. MIT reality mining data set estimation, description in Sect. 4.3.

Table 1. Parameter estimates and test statistics, description in Sects. 4.2–4.3.

Data	N obs	$\hat{\lambda}_1/\hat{\alpha}$	$\hat{\lambda}_2/\hat{\beta}$	$\hat{p}$	MSE	Λ_g p-value	LR-test quantized
sim., $\lambda_2 = 0.01$	27,839	0.25	0.01	0.19	0.000	0.479	9.017
sim., $\lambda_2 = 0.10$	64,144	0.25	0.10	0.19	0.017	0.696	6.485
sim., $\lambda_2 = 0.15$	84,503	0.25	0.15	0.18	0.065	0.807	5.641
MIT, Bernoulli model	1,113	0.41	0.01	0.10	-	*	*
MIT, Poisson model	1,113	0.38	0.01	0.10	-	0.568	5.535

5 Conclusion

We proposed a latent relational point process framework for reconstructing networks from discrete event data. The novel framework comprises the latent relational point process model, an algorithm for fitting the model to data, and goodness-of-fit tests to quantify the model's suitability to the data. The latent relational point process model specialises Newman's model to point processes of latent relations. Using a variational argument, we devised a new, more convenient expectation-maximisation algorithm that uses the log-likelihood function of evolutionary point processes. The goodness-of-fit tests leveraged a new quantization strategy for the multinomial test of the estimated distribution of interaction labels. We empirically evaluated the model's performance in a synthetic data set, demonstrating the effectiveness of our expectation-maximisation algorithm, and showed that our framework agrees with the MIT Reality Sensing data set.

We are now working on several extensions and applications of the latent relational point process framework. The relational point process model can be extended to self-exciting point processes [10], time-varying co-evolution [5], higher dimension spaces, and hypergraphs to capture complex phenomena such as the various social mechanisms underlying human mobility. The expectation-maximisation algorithm can be further improved by incorporating Monte Carlo approximation when closed-form solutions are unavailable and automatic model discovery [13].

Acknowledgement. This research is partially supported by the National Research Foundation, Prime Minister's Office, Singapore, under its Campus for Research Excellence and Technological Enterprise (CREATE) programme as part of the programme DesCartes and by the Ministry of Education, Singapore, under its Academic Research Fund Tier 2 grant call (Award ref: MOE-T2EP50120-0019). Any opinions, findings and conclusions or recommendations expressed in this material are those of the authors and do not reflect the views of the National Research Foundation or of the Ministry of Education, Singapore.

A Appendix

A.1 The Lower Bound of the Posterior

The objective of this section is to derive the lower bound in Eq. 5, using the duality formula from [9]:

$$
\begin{aligned}
&\log \Pr(\theta \mid H_{T^-}) \\
&= \log\left[\int_{\mathbb{E}} \mathcal{L}(H_{T^-} \mid E, \theta)\, d\mu_{\mathcal{E}}(E \mid \theta)\right] + \log \Pr(\theta) - \log \Pr(H_{T^-}) \\
&\geq \int_{\mathbb{E}} \ell(H_{T^-} \mid E, \theta)\, d\hat{\mu}_{\mathcal{E}}(E) - \mathrm{KL}(\hat{\mu}_{\mathcal{E}} \| \mu_{\mathcal{E}}) + \log \Pr(\theta) - \log \Pr(H_{T^-}) \qquad (16)\\
&= \int_{\mathbb{E}} \ell(H_{T^-} \mid E, \theta) \hat{f}_{\mathcal{E}}(E)\, d\iota(\epsilon) - \int_{\mathbb{E}} \log\left(\frac{\hat{f}_{\mathcal{E}}(E)}{f_{\mathcal{E}}(E \mid \theta)}\right) \hat{f}_{\mathcal{E}}(E)\, d\iota(\epsilon) \\
&\quad + \log \Pr(\theta) - \log \Pr(H_{T^-})
\end{aligned}
$$

A.2 The Surrogate is the Posterior Distribution

The objective of this section is to show that the surrogate distribution $\hat{f}_{\mathcal{E}}(E)$ in Eq. 7 is the posterior $\Pr(E \mid H_{T^-}, \theta)$. Applying Bayes theorem to Eq. 7 we have that it is equal to:

$$
\frac{\frac{\Pr(E,\theta \mid H_{T^-}) \Pr(H_{T^-})}{f_{\mathcal{E}}(E \mid \theta) \Pr(\theta)} f_{\mathcal{E}}(E \mid \theta)}{\int_{\mathbb{E}} \frac{\Pr(E,\theta \mid H_{T^-}) \Pr(H_{T^-})}{f_{\mathcal{E}}(E \mid \theta) \Pr(\theta)} f_{\mathcal{E}}(E \mid \theta)\, d\iota(\epsilon)} = \frac{\Pr(E, \theta \mid H_{T^-})}{\Pr(\theta \mid H_{T^-})} = \Pr(E \mid H_{T^-}, \theta) \qquad (17)
$$

A.3 Simple-Pair Model: Gilbert Graph

We obtain the conditional log-likelihood of the observed data by plugging Eqs. 11 and 12 into Eq. 2 and taking the logarithm:

$$
\begin{aligned}
\ell(H_{T^-} \mid A, \theta) &= \sum_{n=1}^{N} [\log \lambda_g + \log(a_n \lambda_1 + (1 - a_n)\lambda_2) - \log \lambda)] - \int_0^T \lambda_g du \\
&= \sum_{i<j} [a_{i,j}(N_{i,j} \log \lambda_1 - T\lambda_1) + (1 - a_{i,j})(N_{i,j} \log \lambda_2 - T\lambda_2)]
\end{aligned} \qquad (18)
$$

Above, we are able to split the log because either $a_n = 1$ or 0. Also note, $\sum_{n=1}^{N} = \sum_{i<j} N_{i,j}$, $|E| = \sum_{i<j} a_{i,j}$ and $|\mathcal{K}_{\mathbb{V}}| - |E| = \sum_{i<j}(1 - a_{i,j})$.

By plugging Eqs. 10 and 18 into Eq. 6 and assuming uniform priors such that $\Pr(\theta)$ is a constatnt, we find expressions for the maximisation step:

$$\begin{cases} \sum_{A}\left[\sum_{i<j} a_{i,j}\left(\frac{N_{i,j}}{\hat{\lambda}_1}-T\right)\right]\hat{f}_{\mathcal{E}}(A)+0=0 \\ \sum_{A}\left[\sum_{i<j}(1-a_{i,j})\left(\frac{N_{i,j}}{\hat{\lambda}_2}-T\right)\right]\hat{f}_{\mathcal{E}}(A)+0=0 \\ \sum_{A}\left[\frac{\sum_{i<j} a_{i,j}}{p}-\frac{\sum_{i<j}(1-a_{i,j})}{1-p}\right]\hat{f}_{\mathcal{E}}(A)+0=0 \end{cases} \tag{19}$$

Let $\hat{f}_{\mathcal{E}}(a_{i,j})=\sum_A a_{i,j}\hat{f}_{\mathcal{E}}(A)=\Pr(a_{i,j}=1 \mid H_{T^-},\theta)$, then we can simplify the first equality in Eq. 19 to find an expression for updating λ_1:

$$\sum_{i<j}\left(\frac{N_{i,j}}{\hat{\lambda}_1}-T\right)\sum_{A} a_{i,j}\hat{f}_{\mathcal{E}}(A)=0 \iff \hat{\lambda}_1=\frac{\sum_{i<j}\hat{f}_{\mathcal{E}}(a_{i,j})N_{i,j}}{T\sum_{i<j}\hat{f}_{\mathcal{E}}(a_{i,j})} \tag{20}$$

Likewise, we find the remaining estimates of Eq. 13—note that $\binom{|\mathbb{V}|}{2}=\sum_{i<j}1$.

Finally, we obtain Eq. 14 by plugging the parameter estimates of Eq. 13, the network prior from Eq. 10 and the likelihood from Eq. 18 into Eq. 7:

$$\begin{aligned} &\hat{f}_{\mathcal{E}}(A)= \\ &=\frac{\prod_{i<j}\left(\hat{\lambda}_1^{N_{i,j}}\exp(-T\hat{\lambda}_1)\,\hat{p}\right)^{a_{i,j}}\left(\hat{\lambda}_2^{N_{i,j}}\exp(-T\hat{\lambda}_2)\,(1-\hat{p})\right)^{(1-a_{i,j})}}{\sum_A\prod_{i<j}\left(\hat{\lambda}_1^{N_{i,j}}\exp(-T\hat{\lambda}_1)\,\hat{p}\right)^{a_{i,j}}\left(\hat{\lambda}_2^{N_{i,j}}\exp(-T\hat{\lambda}_2)\,(1-\hat{p})\right)^{(1-a_{i,j})}} \\ &=\prod_{i<j}\frac{\left(\hat{\lambda}_1^{N_{i,j}}\exp(-T\hat{\lambda}_1)\,\hat{p}\right)^{a_{i,j}}\left(\hat{\lambda}_2^{N_{i,j}}\exp(-T\hat{\lambda}_2)\,(1-\hat{p})\right)^{(1-a_{i,j})}}{\left(\hat{\lambda}_1^{N_{i,j}}\exp(-T\hat{\lambda}_1)\,\hat{p}\right)+\left(\hat{\lambda}_2^{N_{i,j}}\exp(-T\hat{\lambda}_2)\,(1-\hat{p})\right)} \\ &=\prod_{i<j}\left[\hat{f}_{\mathcal{E}}(a_{i,j})\right]^{a_{i,j}}\left[1-\hat{f}_{\mathcal{E}}(a_{i,j})\right]^{(1-a_{i,j})} \end{aligned} \tag{21}$$

References

1. Brugere, I., Gallagher, B., Berger-Wolf, T.Y.: Network structure inference, a survey: motivations, methods, and applications. ACM Comput. Surv. **51**(2), 1–39 (2018). https://doi.org/10.1145/3154524
2. Daley, D.J., Vere-Jones, D.: An Introduction to the Theory of Point Processes: Volume I: Elementary Theory and Methods. Probability and Its Applications, An Introduction to the Theory of Point Processes, 2nd edn. Springer, New York (2003). https://doi.org/10.1007/b97277
3. Daley, D.J., Vere-Jones, D.: An Introduction to the Theory of Point Processes: Volume II: General Theory and Structure, 2nd edn. Springer, New York (2007)

4. Eagle, N., (Sandy) Pentland, A.: Reality mining: sensing complex social systems. Pers. Ubiquitous Comput. **10**(4), 255–268 (2006). https://doi.org/10.1007/s00779-005-0046-3
5. Farajtabar, M., Wang, Y., Gomez-Rodriguez, M., Li, S., Zha, H., Song, L.: COEVOLVE: a joint point process model for information diffusion and network evolution. J. Mach. Learn. Res. **18**(1) (2017)
6. Kolaczyk, E.D.: Statistical Analysis of Network Data: Methods and Models. Springer, New York; London (2009). https://doi.org/10.1007/978-0-387-88146-1
7. Krings, G., Karsai, M., Bernhardsson, S., Blondel, V.D., Saramäki, J.: Effects of time window size and placement on the structure of an aggregated communication network. EPJ Data Sci. **1**(1), 1–16 (2012). https://doi.org/10.1140/epjds4
8. Last, G., Penrose, M.: Lectures on the Poisson Process, 1st edn. Cambridge University Press, Cambridge (2017)
9. Lee, S.Y.: Gibbs sampler and coordinate ascent variational inference: a set-theoretical review. Commun. Stat. Theory Methods **51**(6), 154–1568 (2021). https://doi.org/10.1080/03610926.2021.1921214
10. Linderman, S.W., Wang, Y., Blei, D.M.: Bayesian inference for latent hawkes processes. In: Conference on Neural Information Processing Systems (NIPS 2017) (2017)
11. Lorch, L., et al.: Quantifying the Effects of Contact Tracing, Testing, and Containment Measures in the Presence of Infection Hotspots. arXiv:2004.07641 [physics, q-bio, stat], October 2020. http://arxiv.org/abs/2004.07641
12. Masuda, N., Takaguchi, T., Sato, N., Yano, K.: Self-exciting point process modeling of conversation event sequences. In: Holme, P., Saramäki, J. (eds.) Temporal Networks. Springer, Heidelberg (2013). https://doi.org/10.1007/978-3-642-36461-7_12
13. Mei, H., Eisner, J.M.: The Neural Hawkes Process: A Neurally Self-Modulating Multivariate Point Process (2017)
14. Newman, M.E.J.: Estimating network structure from unreliable measurements. Phys. Rev. E **98**(6), 062321 (2018). https://doi.org/10.1103/PhysRevE.98.062321
15. Newman, M.E.J.: Network structure from rich but noisy data. Nat. Phys. **14**(6), 542–545 (2018). https://doi.org/10.1038/s41567-018-0076-1
16. Pernice, V., Staude, B., Cardanobile, S., Rotter, S.: How structure determines correlations in neuronal networks. PLOS Comput. Biol. **7**(5), e1002059 (2011). https://doi.org/10.1371/journal.pcbi.1002059
17. Ribeiro, B., Perra, N., Baronchelli, A.: Quantifying the effect of temporal resolution on time-varying networks. Sci. Rep. **3**(1), 1–5 (2013). https://doi.org/10.1038/srep03006
18. Robins, G., Pattison, P., Kalish, Y., Lusher, D.: An introduction to exponential random graph (p*) models for social networks. Soc. Netw. **29**(2), 173–191 (2007). https://doi.org/10.1016/j.socnet.2006.08.002
19. Young, J.G., Cantwell, G.T., Newman, M.E.J.: Bayesian inference of network structure from unreliable data. J. Complex Netw. **8**(6), cnaa046 (2021). https://doi.org/10.1093/comnet/cnaa046
20. Young, J.G., Valdovinos, F.S., Newman, M.E.J.: Reconstruction of plant-pollinator networks from observational data. Nat. Commun. **12**(1), 1–12 (2021). https://doi.org/10.1038/s41467-021-24149-x

InTrans: Fast Incremental Transformer for Time Series Data Prediction

Savong Bou[1(✉)], Toshiyuki Amagasa[1], and Hiroyuki Kitagawa[2]

[1] Center for Computational Sciences, University of Tsukuba, Tsukuba, Japan
{savong-hashimoto,amagasa}@cs.tsukuba.ac.jp
[2] International Institute for Integrative Sleep Medicine,
University of Tsukuba, Tsukuba, Japan
kitagawa@cs.tsukuba.ac.jp

Abstract. Predicting time-series data is useful in many applications, such as natural disaster prevention system, weather forecast, traffic control system, etc. Time-series forecasting has been extensively studied. Many existing forecasting models tend to perform well when predicting short sequence time-series. However, their performances greatly degrade when dealing with the long one. Recently, more dedicated research has been done for this direction, and Informer is currently the most efficient predicting model. The main drawback of Informer is the inability to incrementally learn. This paper proposes an incremental Transformer, called *InTrans*, to address the above bottleneck by reducing the training/predicting time of Informer. The time complexities of InTrans comparing to the Informer are: (1) $O(S)$ vs $O(L)$ for positional and temporal embedding, (2) $O((S+k-1)*k)$ vs $O(L*k)$ for value embedding, and (3) $O((S+k-1)*d_{dim})$ vs $O(L*d_{dim})$ for the computation of Query/Key/Value, where L is the length of the input; k is the kernel size; d_{dim} is the number of dimensions; and S is the length of the non-overlapping part of the input that is usually significantly smaller than L. Therefore, InTrans could greatly improve both training and predicting speed over the state-of-the-art model, Informer. Extensive experiments have shown that InTrans is about 26% faster than Informer for both short sequence and long sequence time-series prediction.

Keywords: Incremental learning · Transformer · Time-series forecasting

1 Introduction

Planning for the best of the un-foreseeable future is crucial in business, education, science, and other fields. Time-series forecasting (TF) is one of the core operations in such planning and has been extensively used in many applications, such as weather forecast, traffic control system, disaster prevention system,

S. Bou—Due to name change, Savong Bou is now known as Takehiko Hashimoto.

C. Strauss et al. (Eds.): DEXA 2022, LNCS 13427, pp. 47–61, 2022.
https://doi.org/10.1007/978-3-031-12426-6_4

stock market forecasting, and other planning-related applications [8,10]. TF involves in learning the experience and knowledge from past and current data, then uses the learnt knowledge to predict the future time-series data.

In general, TF methods can be categorized into two types: (1) short sequence time-series forecasting (SSTF) and (2) long sequence time-series forecasting (LSTF) [13]. SSTF is for forecasting the data in the near future, whereas LSTF is for predicting long sequence of data that will happen further away in the future from now. LSTF is usually more useful than SSTF in planning for the long future. Recently, there are strong demands for effective LSTF for long term planning, such as long business strategic planning, future natural disaster prevention systems, and prevention of future pandemics. However, high accurate prediction and fast training/predicting speed are harder to achieve for LSTF than for SSTF. To meet LSTF demand, the predicting model must have: (1) high accurate prediction and (2) fast and efficient learning and predicting capability. This paper focuses on addressing the later problem by attempting to speed up the learning/predicting time by achieving the same predicting accuracy as the state-of-the-art approach, which we will mention below.

LSTF has been extensively studied and there are a lot of different predicting models [1–3,6,7,9,11,13], such as Informer [13], Longformer [3], Reformer [6], and LSTMa [2]. Informer is currently the state-of-the-art model (SOTA). It utilizes probSparse self-attention mechanism to achieve higher predicting accuracy than the existing models. However, its main problem is its inability to train/predict in an incremental manner.

In TF model, the forecasting is usually done in a continuous manner. For example, a weather forecasting system predicts the weather in the next 24 h for every two minutes. Therefore, there are many cases in which both input and output are overlapping with each other from one training/testing sample to another, which is usually prepared by using a sliding window [4,5]. The example is in Fig. 1. We can see that both input and output are overlapping from one training/testing sample to another according to the defined slide, i.e., training sample i and training sample $i+1$. This paper focuses on TF where training/testing samples are overlapping with each other. When learning/predicting from training/testing samples that are overlapping with each other, existing models could not incrementally learn, so all parameters are redundantly and unnecessary computed for every training/testing sample.

The mentioned observation leads to the proposed Incremental Transformer, called *InTrans*. InTrans can incrementally learn/predict from the time-series data by computing and updating only the necessary parameters. The contributions of the paper are as follows: (1) we propose the incremental embedding mechanism for embedding positions, times and values of both input/output; and (2) we propose incremental self-attention by focusing on incrementally computing Query, Key, and Value.

The time complexities of InTrans comparing to the SOTA Informer are: (1) $O(S)$ vs $O(L)$ for positional and temporal embedding, (2) $O((S+k-1)*k)$ vs $O(L*k)$ for value embedding, and (3) $O((S+k-1)*d_{dim})$ vs $O(L*d_{dim})$ for the computation of Query/Key/Value, where L is the length of the input,

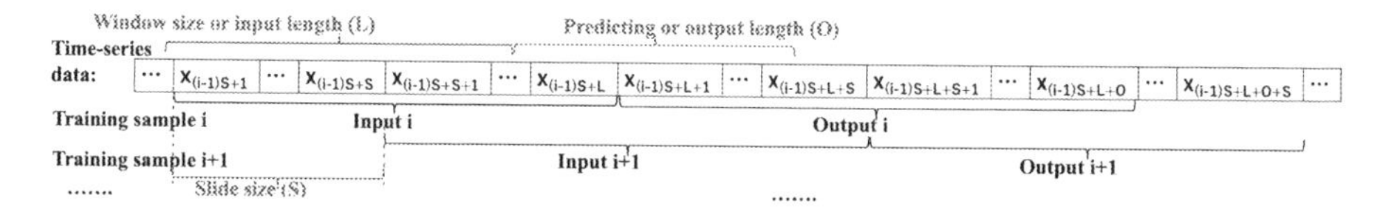

Fig. 1. Training sample prepared by using sliding window (Input length: L, Output length: O, Slide size: S).

k is the kernel size, d_{dim} is the number of dimensions and S is the length of the non-overlapping part of the input that is usually significantly smaller than L. Hence, InTrans significantly needs much less training/testing time than the SOTA Informer. The experimental results show that InTrans could reduce the learning/predicting time by about 26% compared to the SOTA Informer while achieving the same predicting accuracy.

2 Preliminaries

2.1 Time-Series Data

In this section, some definitions related to TF are explained. Time-series data (known as time-series in short form) is defined as follow:

Definition 1. *Time-series is a totally ordered sequence of data points* $< x_1, x_2, ..., x_t >$, $x_t \in \mathbb{R}^{d_x}$ *where* d_x *is the number of attributes and* $x_i \prec x_{i+1}$ *where* $\prec$ *is order relation.*

Definition 2. *Time-series forecasting (TF) is the use of model to predict future time-series based on the previously and currently observed time-series. Let L, O, and S be the size of input, output, and slide. The input and output for TF are defined as follows:*

- *Input i is* $X_i = x_{(i-1)S+1}, x_{(i-1)S+2}, ..., x_{(i-1)S+L},$
- *Output i is* $Y_i = x_{(i-1)S+L+1}, x_{(i-1)S+L+2}, ..., x_{(i-1)S+L+O},$

Definition 3. *Short sequence time-series forecasting (SSTF) and long sequence time-series forecasting (LSTF) are the TF problems where the lengths of output are short and long respectively.*

2.2 Informer

Informer [13] is implemented on top of Transformer [12], which is one of the most successful deep learning model. It adopts the encoder-decoder architecture to transform the input and output into the corresponding hidden state representations.

In Informer, time-series is embedded by using the three combined embeddings: (1) temporal, (2) positional, and (3) value embeddings non-incrementally.

Given an input i (or output), X_i (an $L * d_x$-matrix), its embedding is computed in Eq. 1.

$$E_i = T_i + P_i + V_i \tag{1}$$

Note here that T_i, P_i, and V_i are respectively the $L{*}d_{dim}$-matrix of the temporal, positional, and value embeddings of X_i where d_{dim} is the number of dimensions.

P_i is computed using Eq. 2 where pos is a positional information and $j \in 1, 2, ..., d_{dim}$. For example, $pos \in 1, 2, ..., L$ for the input i in Definition 2.

$$P = \begin{cases} sin(pos/10000^{\frac{2j}{d_{dim}}}), & \text{if } j\%2 = 0 \\ cos(pos/10000^{\frac{2j}{d_{dim}}}), & \text{if } j\%2 = 1 \end{cases} \tag{2}$$

Similarly, T_i is computed by interpreting the global temporal information, such as week, month, year, and holiday, from the local time stamp (Fig. 2). Then, such global temporal information is embedded using Eq. 2 to create T_i. Finally, V_i is computed using a convolutional layer Conv1d [13].

3 Incremental Transformer: InTrans

3.1 Motivation

In TF, the prediction is usually done in a continuous manner. Therefore, the input/output from one training/testing sample usually overlaps with the input/output from the subsequent training/testing samples. Look at the training samples i and $i+1$ in Fig. 1. The overlapping parts of both training samples consist of time-series $x_{(i-1)S+S+1}, ..., x_{(i-1)S+L}$ for the input and time-series $x_{(i-1)S+L+S+1}, ..., x_{(i-1)S+L+O}$ for the output. For the existing models for TF, after training/testing the model by the training/testing sample i, all parameters are computed. When training/testing the model by the subsequent training sample $i+1$, all parameters associated with the overlapping part of the training sample will be computed again. Such unnecessary processing is not useful but a big burden for the training/predicting process.

This observation leads us to propose the incremental Transformer, called *InTrans*, for efficient incremental learning/testing. Specifically, InTrans does not compute all related parameters associated with the overlapping part of the training/testing sample. Instead, the related parameters from the preceding training/testing sample i can be reused for the current training/testing sample $i+1$.

This paper focuses on TF where training/testing samples are overlapping with each other as explained in Fig. 1. The rest of the paper discusses about the input of the training sample, but same discussion also applies to: (1) the output of the training sample and (2) the input and output of the testing sample.

3.2 Overview

InTrans is based on the most recent Transformer-based model, Informer [13]. The overview of InTrans is in Fig. 3. For each training sample, we have input

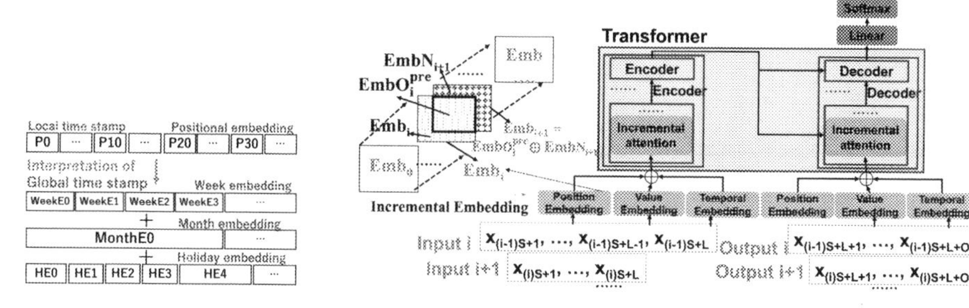

Fig. 2. Global temporal information.

Fig. 3. Overview of the proposed InTrans

and output. Both input and output are prepared by using a sliding window so that the training is done in a continuous manner. The input and output can be regarded as windows, and the slide indicates the time gap between each training sample. The input and output are embedded using position, value, and temporal embedding following the same procedure in the Informer paper [13]. Then, the embedding of the input and output is sent to the encoder and decoder respectively, and goes through other processes similar to those of the original Informer. We take the best advantages from the overlapping among the training data, and the proposal is on two points:

- Incremental Embedding: Efficient mechanism to incrementally compute the positional, temporal, and value embeddings of the input/output. Specifically, the positional, temporal, and value embeddings corresponding to the overlapping part are not recomputed and reused when embedding the subsequent input/output, which is explained in Sects. 3.3, 3.4, and 3.5.
- Incremental Self-attention: Query, Key, and Value are the key components for computing the self-attention. Similar to the incremental embedding, we propose efficient mechanism for incrementally computing Query, Key, and Value in Sect. 3.6.

3.3 Incremental Positional Embedding

Positions of records can be categorized into two groups: (1) absolute and (2) relative positions. For absolute position, each record has the same positional information regardless of residing in different training samples. The positional information can either be interpreted as position information or local time stamp.

The training samples are overlapping with each other. We divide the positional embedding into: (1) Po: Positional embedding of the overlapping part, and (2) Pn: Positional embedding of the non-overlapping part between two consecutive input. The example is in Fig. 4a. Positional embedding of input $i+1$ is computed using Eq. 3, where $\oplus$ is a concatenation operation.

$$P_{i+1} = Po_{i+1} \oplus Pn_{i+1} \quad (3)$$

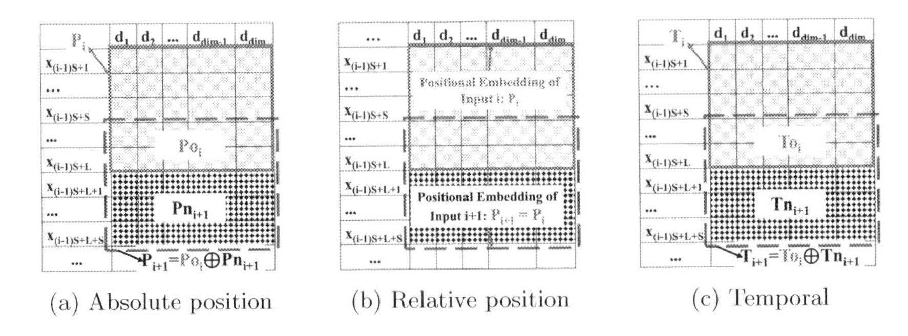

(a) Absolute position (b) Relative position (c) Temporal

Fig. 4. Incremental positional and temporal embedding.

Po_{i+1} and Pn_{i+1} are respectively the positional embeddings of input $i+1$ at the overlapping and non-overlapping parts between inputs i and $i+1$. To make the positional embedding incremental, Eq. 3 is rewritten to Eq. 4.

$$P_{i+1} = Po_i \oplus Pn_{i+1} \tag{4}$$

Po_i is the positional embedding of the overlapping part between inputs i and $i+1$ computed when training input i. We have Theorem 1.

Theorem 1. *P_{i+1} computed using Eqs. 4 and 3 are equal and are the same as the positional embedding fully computed using the input $i+1$.*

Proof: See Appendix A.

For relative position, we need a more complicated technique to compute. The positions of records residing on the overlapping part between two consecutive inputs are different. Look at Fig. 4b. The position of record $x_{(i-1)S+S+1}$ is $S+1$ in input i, but its position in input $i+1$ is 1. Therefore, conventional incremental computation cannot be applied to positional embedding. We observe that the position always starts from 1 to the length of each input, which has the same length. We propose Theorem 2, which means positional embedding is computed only once for the first training sample and can be reused throughout the whole training process.

Theorem 2. *For relative position, positional embeddings of all inputs are the same. We have $P_1 = P_2 = ... = P_i = P_{i+1} = ...$*

Proof: See Appendix B.

3.4 Incremental Temporal Embedding

Global temporal information such as week, month, year, and holiday can be interpreted from the local time stamps of the records (Fig. 2). Temporal embedding is done by using such global temporal information.

The example is shown in Fig. 4c. Therefore, temporal embedding can be incrementally computed using Eq. 5 in similar manner to positional embedding.

$$T_{i+1} = To_i \oplus Tn_{i+1} \tag{5}$$

To_i is the temporal embedding of the overlapping part between inputs i and $i+1$ computed when training input i and Tn_{i+1} is the temporal embedding of input $i+1$ at the non-overlapping part between inputs i and $i+1$. We have Theorem 3.

Theorem 3. *T_{i+1} computed using Eq. 5 is equal to temporal embedding fully computed using the input $i+1$.*

Proof: See Appendix C.

3.5 Incremental Value Embedding

The value embedding in Informer adopts Conv1d [13]. In order to make value embedding incremental, we need to make Conv1d incremental. Figure 5 shows the proposed incremental Conv1d. This example sets stride value to one and padding values are set so that input and its value embedding have the same length. However, similar discussion also applies to general cases with different stride and padding values. Note, Informer sets kernel width to three and stride to one by default. The values of the input are convolved by the kernel (e.g., width k), so the last $k-1$ records in the input are convolved using padding values, which cannot be reused for value embedding of the next input. The proposed incremental value embedding is computed using Eq. 6.

$$V_{i+1} = Vr_i \oplus Vc_{i+1} \tag{6}$$

Vr_i is a reusable subset of value embedding of the overlapping part, excluding the value embedding of the last $k-1$ records, between inputs i and $i+1$ after computing value embedding of input i. Vc_{i+1} is the value embedding that needs to fully compute for the input $i+1$. Vc_{i+1} contains values embedding of the last $k-1$ records of the overlapping part and the value embedding of the non-overlapping part between inputs i and $i+1$. We have Theorem 4.

Theorem 4. *V_{i+1} computed using Eq. 6 is equal to the value embedding fully computed using the input $i+1$.*

Proof: See Appendix D.

3.6 Incremental Self-Attention

Incremental Query, Key, and Value. In Transformer, Query, Key, and Value are computed as in Eq. 7 where W_{Q_i}, W_{K_i}, and W_{V_i} are their respective weights with the same size $(d_{dim} * d_{dim})$ and $Q_i/K_i/V_i$ is a $L * d_{dim}$-matrix. The example of incremental Query creation is in Fig. 6. Mini Batch Gradient Descent is

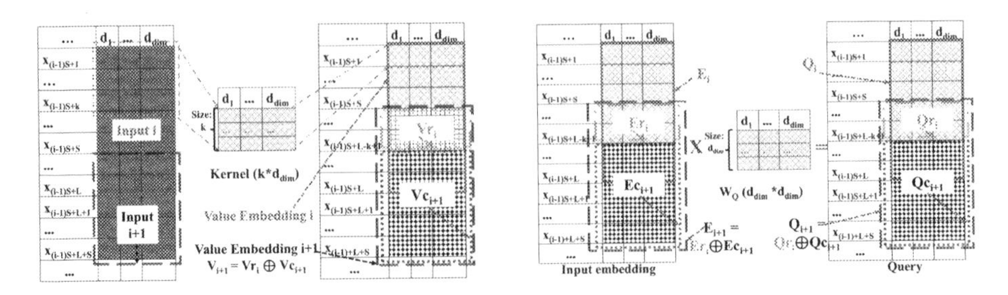

Fig. 5. Incremental value embedding

Fig. 6. Incremental query

adopted by Informer, so W_{Q_i} of all inputs in the same batch are the same, which is also true for W_{K_i} and W_{V_i}.

$$Q_i = E_i * W_{Q_i}, K_i = E_i * W_{K_i}, V_i = E_i * W_{V_i}. \tag{7}$$

The creation of Query is incrementally computed using Eq. 8 where Qr_i is the reusable computation from Query i at the overlapping part between Query i and $i+1$ excluding the last $k-1$ records and Qc_{i+1} are the remaining values that need to compute for Query $i+1$. Same process applies to incrementally create Key and Value. Theorem 5 is proposed.

$$Q_{i+1} = Qr_i \oplus Qc_{i+1} \tag{8}$$

Theorem 5. *Q_{i+1} computed using Eq. 8 is equal to the Query fully computed using the input embedding E_{i+1}. The same is also true to K_{i+1} and V_{i+1}.*

Proof: See Appendix E.

3.7 Prediction Accuracy

InTrans is implemented on top of Informer by adopting the proposed incremental computation of the temporal/positional/value embeddings and Query/Key/Value. We show that the prediction accuracy of InTrans is the same as that of Informer. We have Theorem 6.

Theorem 6. *Given the same training/validating/testing data and the same parameter settings, InTrans and Informer have the same prediction accuracy.*

Proof: See Appendix F.

3.8 Complexity Analysis

The time complexity of the proposed and other existing approaches are in Table 1 for each input. The time complexity for the output can be similarly obtained by replacing L by O in Table 1. The proofs of the time complexity of existing models can be referred to the original papers or Informer [13]. The complexity of InTrans is defined in Theorem 7.

Table 1. Computational complexity analysis

Approach	Embedding			Query/Key/Value
	Positional	Temporal	Value	
InTrans	$O(S)$	$O(S)$	$O((S+k-1)*k)$	$O((S+k-1)*d_{dim})$
Informer	$O(L)$	$O(L)$	$O(L*k)$	$O(L*d_{dim})$
Transformer	$O(L)$	$O(L)$	$O(L*k)$	$O(L*d_{dim})$
LogTrans	$O(L)$	$O(L)$	$O(L*k)$	$O(L*d_{dim})$
Reformer	$O(L)$	$O(L)$	$O(L*k)$	$O(L*d_{dim})$

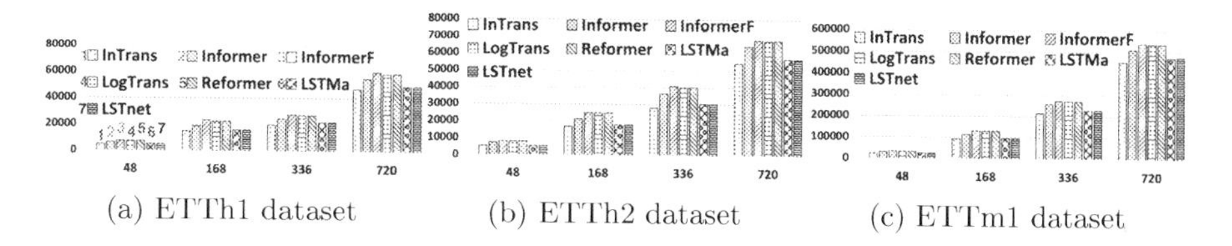

(a) ETTh1 dataset (b) ETTh2 dataset (c) ETTm1 dataset

Fig. 7. Training time (second) when increasing the length of input.

Theorem 7. *The time complexity of InTrans for each input (or output) is:*

- $O(S)$*: For positional or temporal embedding.*
- $O((S+k-1)*k)$*: For value embedding.*
- $O((S+k-1)*d_{dim})$*: For computing the Query or Key or Value.*

Where S*,* L*,* k*, and* d_{dim} *are the slide size, the window size or length of input sequence, the size of the kernel, and the number of dimensions respectively.*

Proof: See Appendix G. S is usually much smaller than L in many practices, so InTrans significantly needs less training time than Informer.

4 Experimental Evaluation

4.1 Dataset and Settings

The five publicly-released datasets from the Informer [13] paper were used: (1) ETTh1, (2) ETTh2, (3) ETThm1, (4) Weather, and (5) ECL.

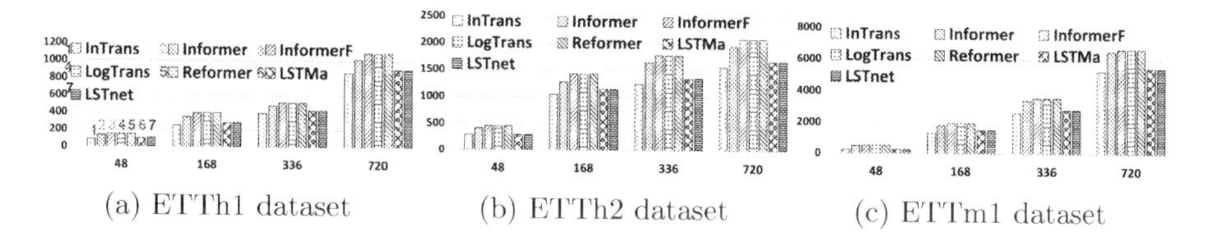

(a) ETTh1 dataset (b) ETTh2 dataset (c) ETTm1 dataset

Fig. 8. Testing time (second) when increasing the length of input.

ETT (Electricity Transformer Temperature)[1]: is the Electricity consumption in two villages in China for two years. Two separated datasets were prepared for different granularity: (1) 1-h level: (a) ETTh1, and (b) ETTh2; and (2) 15-min level: ETTm1. There are seven attributes describing oil temperature and power load features. The dataset was divided into 12 months for training, 4 months for validating, and 4 months for testing.

ECL (Electricity Consuming Load)[2]: consists of electricity consumption of 321 households. The dataset was converted into hourly two-year period. The ratio of training, validating, and testing is 15/3/4 months respectively.

Weather[3]: Contains data about weather from 1,600 locations in the U.S. for four years from 2010 to 2013. Each data point has 12 values: (1) The target value: Wet bulb, and (2) Eleven features. The gap between each data point is one hour. The dataset was divided into 28/10/10 months of training, validating, and testing respectively.

Section 3.7 has shown that InTrans could achieve the same predicting accuracy as the SOTA Informer. Since the contribution of this paper is to reduce the training/testing time of Informer, only experimental results about training/testing time are presented. The accuracy can be referred to the Informer [13].

4.2 Training and Testing Time Without GPU

This section measures the CPU running time of all approaches by running on an iMac PC (3.3 GHz 6 core Intel Core i5) without any GPU. The length of the input is increased from 48 to 720 records. The fixed parameters are batch size (32), Iteration (5), Epoch (6), and length of the output (168 records). The results for training time are in Figs. 8a, 8b, and 8c for ETTh1, ETTh2, and ETTm1 respectively. The results for testing time are in Figs. 8a, 8b, and 8c for ETTh1, ETTh2, and ETTm1 respectively. When the length of the input is increased, the training/testing time of all approaches also increases for all datasets due to heavier workload is added to the training/predicting process.

The proposed InTrans needs significantly less training/testing time than the SOTA Informer and other existing approaches due to the following reasons: (1) the training/testing sample is highly overlapping with each other; (2) Informer and other existing approaches could not incrementally compute the embedding and Query/Key/Value from the overlapping training/testing sample, so the temporal/positional/value embeddings and Query/Key/Value of the training/testing samples are redundantly and unnecessarily computed; and (3) the proposed InTrans makes use of the proposed incremental temporal/positional/value embeddings and Query/Key/Value for incremental computation from the overlapping training/testing sample. InTrans guarantees that the computation corresponding to the overlapping part of the training/testing samples can be totally reused for the subsequent training/testing. InTrans is implemented on top of Informer and is about 26% faster than Informer.

[1] https://github.com/zhouhaoyi/ETDataset.
[2] https://archive.ics.uci.edu/ml/datasets/ElectricityLoadDiagrams20112014.
[3] https://www.ncei.noaa.gov/data/local-climatological-data/.

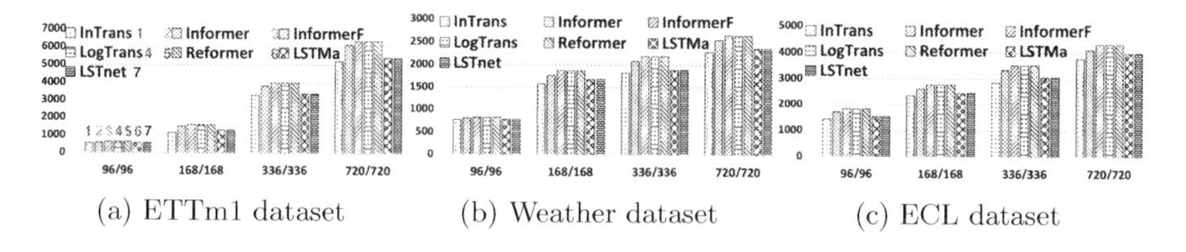

(a) ETTm1 dataset (b) Weather dataset (c) ECL dataset

Fig. 9. GPU: Training time (second) when increasing input/output length.

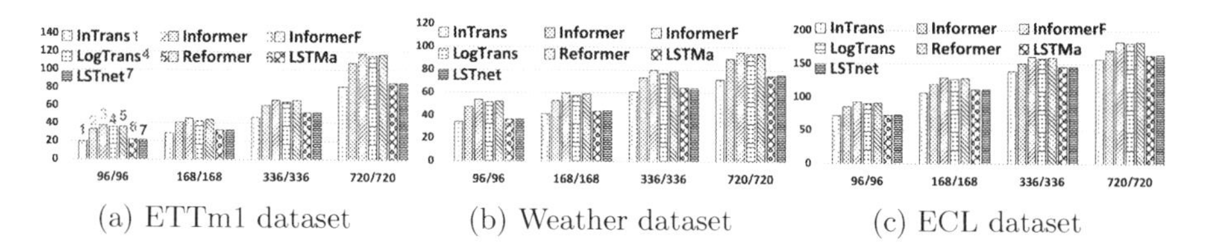

(a) ETTm1 dataset (b) Weather dataset (c) ECL dataset

Fig. 10. GPU: Testing time (second) when increasing input/output length.

4.3 Training and Testing Time Using a GPU

The experiment was done on an Ubuntu 20.04 server with a GPU (RTX A5000). The parameter setting is the same as that in Sect. 4.2 with the exception that the lengths of both input and output are increased at the same time. The results are in Figs. 9 and 10 for training and testing respectively. When the lengths of both input and output are increased, the training/testing time of all approaches also increases for all datasets due to heavier workloads. InTrans significantly uses less training/testing time than the SOTA Informer and other existing approaches for all datasets and parameters with similar reasons as explained in Sect. 4.2.

5 Related Works

TF has long been studied. Informer [13] is the current state-of-the-art predicting model for TF. Its main focus is LSTF. It uses encoder-decoder-based transformer. There are three main characteristics: (1) a ProbSparse self-attention, (2) the self-attention distilling, and (3) the generative style decoder. The ProbSparse self-attention feature allows each key to only attend to the dominant queries. The query is a sparse matrix and only contains the top-u dominant queries under the sparsity measurement. Informer uses Con1d to reduce the redundant combination of the values by privileging the superior ones with dominating features and making a focused self-attention feature map in the next layer. Thereby, the size of the output of each encoder and decoder layer is reduced by half. In addition, Informer adopts generative inference by concatenating the earlier set of time-series to the output to boost the learning capability of the model. InformerF is a variant of Informer by incorporating the canonical self-attention [12].

LogTrans [3] proposed to set the limit on the boundary of the self-attention of each token. The computation of self-attention is done against the defined subset of the whole sequence. LogTrans could speed up the time complexity of the model by sacrificing the accuracy rate. To achieve its best accuracy, the limit on the boundary of self-attention is not set in the experiment of this paper.

Reformer [6] introduces two techniques to improve the efficiency of Transformers. It introduces locality-sensitive hashing to improve the dot-product attention. Its argument is Softmax is dominated by the largest elements, so it only needs to focus on the keys that are closest to the queries. It uses locality-sensitive hashing to find the nearest neighbors among the keys.

Prophet [11] is part of Facebook forecasting at scale. Its prediction is done using an additive model that is based on regression with interpretable parameters. It can be adjusted by users with the domain knowledge about the time-series. The temporal trends are classified by year, week, day, season, and holiday.

ARIMA [1] and DeepAR [9] are auto-regressive recurrent network model. Other models are LSTnet [7] and LSTMa [2] that are based on Convolution Neural Network (CNN) and the Recurrent Neural Network (RNN) to learn the short term dependencies in time-series to discover the long term dependencies.

6 Conclusion

This paper proposes an incremental Transformer, called *InTrans*, that can efficiently learn/predict from overlapping training/testing sample. Two main features of InTrans are the incremental computation of: (1) Embedding and (2) Query/Key/Value. The proposed InTrans guarantees that the computation corresponding to the overlapping part of the training/testing samples can be totally reused for the subsequent training/testing. The time complexities of InTrans comparing to the SOTA Informer are: (1) $O(S)$ vs $O(L)$ for positional and temporal embedding, (2) $O((S+k-1)*k)$ vs $O(L*k)$ for value embedding, and (3) $O((S+k-1)*d_{dim})$ vs $O(L*d_{dim})$ for the computation of Query/Key/Value. S is usually much smaller than L in many practices, so InTrans significantly needs much less training/testing time than the SOTA Informer. Extensive experiments on various datasets show that the training/testing speed of InTrans is about 26% less than training/testing time of the SOTA Informer. For future works, we would explore the possibility to apply the incremental computation to other parts of Informer or Transformer at both forward and backward propagation.

Acknowledgements. This work was supported by University of Tsukuba Basic Research Support Program Type A, Japan Society for the Promotion of Science (JSPS) KAKENHI under Grant Number JP19H04114 and JP22H03694, the New Energy and Industrial Technology Development Organization (NEDO) Grant Number JPNP20006, and Japan Agency for Medical Research and Development (AMED) Grant Number JP21zf0127005.

Appendix

A Proof of Theorem 1

The inputs to the Eq. 2 are positional and dimensional information. The dimensional information is the same for all records in the training data. Assuming that we have two consecutive input i and $i+1$ of L records. The gap between the beginning of input i and $i+1$ is S records. The positional information of input i is $POS_i = [pos_{(i-1)S+1}, pos_{(i-1)S+2}, ..., pos_{(i-1)S+S}, pos_{(i-1)S+S+1}, ..., pos_{(i-1)S+L}]$, and the positional information of input $i+1$ is $POS_{i+1} = [pos_{(i-1)S+S+1}, pos_{(i-1)S+S+2}, ..., pos_{(i-1)S+L}, pos_{(i-1)S+L+1}, ..., pos_{(i-1)S+L+S}]$. We have

- $POS_i = POSn_i \oplus POSo_i$, and
- $POS_{i+1} = POSo_{i+1} \oplus POSn_{i+1}$, where
  - $POSn_i = [pos_{(i-1)S+1}, pos_{(i-1)S+2}, ..., pos_{(i-1)S+S}]$, and
  - $POSo_i = POSo_{i+1} = [pos_{(i-1)S+S+1}, ..., pos_{(i-1)S+L}]$
  - $POSn_{i+1} = [pos_{(i-1)S+L+1}, ..., pos_{(i-1)S+L+S}]$

$PosEncoding(POS)$ represents encoding the POS by Eq. 2. We have:

- $Pn_i = PosEncoding(POSn_i)$, and
- $Po_i = PosEncoding(POSo_i)$, so
- $P_i = Pn_i \oplus Po_i$.

Because $POSo_i = POSo_{i+1}$, then we have:

- $Pn_{i+1} = PosEncoding(POSn_{i+1})$, therefore
- $P_{i+1} = Po_{i+1} \oplus Pn_{i+1} = Po_i \oplus Pn_{i+1}$.

Theorem 1 is proven.

B Proof of Theorem 2

The notations in Appendix A are also used in this Section. For relative position, the positional information of all input/output is the same, so $POS_1 = POS_2 = ... = POS_i = POS_{i+1} = [pos_1, pos_2, ..., pos_L]$. Therefore, we have $P_1 = P_2 = ... = P_i = P_{i+1} = PosEncoding(POS_1)$. Theorem 2 is proven.

C Proof of Theorem 3

The temporal embedding takes temporal information, such as weak, month, and holiday, of the records as a basis for embedding. The temporal information of the same record does not change wrt different training samples. Therefore, the temporal embedding of all records belongs to the overlapping part between input i and $i+1$ is the same. The proof is similar to that of absolute positional embedding in Appendix A. Thus, $To_{i+1} = To_i$, so Theorem 3 is proved.

D Proof of Theorem 4

The notations in Appendix A are also used in this Section. We need to prove that $Vr_i = Vr_{i+1}$. When computing the embedding value of input i, the input i is convolved by a kernel (e.g., width k). Since the default stride is set to one, the resulting embedding values of records $x_{(i-1)S+1}$ to $x_{(i-1)S+L-(k-1)}$ are fully convolved without including the padding values. Similarly, the embedding values of records $x_{(i-1)S+S+1}$ to $x_{(i-1)S+L-(k-1)}$ are fully convolved. Therefore, the embedding values of records $x_{(i-1)S+S+1}$ to $x_{(i-1)S+L-(k-1)}$ are the same for both input i and $i+1$. Therefore, $Vr_i = Vr_{i+1}$, which proves Theorem 4.

E Proof of Theorem 5

Query, Key, and Value are computed as in Eq. 7. Such multiplication does not change the value distribution from the original input embedding. Therefore, Query, Key, and Value can be incrementally computed in similar manner to that of input embedding, which is proved in Theorems 3, 1, and 4. Therefore, Theorem 5 is proven.

F Proof of Theorem 6

InTrans is implemented on top of Informer by adopting the incremental computation of the temporal/positional/value embeddings and Query/Key/Value of the training sample. To prove that InTrans has the same predicting accuracy as that of Informer, we have to prove that the temporal/positional/value embeddings and Query/Key/Value incrementally computed by InTrans is the same as the temporal/positional/value embeddings and Query/Key/Value computed by Informer. Theorems 1, 2, 3, 4, 5 have proved that the embedding of the input and the Query/Key/Value incrementally computed by InTrans are the same as those non-incrementally computed by Informer. Theorem 3.7 is proven.

G Proof of Theorem 7

Equations 4 and 5 suggests that positional and temporal embedding can be incrementally computed, which is proved in Theorems 1 and 3. For each input $i+1$, the positional embedding (Po_i) and the temporal embedding (To_i) corresponding to the overlapping part between inputs i and $i+1$ computed when embedding the input i can be reused for input $i+1$. Therefore, only the positional embedding (Pn_{i+1}) and the temporal embedding (Tn_{i+1}) corresponding to the non-overlapping part between input i and $i+1$ need to be computed when embedding input $i+1$. The size of the non-overlapping part between input i and $i+1$ is represented by S, so the time complexity to compute the positional and temporal embedding of each input is $O(S)$.

Similarly, value embedding is done by using Conv1d. Equation 6 and Theorem 4 suggest that the value embedding of the overlapping part, excluding the value embedding of the last $k-1$ records, between input i and $i+1$ after computing value embedding of input i can be reused for embedding the input $i+1$. Therefore, the time complexity to compute the value embedding of each input is $O((S+k-1)*k)$.

Similar to value embedding, Eq. 6 and Theorem 4 suggest that the time complexity to compute Query or Key or Value is $O((S+k-1)*d_{dim})$. Theorem 7 is proved.

References

1. Ariyo, A.A., Adewumi, A.O., Ayo, C.K.: Stock price prediction using the arima model. In: 2014 UKSim-AMSS 16th International Conference on Computer Modelling and Simulation, pp. 106–112 (2014)
2. Bahdanau, D., Cho, K., Bengio, Y.: Neural machine translation by jointly learning to align and translate (2016)
3. Beltagy, I., Peters, M.E., Cohan, A.: Longformer: the long-document transformer. arXiv:2004.05150 (2020)
4. Bou, S., Kitagawa, H., Amagasa, T.: L-Bix: incremental sliding-window aggregation over data streams using linear bidirectional aggregating indexes. Knowl. Inf. Syst. **62**(8), 3107–3131 (2020)
5. Bou, S., Kitagawa, H., Amagasa, T.: Cpix: real-time analytics over out-of-order data streams by incremental sliding-window aggregation. IEEE Trans. Knowl. Data Eng. (2021). https://doi.org/10.1109/TKDE.2021.3054898
6. Kitaev, N., Kaiser, L., Levskaya, A.: Reformer: the efficient transformer. In: International Conference on Learning Representations (2020)
7. Lai, G., Chang, W.C., Yang, Y., Liu, H.: Modeling long- and short-term temporal patterns with deep neural networks (2018)
8. Park, H.J., Kim, Y., Kim, H.Y.: Stock market forecasting using a multi-task approach integrating long short-term memory and the random forest framework. Appl. Soft Comput. **114**, 108106 (2022)
9. Salinas, D., Flunkert, V., Gasthaus, J., Januschowski, T.: Deepar: probabilistic forecasting with autoregressive recurrent networks. Int. J. Forecast. **36**(3), 1181–1191 (2020)
10. Su, T., Pan, T., Chang, Y., Lin, S., Hao, M.: A hybrid fuzzy and k-nearest neighbor approach for debris flow disaster prevention. IEEE Access **10**, 21787–21797 (2022). https://doi.org/10.1109/ACCESS.2022.3152906
11. Taylor, S., Letham, B.: Forecasting at scale. Am. Stat. **72**, 37–45 (2018)
12. Vaswani, A., et al.: Attention is all you need. In: Guyon, I., et al. (eds.) Advances in Neural Information Processing Systems, vol. 30. Curran Associates, Inc. (2017)
13. Zhou, H., et al.: Informer: beyond efficient transformer for long sequence time-series forecasting. In: The Thirty-Fifth AAAI Conference on Artificial Intelligence, AAAI 2021, Virtual Conference, vol. 35, pp. 11106–11115. AAAI Press (2021)

A Knowledge-Driven Business Process Analysis Methodology

Michele Missikoff(✉)

Institute of Analysis of Systems and Informatics (IASI), CNR, Rome, Italy
Michele.missikoff@iasi.cnr.it

Abstract. Business Process Analysis (BPA) is a strategic activity, necessary for enterprises to model their business operations, especially in the context of information system development. This paper proposes a knowledge framework, referred to as BPA Canvas, primarily conceived to guide business people in building a BPA knowledge base. The resulting knowledge base is organized into eight sections where only the last one, the BPA ontology, requires specialist competences.

Keywords: Information system · Business process analysis · Business model canvas · Knowledge representation · Ontology

1 Introduction

Business process analysis (BPA) requires a thorough understanding of the enterprise reality and its effective modeling. The produced enterprise models represent a solid basis for Information system development, but also enterprise management, business process reengineering, and, among others, digital transformation. Nowadays, the instability of markets, supply chains, commodities prices, but also the continuous evolution of digital technologies, requires continuous transformations for enterprise to cope with the uncertain operating scenarios. Then, BPA, enterprise modeling and knowledge management are increasingly becoming strategic for the success, or even the survival, of enterprises.

In this paper we present a systematic framework for business process analysis, BPA Canvas, aimed at building an enterprise knowledge base. It has been designed to guide business people, without specific competences in knowledge management, to progressively construct various business models (Knowledge Artefacts), grouped into eight sections that form an enterprise Body of Knowledge (BOK). The results presented here start from, and represent an extension and improvement of, the work reported in [1] and are more extensively elaborated in [12].

2 Business Process Analysis Canvas

The proposed methodology is characterised by the following points: (i) ease of use for the business experts who are moved at the center of the stage; (ii) well defined guidelines,

C. Strauss et al. (Eds.): DEXA 2022, LNCS 13427, pp. 62–67, 2022.
https://doi.org/10.1007/978-3-031-12426-6_5

with a progressive business modeling, from informal to formal (inspired by some early ideas [2]); (iii) knowledge-driven approach, based on a solid formal grounding [3]. The points may sound contradictory, since rigid formality reduces the acceptance of business people. Our proposal tries to solve this contradiction.

The BPA Canvas is organised in eight knowledge sections that hold different kinds of knowledge artefacts, i.e., models that assume various forms. In particular, we have: (i) *free-form text*; (ii) *structured text* (bullet points, numbered lists, etc.); (iii) *tables*; (iv) *diagrams*; (v) a formal representation of the business domain by means of a *BP Ontology*. Figure 1 shows the layout of the BPA Canvas with its eight sections.

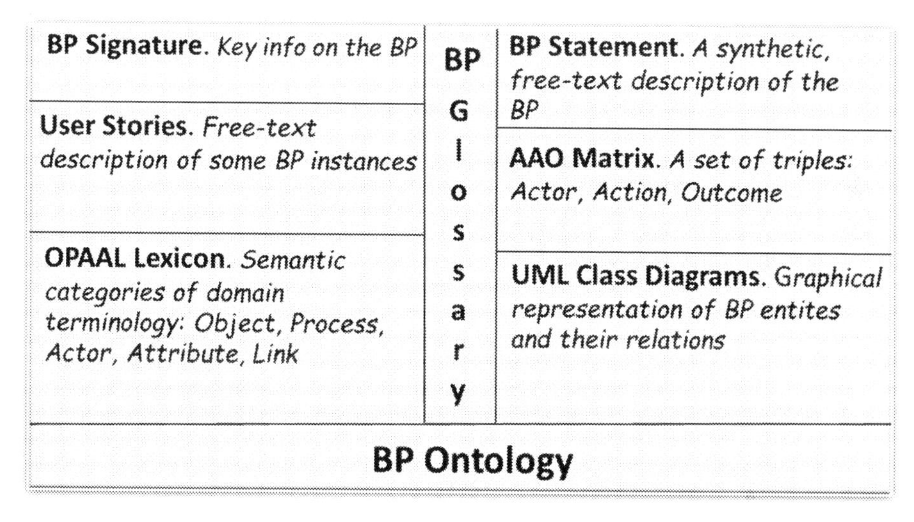

Fig. 1. BPA canvas layout

The methodology, illustrated in the next section using a simple example, suggests to start specifying the *BP Signature*, and then continue with *the BP Statement* and a number of *User Stories* (one for each business case). These KAs are in the form of plain text descriptions, easily provided by domain experts. Then, the *AAO Matrix* requires a first linguistic analysis of the collected knowledge, extracting simple triples: *<subject, verb, direct/indirect object>* from the given texts. In parallel, we start to build the *BP Glossary* that contains all the terms used in above KA, together with their descriptions. The Glossary represents a solid reference point for the business terminology: it is very useful when the picture gets large and complicated. Then we have the first semantic step: where the terminology is classified according to the first four categories of the *OPAAL (Object, Process, Actor, Attribute, Link) Lexicon*. The next step consists in creating the *UML Class Diagrams*, starting from the content of the Lexicon, and finally the BPA Ontology that formalises the whole picture. We presented the BPA Canvas sections in a sequence, but in carrying out the analysis we would rather proceed in a spiral way, progressively enriching the first models, then going back and forth to keep aligned and consistent the various KAs in the different sections.

The above KA are sufficiently intuitive and can be built by business experts without specific technical competences (and, after a suitable training, without the help of knowledge specialists). Only the last section of the methodology, the *BPA Ontology*, i.e., a formal representation of the business scenario (by using an ontology language, such as OWL), requires the intervention of an ontology engineer.

Please note that the BPA Canvas focuses on the structural elements of the BP, where tasks, activities, operations are considered as entities to be linked with the other business elements (document, actors, etc.). According the philosophy of an incremental knowledge modeling, the intricacy of the business logic and the formal modeling of the temporal sequencing tasks is postponed to another stage and then falls outside of the scope of this paper.

3 Applying the BPA Canvas: A Running Example

The example chosen to illustrate the BPA Canvas consists in a home delivery pizza business. The aim is to show the progression in building the various knowledge artefacts in a stepwise fashion to tame the complexity of the knowledge management endeavour, until the BPA ontology is eventually produced.

BP Signature:	**BP Statement**: *My business, PizzaPazza, is a home delivery pizza shop. The customer fills in the order and then submits it to the shop, with the payment, by using our Web site. Making good pizzas requires good quality dough, produced in-house, and a careful baking of the pizza. To make clients happy, we need to quickly fulfil the order and the delivery boy needs to know streets and how to speedily reach the customer's address*	**User Story**: *Mary connects to the PizzaPazza Web site and places her order of two Napoli pizzas, providing also the payment. On the arrival of Mary's order at PizzaPazza, John, the cook, puts the order on the worklist. When the Mary's turn arrives, John prepares the ordered pizzas, cooks them, and then alerts the delivery boy Ed to come and pick up the pizzas. Thus, Ed collects the pizzas and starts his delivery trip, eventually achieving the delivery to Mary's home*
BP Name: Home Pizza Delivery **Trigger**: Order Arrived **Key Actors**: Customer, Cook, Delivery Boy **Key objects**: Order, Dough, Pizza, Delivery Vehicle **Input**: Order Objective: Cook and deliver pizzas to customers **Output**: Pizzas Delivered, Customer happy		

The semantic analysis starts from the free-form text to extract a first structured knowledge artefact: *AAO Matrix*, consisting of trigrams representing the key knowledge about who (Actor) is doing what (Actions) yielding what results (Outcome).

AAO MAtrix			BPA Glossary	
Actor	**Action**	**Outcome**	**Term**	**Description**
Customer	filling	order	*Customer*:	One who buys goods or services from a store or business.
	submitting	order	*Cooking*:	To cook food with dry heat, especially in an oven.
PizzaShop	receiving	order	*DeliveryBoy*:	One that performs the act of conveying or delivering.
	making	pizza	*Order*:	A request made by a customer at a pizza shop for food
	producing	dough	*PizzaKind*:	Different types of pizza the customer can chose to order
	baking	pizza		
DeliveryBoy	collecting	pizza		
	delivering	pizza		
Customer	appraising	service		

Then, the *BPA Glossary* that gathers all the terms, with their descriptions, characterising the business domain. In parallel, the terms are then organised into a Lexicon, introducing their categorization according to the OPAAL scheme. Note that the *Link* category includes pairs of semantically related [4] terms, subdivided into *Structural* and *Functional* ones.

OPAAL Lexicon

Object	Order, Pizza, Margherita, Base, Topping, ...
Process	Cooking, MakingDough, PlacingOrder, AcceptingOrder, DeliveringPizza, ReceivingPizza, ...
Actor	PizzaShop, Customer, Cook, DeliveryBoy, ...
Attribute	Price, Quantity, Calories, PizzaKind, Address, ...
Link	*Structural*: Order-Pizza, Customer-Address, Pizza-Margherita,...
	Functional: Customer-Order, DeliveryBoy-Pizza, PizzaShop-Order, Customer-Pizza, PizzaShop-Pizza,...

4 Building Class Diagrams and the BPA Ontology

Starting from the above knowledge artefacts, and in particular from the OPAAL Lexicon, the next two artefacts consist in the *UML-Class Diagrams* (CD) [5] and the *BPA Ontology* of the Pizza shop. As anticipate, the eight sections of the BPA Canvas have been listed in a sequence, but their construction does not take place sequentially. In particular, in this section we carry out the building of the last two BOK sections (diagrams and ontology) in parallel.

To build the CD we start from the OPAAL Lexicon applying a few rules (not reported for sake of brevity). Essentially, *objects* and *actors* become CD classes, where *attributes* are reported in the corresponding class boxes. The *process* concepts become arcs in the functional dimension, while *ISA*, *PartOf*, etc. become the arcs in the structural dimension. To tackle the overall complexity, instead of building a single large CD diagram, we partition it in subdiagrams, adopting a partitioning criterion based on the different kinds of links. Diagram partitioning, based on well-known techniques rooted in Graph Theory, is not an easy job. In particular, it presents a number of problems when reconstructing a coherent global graph, especially if the semantics of edges and nodes is involved. We address the problems with ontology merging techniques [6].

For sake of space, we report only one functional CD, having operational links, together with the corresponding ontology fragment (using a simplified Turtle syntax, and omitting namespaces) (Fig. 2).

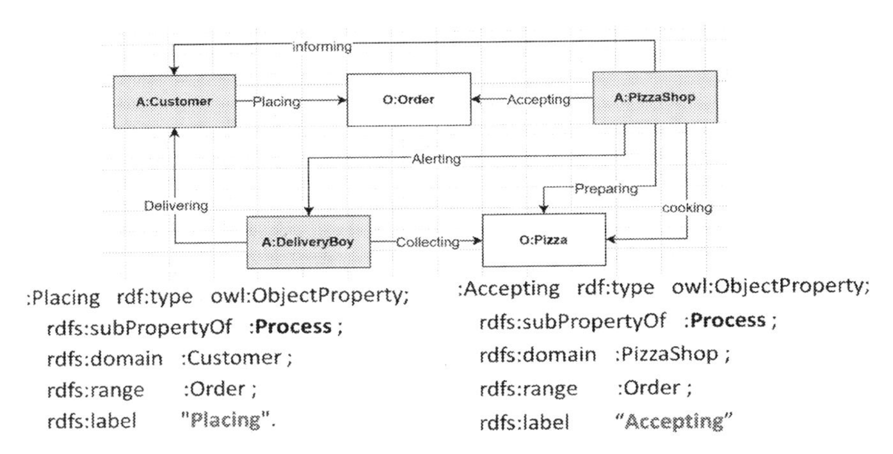
informing
A:Customer
Placing
O:Order
Accepting
A:PizzaShop
Alerting
Preparing
Delivering
cooking
A:DeliveryBoy
Collecting
O:Pizza
:Placing rdf:type owl:ObjectProperty;
rdfs:subPropertyOf :Process ;
rdfs:domain :Customer ;
rdfs:range :Order ;
rdfs:label "Placing".
:Accepting rdf:type owl:ObjectProperty;
rdfs:subPropertyOf :Process ;
rdfs:domain :PizzaShop ;
rdfs:range :Order ;
rdfs:label "Accepting"

Fig. 2. Excerpt of a functional CD and ontology fragment

A formal representation of the BPA offers various advantages, from the possibility of querying the BOK (e.g., to discover which actors perform what actions) to the possibility to apply a reasoner (we adopted Protégé) to prove the absence of (formal) inconsistencies (to this end, and to improve the models, constraints are added).

As anticipated, we presented the knowledge artefacts in a sequence, but in building them we proceed in a spiral way. Periodically, according to the Agile philosophy [7], the produced artefacts are released and shared with end-users and stakeholders for a validation. Then, comments and observations are used to improve the models and release a new version of the PBA Canvas knowledge base.

5 Related Work and Conclusions

Firstly, we need to mention Business Model Canvas [8] that inspired the BPA Canvas layout. The former addresses a high level, enterprise space with respect to our proposal that is focussed on business processes. Furthermore, the former remains at an informal level and lacks of a systematic approach for the modeling practices and the produced documents. Along the line of ontology-based business analysis there are various proposal [9], among others, COBRA, a Core Ontology for Business pRocess Analysis [10] that is based on a Time Ontology. Another research line, with a wider scope, is represented by the adoption of ontologies and semantic web services for BP management, such as

Semantic Business Process Management (SBPM) [11]. Such proposals differ from BPA Canvas since they are more inclined towards the formal aspects than the ease of use for business experts.

BPA Canvas methodology is currently being tested in two field applications, in the area of SMEs (a fashion atelier) and Public Administration (Italian Ministry of Economy and Finance). The first feedback is very encouraging.

References

1. Missikoff, M.: A knowledge-based approach to business analysis for process innovation. In: Proceedings of Italian Symposium on Advanced Database Systems, SEBD 2021 in CEUR Workshop Proceedings (2021)
2. Rolland, C.: A contextual approach for the requirements engineering process. In: SEKE 1994, The 6th International Conference on Software Engineering and Knowledge Engineering (1994)
3. Holsapple, C.W., Joshi, K.D.: A formal knowledge management ontology: conduct, activities, resources, and influences. J. Am. Soc. Inf. Sci. Technol. **55**, 7 (2007)
4. Bagheri, E., Feng, Y.: Methods and resources for computing semantic relatedness. Encyclopedia with Semantic Computing and Robotic Intelligence, vol. 01, no. 01. World Scientific (2016)
5. Franzone, J.: A pimer on UML class diagrams. In: Proceedings of the 2003 American Society for Engineering Education Annual Conference and Exposition, American Society for Engineering Education (2003)
6. Chatterjee, N., Kaushik, N., Gupta, D., Bhatia, R.: Ontology merging: a practical perspective. In: Satapathy, S.C., Joshi, A. (eds.) ICTIS 2017. SIST, vol. 84, pp. 136–145. Springer, Cham (2018). https://doi.org/10.1007/978-3-319-63645-0_15
7. Badakhshan, P., Conboy, K., Grisold, T., vom Brocke, J.: Agile business process management: a systematic literature review and an integrated framework. Bus. Process. Manag. J. **26**(6), 1505–1523 (2020)
8. Pigneur, Y., Osterwalder, A.: Business Model Generation: A Handbook for Visionaries, Game Changers and Challengers. Wiley, Hoboken, New Jersey (2010)
9. Andersson, B., et al.: Towards a reference ontology for business models. In: Embley, D.W., Olivé, A., Ram, S. (eds.) ER 2006. LNCS, vol. 4215, pp. 482–496. Springer, Heidelberg (2006). https://doi.org/10.1007/11901181_36
10. Pedrinaci, C., Domingue, J., Alves de Medeiros, A.K.: A core ontology for business process analysis. In: Bechhofer, S., Hauswirth, M., Hoffmann, J., Koubarakis, M. (eds.) ESWC 2008. LNCS, vol. 5021, pp. 49–64. Springer, Heidelberg (2008). https://doi.org/10.1007/978-3-540-68234-9_7
11. Hepp, M., Leymann, F., Domingue, J., Wahler, A., Fensel, D.: Semantic business process management: a vision towards using semantic Web services for business process management. In: IEEE International Conference on e-Business Engineering (ICEBE 2005), pp. 535–540 (2005)
12. Missikoff, M.: A Knowledge-driven Business Process Analysis Canvas. Arxiv arxiv:2201.06860 (2022)

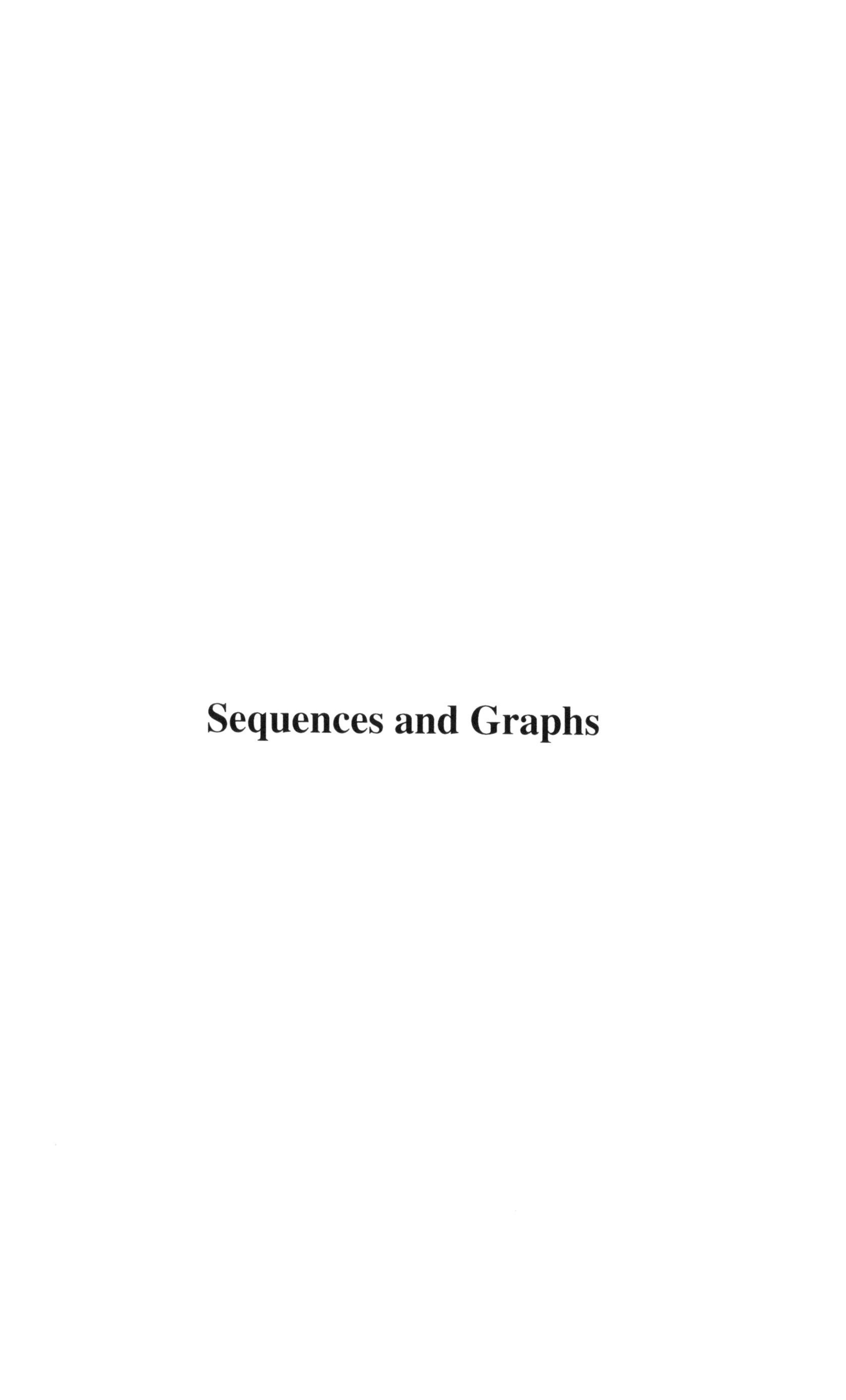

Sequences and Graphs

Extending Authorization Capabilities of Object Relational/Graph Mappers by Request Manipulation

Daniel Hofer[1,2](✉), Stefan Nadschläger[1], Aya Mohamed[1,2], and Josef Küng[1,2]

[1] Institute for Application-Oriented Knowledge Processing (FAW), Faculty of Engineering and Natural Sciences (TNF), Johannes Kepler University (JKU) Linz, Linz, Austria
{dhofer,snadschlaeger,amohamed,jkueng}@faw.jku.at

[2] LIT Secure and Correct Systems Lab, Linz Institute of Technology (LIT), Johannes Kepler University (JKU) Linz, Linz, Austria
{daniel.hofer,aya.mohamed,josef.kueng}@jku.at

Abstract. Enforcing authorization for web applications must be done on the server side. Thus, either the backend or the persistent storage are suitable layers. From a developer's point of view, we want to use a framework to automate creating persistent storage models and to map the entities between storage and backend. However, not all such frameworks offer sufficient authorization support. From a scientist's perspective, we want to generally combine the filtering capabilities of the persistent storage with the advantages of using a mapper framework. Therefore, we propose to intercept the communication between the backend and the mapper framework and thus provide a central point of authorization. This offers the advantage that developers are unlikely to inadvertently introduce security vulnerabilities. The request is modified by adding a filter to return only authorized entities. Filtering directly in the storage saves performance and bandwidth besides reducing development and maintenance effort.

Keywords: Information security · Web application · Query rewriting · Object-graph mapper · Aspect-oriented programming

1 Introduction

In object-oriented software development, we use frameworks to speed up development and reduce repetitive boilerplate code. To persist data (e.g. objects) of an object-oriented system in a database, frameworks exist, which translate these objects to data store specific formats. Some of these frameworks automatically extract the entity's metadata like the names and types of its fields while others have to be configured manually. In any case, they use their abstraction of the data model to build a suitable one in the target storage.

C. Strauss et al. (Eds.): DEXA 2022, LNCS 13427, pp. 71–83, 2022.
https://doi.org/10.1007/978-3-031-12426-6_6

Let's consider an example for an automated framework: *Store a Java object in a relational database*. The framework uses reflection and annotations to get metadata about the object. It creates a table in the database with the name of the object's class. Further, it maps all field names to attribute names and their data types are defined accordingly in the database table. In that case, the framework is an *Object-relational Mapper* (ORM). The same type of framework also exists for graph databases like Neo4j and is called *object-graph mapper* (OGM).

Such a framework takes care of the whole database access, so developers of an application do not need to create tables for each class on their own. More importantly, a set of queries like the standard CRUD (Create Read Update Delete) operations are predefined and provided by the framework and do not need manual maintenance. While this is very useful to avoid errors and to speed up development, these frameworks do exactly this task and nothing more. So what is the problem? The short version: We want to enforce authorization automatically by modifying a request in a way that the persistent storage filters out unauthorized entities. The long version is described in Sect. 2.

Our main contributions to solve this problem are:

- Identify a single location each storage access passes through. This works against code duplication and inadvertent developer mistakes.
- Design the artifact so that it can be added to an existing system without intrusive changes.
- Propose using the filter capabilities of the storage to save resources.

While there are already solutions like *Spring Security*[1], a major distinction to our work is, that *Spring Security*, greatly shortened, assures that no information is leaving the backend without permission. However, we do not even load data to which access was denied from the storage into the backend. Thus, we aim at making it harder to let sensitive information leak out of the system. However, we are not able to protect the system against a developer intentionally breaking access control.

The remainder of the paper is organized as follows: In Sect. 2, we describe the stakeholder goals, the problem context, the requirements and the resulting research questions we want to answer. Section 3 shows our proposed solution followed by a brief assessment in Sect. 4. We then present related work in Sect. 5 and conclude our paper and give an outlook to future work in Sect. 6.

2 Outline of the Problem

We loosely follow the *design science* research method. According to Wieringa [21], *design science* contains an artifact interacting with its context attempting to solve a certain problem. In this section, we describe the problem in more detail, i.e., the problem context, the stakeholder goals, the requirements and the resulting research questions.

[1] https://spring.io/projects/spring-security.

2.1 Stakeholder Goals

The stakeholders in our application scenario define the following goals for adding access control to the existing application:

- Prevent unauthorized access to the entities in the persistent storage.
- Resilience of the authorization extension, i.e., how and where to access a protected entity must not have side effects on the access control.
- No changes to the current software architecture.
- Minimal effects for software developers.
- Minimal changes in the implementation.
- Follow a general, implementation-independent approach.
- Efficient access control processing.

2.2 Problem Context

The artifact we study is part of a web application with a backend using some kind of object to storage mapper. The mapper is loosely coupled to the remainder of the backend and accessed through a dedicated and well-defined interface.

The web application is built upon *Spring Boot*[2]. Its data is stored in the graph database *Neo4j*[3]. Therefore, the application uses *Neo4j OGM*[4]. This is the database's specific object-graph mapper implementation. The query language for this database is Cypher[5].

The overall application handles a potentially huge amount of building blocks for physical simulations. These components are either freely available or proprietary, in which case not even their existence shall be disclosed. Furthermore, there might be more building blocks in the database than can be handled in a machine's main memory.

Neo4j's Built-in Authorization. According to the documentation [14], Neo4j supports fine-grained access control. It provides role-based access control and permissions down to the attribute level. An introduction of the role-based access control model and a comparison with other models can be found in [9]. However this built-in support is not sufficient due to the following reasons:

- Only the *Enterprise Edition* supports built-in access control, but not the *Community Edition.*
- A general, implementation-independent approach is needed.
- The granularity level of the built-in access control is not sufficient, as for example the presence or absence of a relationship cannot be used for permission decisions on individual nodes.
  e.g.: Users can write instances of an entity type only if there is a path

[2] https://spring.io/projects/spring-boot.
[3] https://neo4j.com/product/neo4j-graph-database/.
[4] https://neo4j.com/developer/neo4j-ogm/.
[5] https://neo4j.com/developer/cypher/.

`(:EntityType)-[:HAS_OWNER]->(:User)`.
With the built-in access control support, it is only possible to grant a permission on all nodes of a certain *EntityType*.

2.3 Artifact

The artifact is all components necessary to protect the sensitive data, i.e., the policy schema, its representation in the persistent storage as well as policy enforcement.

Requirements. Based on the stakeholder goals and the context of the artifact, we have to consider the following requirements:

- **Use a single point of enforcement**: The application returns entities in multiple, different locations. Each of them must be secure. Consequently, a single point of enforcement is needed to prevent vulnerabilities because, for instance, developers forget to filter for only allowed entities in some place.
- **No breaking changes, keep the existing architecture**: The application's overall architecture must be maintained when adding access control. Especially, no breaking changes are allowed.
- **Activate and deactivate access control**: It should be possible to easily activate and deactivate access control for the overall system (at configurationtime and not runtime).
- **Transparent to software developers**: For loose coupling and unchanged business logic, the artifact must be transparent to the developers.
- **Reduce development complexity**: Filtering and re-fetching of items require custom algorithms.
- **Do not load unnecessary items**: If we fetch entities the user is not allowed to access, we basically waste time, memory and bandwidth.

Research Questions. From these requirements for the artifact in the given problem context, we can extract the following research questions:

RQ1 How to add a loosely coupled authorization layer without changing the application's architecture and with only minimal effects on its implementation and developers?
RQ2 Where to place the access control layer to provide a single point of access control?
RQ3 How to define protected entities and store the policy definitions.

3 Our Solution

As mapper frameworks are dedicated projects, they are loosely coupled to the rest of the system and shall be only accessed through a well defined interface.

For *JPA*[6] this Java interface is called *EntityManager* and for *Neo4j OGM* it is called *Session*. In either case, we can provide our own implementation which is wrapped around the original one.

An example is shown in Fig. 1 where the *Business Logic* requests all objects of the entity type *Component*. The method call is intercepted by our extension in the wrapper. Inside the extension, the original method call is mapped to another method which allows usage of filters and is provided by the original *Mapper*. The Mapper eventually translates the method call to the query language used by the database, containing the filter specified by the extension.

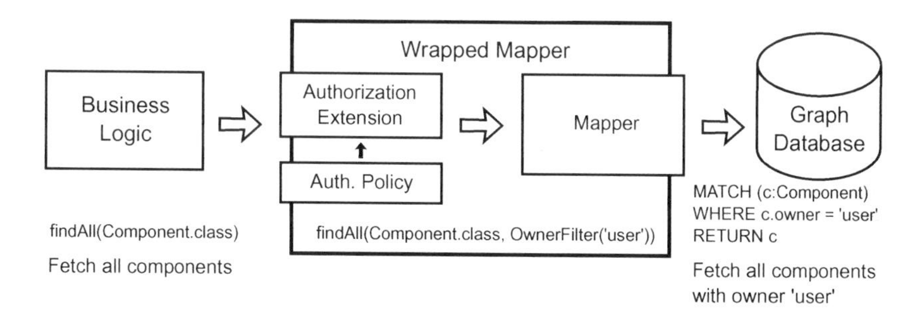

Fig. 1. Example of how a request for all objects of an entity type is changed.

Additionally, we have to indicate the protected entities. Therefore, we provide a marker interface without methods. At runtime, it can be checked whether an object implements it or not.

3.1 Interception

To maintain loose coupling, we do not want to change the existing system by explicitly adding the wrapper to the ORM/OGM interface. *Aspect-oriented programming* (AOP) [5,6] allows to add behaviour at runtime.

In our case, we monitor whenever a new instance of the ORM/OGM interface is requested and intercept this request. We then get the instance and put it inside a new instance of our wrapper. As this implements the same interface, we can pass it back to the calling part of the system. By this, we can apply our logic before each request to the OGM/ORM framework without modifying the existing source code. Whenever the access control extension has to be disabled, we simply deactivate the monitoring for ORM/OGM interface requests.

So far, we described the outer part of the wrapper. The next section describes its inner functionality dealing with the control flow of a data request.

[6] https://www.oracle.com/java/technologies/persistence-jsp.html.

3.2 Responsibility of the Wrapper

The wrapper controls each information and data flow from the backend through the mapping framework to the storage and vice versa. Basically, we can categorize the methods by one of the following four groups:

1. **No security relevance**: These methods do not need special attention and every call can be passed along without further actions.
2. **Fixed entity type access**: Most of the methods require the entity type to be specified. In this case, only the applicable filter needs to be added to the request before passing it along.
3. **Free-text query**: As the interface also supports storage native queries, the authorization extension has to process them by extracting the intentions of a query and rewriting it to honor authorization aspects.
4. **Administrator only methods**: These are methods which must not be invoked by any user, e.g., dropping the entire database. Therefore, these method calls must only be issued by administrators.

The wrapper uses an implementation specialized to the method's category, but independent from the handled entity type. Listing 1 for example shows the general template for handling case 2.

```
T loadEntityById(EntityType<T> type, ID id) {
    if(type implements AccessControlled) {
        User user = getCurrentUser()
        Filter permissionFilter = new Filter(user, "CAN_READ")
        Filter idFilter = new Filter("id", "EQUALS", id)
        Filters composition = idFilter.and(permissionFilter)
        return originalMapperInterface.loadEntity(composition)
    }
    return originalMapperInterface.loadEntityById(type, id)
}
```

Listing 1: Pseudocode for general access handling on an entity type.

We need additional information about the entity types to decide if a request must be intercepted and altered or whether it can be passed along in its original form.

3.3 Encoding the Authorization Properties

To generate the filtering statements, the authorization extension needs two inputs: (1) the information whether an entity type is protected and (2) the structure of the policy encoded in the persistent storage.

Marking Access-Controlled Entity Types. During processing, our authorization extension requires some contextual information. First of all, we need to distinguish access-controlled entity types from those everyone can access. For this, we are using a marker interface. It has no methods and is only used to indicate the access control status by taking advantage of the programming language's type system. To meet our requirements, this interface is only used by the access control system and does not change any behaviour outside the artifact. In the example in Listing 2, the marker interface is called *AccessControlled*.

Providing Authorization Data to the Persistent Storage. The encoding of the access permissions must be stored in the persistent storage to be available during evaluation of the modified queries. We propose to introduce a new entity type called *Permission*. By using a dedicated entity type for this purpose, we enforce the same form of permission encoding for all protected entity types. Another advantage is that *OGM* uses a default query depth of 1 and therefore, we are not loading the permission encoding when fetching an entity. This information is rarely required by the user and should only be fetched if needed to avoid overhead. However, this query depth is implementation specific. By adding one property encapsulating the required information, we achieve looser coupling as we only have to change *Permission* and not every entity type implementing an interface.

Without a dedicated entity type for policy data, we had to include them directly in the protected entity type. However, such an approach would reduce loose coupling, because properties with different semantics and scopes would be mixed in one entity type. This increases the risk for name collisions and reduces maintainability.

```
public class Entity implements AccessControlled {
    private long id;
    private String name;
    private Permission permission;
}
```

Listing 2: To remove the authorization system data, delete field *permission*.

Other entity types serving as either subject or resource in the authorization process receive their own adjacent instance of the *Permission* entity type. By this, we have common capabilities among all entity types participating in the system while at the same time, changes on existing entity types are kept to a minimum. Actual access rights are then encoded by references between the *Permission* entity types with additional information like the permitted operation, e.g., *read* or *write*.

In the requirements in Sect. 2.3, we stated that we should not make intrusive changes. We are changing the entity types here by adding new information, however, this only affects the information storage and not the overall system. The authorization part can be disabled by deactivating the wrapper of the

ORM/OGM interface. For removing the artifact altogether, it is sufficient to delete it, no replacement is needed. Listing 2 shows how an authorization system can be removed by deleting the field *permission*.

An example which requires a replacement is given in Listing 3. The *return* statement cannot be simply deleted, as the original instance needs to be provided

```
public void MapperInterface getMapperInterface() {
    MapperInterface instance = MapperFactory.getInstance();
    return new AccessControllWrapper(instance);
}
```

Listing 3: Remove the authorization system by returning the original object.

3.4 Summary

We use aspect-oriented programming to add a wrapper around the public interface of the mapper. This wrapper implements the same interface as its original counterpart so they are interchangeable. Its purpose is to redirect data access to our custom logic to classify and modify a request so access control is enforced.

To enable this, all entity types that require authorization must be reported to the system. We do this by providing a marker interface. Such an interface can be implemented by a class but does not have any methods. The interface can be used during type-checks. This information together with the requested method determines the intentions of a method call and modifies the passed request accordingly.

4 Assessing the Prototype

We checked the prototypical implementation of our solution concept if it solves the problem and the performance overhead of the wrapping.

4.1 Applicability of the Solution Concept

Experiments on our reference implementation showed, we catched all method calls originating from the backend to the mapper framework *Neo4j OGM*. This framework is meant to be accessed through *Session* objects which are created by a *SessionFactory* [13]. Therefore, the call of *SessionFactory.openSession()* must always be present to get a *Session* object.

Consequently, we have to ensure that every call of *SessionFactory.openSession()* is intercepted. As long as the wrapping of the *Session* instance is guaranteed and no developer tries to work with undocumented internal parts of the mapper, accessing the storage without passing the interception point is not possible.

If developers wanted to evade the security interception on purpose, they could to this by various ways as we aim at preventing accidental vulnerabilities. Our artifact is not designed to defend against rogue developers like in the *Linux incident* of the University of Minnesota [19].

4.2 Performance

We verify that the interception alone does not have an unexpected and unjustifiable performance impact. In detail, this means we check, that detecting and wrapping new *Session* instances obtained by *SessionFactory.openSession()* does not significantly increase the run time. Therefore, we selected the method *loadAll(Class¡T¿ type)* from the *Session* interface and measured the time it took for the method to complete. Additionally, we used an empty database so we were able to only measure the changes in timing by the wrapping of *Session* and keep the database execution time as constant as possible.

For each of these configurations, we executed the method call as a regular user and for comparison as an administrative one. Our reference implementation bypasses all access control checks for the latter to avoid breaking the system in case of a configuration error.

For each combination, we measured the method 100 000 times for warm-up, followed by 1 000 000 executions. We then used only the latter executions for our statistics shown in Table 1. As the numbers indicate, the difference between access with and without access control in place is smaller than the standard deviation. Also the difference between access by a user and by an administrator appears to be neglectable. Therefore, we conclude that the performance impact of the interception is smaller than variations of the execution time caused by external factors.

Table 1. Performance measurement results in milliseconds (1 000 000 executions).

		AC activated	AC deactivated	Δ
Mean	Admin	0.317 875	0.312 553	0.005 322
	User	0.315 809	0.310 996	0.004 813
Standard deviation	Admin	0.130 436	0.112 674	
	User	0.147 702	0.148 250	

5 Related Work

5.1 Query Rewriting and Interception

The idea of *authorization-transparent access control* was published by Motro [12] and implemented using views by Rosenthal [17,18]. Jarman et al. [3] implement Role-Based Access Control by modifying the *WHERE* clause of SQL queries based on policies. Their framework does query rewriting just before the database. Rizvi et al. [16] investigated the equivalence of queries with regard to security and authorization-transparent access control like we propose here, however on a database level. Their approach was then extended to support XML [4].

To add query rewriting to an existing system, aspect-oriented programming is used by [11] for dynamic message routing in an non-intrusive manner. Some aspects of the *Spring Security* architecture and which authentication and authorization patterns can be implemented are described in [2]. A major distinction between this project and ours is, that *Spring Security* works at the interface between backend and frontend, taking care that no unauthorized information is leaving the backend. In contrast, we work between storage and backend and not even load protected data, reducing risk of backend bugs leading to information leaks.

We use an architecture based on dependency injection. This lowers maintenance effort [15]. In the semantic web domain, a similar overall architecture is proposed by [20] to enforce security but also use it for reusable auditing and logging components. There is also a GitHub project called *strategy-spring-security-acl* [8] trying to solve our goal. They extend the functionality of *Spring Data JPA*'s repositories by deriving from them. However, this project might be abandoned since it has not been updated since 2016. Furthermore, it changed parts of the existing *Spring* source code which requires to readapt each new version of the underlying *Spring Data* project. In contrast, we aim at wrapping around an existing mapper without changing it. Therefore, we expect that our project only needs updates if the behaviour of the mapper changes.

5.2 Authorization in Databases

Leao et al. [7] describe how modifying the behaviour of Spring component by using a subclass of it can be used to add additional context to a request for further processing it in the database for the purpose of access control. In Sect. 2.2 we already mentioned the documentation of access control features for Neo4j Enterprise Edition [14], their limitations and our goal to extend that.

5.3 Common Access Control Approaches

Based on the related work and intuitively straightforward solutions, we propose some approaches we investigated but found to not satisfy our requirements.

Explicitly Adding the Filter to the Query. Mapping frameworks allow to define custom queries. We could redefine every query to already contain the filtering part enforcing the access control policy. However this means we also have to explicitly define queries for CRUD operations already provided by the framework. In case the structure of this information is changed, all queries for all entities have to be updated manually. If errors are made there, information leaks may occur if not tested vigorously.

Using Access Control of the Persistent Storage. In particular when using relational databases this is a very good option since they typically offer highly sophisticated access control mechanisms. Mostly a role-based access control is implemented. If additional authorization functionalities are required, e.g. an attribute-based access control, we still have to provide a custom solution. When using other storage systems, e.g. when using a graph-database, authorization mechanisms are not yet that mature. For instance, the current access control models in graph databases only focus on the nodes and not on paths [1,10]. An additional drawback is that offered capabilities may vary across storage systems and replacing one by another might not be possible.

Filtering Entities in the Backend. This can work around the previous two problems, but introduces new ones. As previously described in Sect. 2.3, entities can be displayed paginated. If the users need more elements, they request the next page and so on. The resulting problem is, that we have to fetch enough elements to fill a page and remove the entities the user is not supposed to see. As we do not have enough items to fill our page any more, we need to fetch additional entities and again filter them. We have to repeat this process, until our first page is full. The next problem arises when the user requests to see the next page. Additionally to the fetch and filter cycle, we also have to keep track of those elements we already included in the previous page. Also, infinite scrolling wouldn't work as we still have to recover from connectivity problems and therefore must track the displayed items per user. If we modified the query to take care of such problems, we suffer from the problems of *Explicitly adding the filter to the query* again.

6 Conclusion and Future Work

To solve the problem of adding access control to an existing web-based application, we propose to wrap the *object-relational/graph mappers* with access control capabilities using an *aspect-oriented programming*-based approach and automatically modified requests. We sum-up the most important results achieved so far with respect to our research questions stated in Sect. 2.3. To allow for a loosely-coupled authorization layer while keeping the architecture of our existing application (see RQ1), we decided for an *aspect-oriented programming*-based approach and intercept each request to the object-graph mapper interface. It catches the creation of the mapper's entry point and wraps it with our custom logic. Thus, no existing code needs to be changed and access control can be easily enabled or disabled. The request is forwarded to an instance of our wrapper for the object-graph mapper, which is the place to decide if some authorization needs to be enforced (see RQ2). Since the modification of the request is automatic and centralized, the software developers are unlikely to inadvertently introduce security vulnerabilities. To define protected entities (see RQ3), we use an interface-based approach. Each entity type (i.e. the respective class) which needs to be protected implements the marker interface. The policies applied on the entities are stored in persistent storage (see RQ3), organized according to the needs of the policy definitions.

The applicability of our approach is assessed in Sect. 4 and shows that adding the core wrapper is a promising approach. A first prototype only implements the wrapper concept and does not manipulate the requests. This prototype is implemented using Neo4j and its object-graph mapper. First tests show that the wrapping does not introduce significant performance overhead.

In the future, we will work on how to manipulate the requests to the storage and how we can handle free-text queries.

Acknowledgements. The research reported in this paper has been partly supported by the LIT Secure and Correct Systems Lab funded by the State of Upper Austria. This work has been supported by the COMET-K2 Center of the Linz Center of Mechatronics (LCM) funded by the Austrian federal government and the federal state of Upper Austria.

References

1. Bogaerts, J., Decat, M., Lagaisse, B., Joosen, W.: Entity-based access control: supporting more expressive access control policies. In: Proceedings of the 31st Annual Computer Security Applications Conference, pp. 291–300 (2015)
2. Dikanski, A., Steinegger, R., Abeck, S.: Identification and implementation of authentication and authorization patterns in the spring security framework. In: The Sixth International Conference on Emerging Security Information, Systems and Technologies (SECURWARE 2012), pp. 14–30 (2012)
3. Jarman, J., McCart, J.A., Berndt, D., Ligatti, J.: A dynamic query-rewriting mechanism for role-based access control in databases. In: AMCIS 2008 Proceedings (2008)
4. Kanza, Y., Mendelzon, A.O., Miller, R.J., Zhang, Z.: Authorization-transparent access control for XML under the non-truman model. In: Ioannidis, Y., et al. (eds.) EDBT 2006. LNCS, vol. 3896, pp. 222–239. Springer, Heidelberg (2006). https://doi.org/10.1007/11687238_16
5. Kiczales, G., Hilsdale, E., Hugunin, J., Kersten, M., Palm, J., Griswold, W.G.: An overview of AspectJ. In: Knudsen, J.L. (ed.) ECOOP 2001. LNCS, vol. 2072, pp. 327–354. Springer, Heidelberg (2001). https://doi.org/10.1007/3-540-45337-7_18
6. Kiczales, G., et al.: Aspect-oriented programming. In: Akşit, M., Matsuoka, S. (eds.) ECOOP 1997. LNCS, vol. 1241, pp. 220–242. Springer, Heidelberg (1997). https://doi.org/10.1007/BFb0053381
7. Leão, F., Azevedo, L.G., Baião, F., Cappelli, C.: Enforcing authorization rules in information systems. In: IADIS International Conference Applied Computing (2011)
8. Lecomte, F.: strategy-spring-security-acl (2016). https://github.com/lordlothar99/strategy-spring-security-acl
9. Mohamed, A., Auer, D., Hofer, D., Küng, J.: Authorization strategies and classification of access control models. In: Dang, T.K., Küng, J., Chung, T.M., Takizawa, M. (eds.) FDSE 2021. LNCS, vol. 13076, pp. 155–174. Springer, Cham (2021). https://doi.org/10.1007/978-3-030-91387-8_11
10. Mohamed, A., Auer, D., Hofer, D., Küng, J.: Extended authorization policy for graph-structured data. SN Comput. Sci. **2**(5), 1–18 (2021)

11. Moser, O., Rosenberg, F., Dustdar, S.: Non-intrusive monitoring and service adaptation for WS-BPEL. In: Proceedings of the 17th International Conference on World Wide Web, WWW 2008, pp. 815–824. Association for Computing Machinery, New York (2008). https://doi.org/10.1145/1367497.1367607
12. Motro, A.: An access authorization model for relational databases based on algebraic manipulation of view definitions. In: Proceedings of Fifth International Conference on Data Engineering, pp. 339–340. IEEE Computer Society (1989)
13. Neo4j Inc: Tutorial - OGM Library (2021). https://neo4j.com/docs/ogm-manual/current/tutorial/. Accessed 21 Dec 2021
14. Neo4j Inc: Fine-grained access control (2022). https://neo4j.com/docs/operations-manual/current/authentication-authorization/access-control/. Accessed 19 Jan 2022
15. Razina, E., Janzen, D.S.: Effects of dependency injection on maintainability. In: Proceedings of the 11th IASTED International Conference on Software Engineering and Applications, Cambridge, MA, p. 7 (2007)
16. Rizvi, S., Mendelzon, A., Sudarshan, S., Roy, P.: Extending query rewriting techniques for fine-grained access control. In: Proceedings of the 2004 ACM SIGMOD International Conference on Management of Data, SIGMOD 2004, pp. 551–562. Association for Computing Machinery, New York (2004). https://doi.org/10.1145/1007568.1007631
17. Rosenthal, A., Sciore, E.: View security as the basis for data warehouse security. In: DMDW, p. 8 (2000)
18. Rosenthal, A., Sciore, E.: Administering permissions for distributed data: factoring and automated inference. In: Olivier, M.S., Spooner, D.L. (eds.) Database and Application Security XV. ITIFIP, vol. 87, pp. 91–104. Springer, Boston, MA (2002). https://doi.org/10.1007/978-0-387-35587-0_7
19. The Linux Foundation: Linux incident (2021). https://cse.umn.edu/cs/linux-incident. Accessed 21 Dec 2021
20. Volz, R., Oberle, D., Staab, S., Motik, B.: Kaon server-a semantic web management system. In: WWW (Alternate Paper Tracks). Citeseer (2003)
21. Wieringa, R.J.: Design Science Methodology for Information Systems and Software Engineering. Springer, Heidelberg (2014). https://doi.org/10.1007/978-3-662-43839-8

Sequence Recommendation Model with Double-Layer Attention Net

Weilun Li(✉), Peng Yang, Xiaoye Wang, and Yingyuan Xiao

Tianjin University of Technology, Tianjin 300384, China
Vlenlee131@gmail.com, yyxiao@tjut.edu.cn

Abstract. Personalized prediction for users based on their historical behavior sequences is a challenging problem in the field of recommender systems. The development of recurrent neural networks enables systems to better process sequence information to capture users' long-term preferences. While it cannot effectively utilize both long-term and short-term preferences. In this paper, we propose a novel double-layer attention mechanism mode, which not only increases the weight of short-term preferences in the model, but also assigns separate weights to users' recent behaviors, and then effectively preventing the prediction bias caused by mis-click. The proposed model is experimented on MovieLens, Amazon video game and Amazon digital music datasets, and the results show that our model achieves the best performance in all datasets.

Keywords: Sequence recommendation · User preferences · Attention net

1 Introduction

Nowadays, recommendation systems are an essential part of the Internet. Facing the problem of information overload, personalized recommendations to users through their historical behaviors can greatly improve users' browsing experience and even can enable platforms to better promote goods. Recurrent neural networks (RNN) have made significant progress in recent research, and the application of RNNs in recommender systems is mainly about modeling the influence of sequences between data, thus helping to obtain more effective hidden representations of users and items.

However, the traditional RNN structures do not perform well for capturing both long-term and short-term preferences of users, and they cannot explicitly capture interactions between items throughout the user history. To solve this problem, Liu and Zeng et al. improved the algorithm based on the Short-Term Memory Priority model (STMP) and proposed an optimized Short-Term Attention/Memory Priority Model (STAMP) [1]. The model uses an attention layer to weight users' long-term and short-term preferences after extracting them, which can assign different weights for the long-term and short-term preferences respectively, effectively solving the problem of traditional RNNs and preventing users from drifting their preferences in long sessions.

However, the STAMP model only uses the user's last click to extract their short-term preferences, so once the user has a wrong click in the instant recommendation,

C. Strauss et al. (Eds.): DEXA 2022, LNCS 13427, pp. 84–96, 2022.
https://doi.org/10.1007/978-3-031-12426-6_7

the system will recommend products that do not meet the user's expectations. Thus the recommendation will lose accuracy and this negative phenomenon will continue. Therefore, we propose a sequential recommendation model (DAttRec) with a double-level attention mechanism, which can achieve the combination of long and short term interests of users. Effectively prevent system errors caused by the user's wrong click. And we introduce a new self-attention module which effectively prevents the user's short-term interest drift. To the best of our knowledge, this is the first attempt to use a Double-layer Attention mechanism in sequential Recommendation (DAttRec). Our contributions are mainly as follows:

- We optimize the Short-Term Attention/Memory Priority Model to expand the number of clicks that capture short-term preferences, which effectively prevent the inaccuracy of the recommendation list due to users' wrong click.
- A double-layer attention net is proposed, introducing a new self-attention module that assigns different weights to the last L clicks behaviors, effectively prevent capturing the interest drift from short-term preference expansion.
- Our proposed model was tested on MovieLens dataset, Amazon video game and digital music dataset respectively, and all achieved the best results, where the double-layer attention net played a significant role. The experimental results confirmed the superiority of our proposed model.

2 Related Work

In this section, we introduce traditional sequential recommendation models, deep neural network models, and attention-based models to review related work, respectively.

2.1 Traditional Sequential Recommendation Models

Based on the recommended systems of Markov chain, we adopt Markov chain model to model the transformation of user-item interaction in sequence to predict the next interaction. The FPMC model proposed by Rendle S et al. [2] is an improved model of Markov chains. There are also improvements to FPMC, such as the HRM model proposed by Hidasi B et al. [3]. which essentially adds nonlinear transformations to FPMC. The TransRec model proposed by He R et al. [4] is also a Markov chain model.

All the above Markov chain-based models have significant drawbacks, one being that they ignore long-term dependencies; and two, they ignore the collective dependencies of user-item interactions.

2.2 Deep Neural Network Models

The rise of neural network has brought new solutions to sequential recommendation, such as convolutional neural network [5–7], RNNs [8–15] and graph neural network [16–24] models. RNN-based recommender systems include GRU4Rec and GRU4Rec proposed by Hidasi B [9, 10] et al. Unlike the RNN-based models, the CNN-based Caser model proposed by Tang J [6] obtains a short-term representation of the user by convolutional operations. The GNN-based SR-GNN proposed by Wu S [19] et al. learning about the Embedding of users or items on the graph, we embed more complex relationships on the entire graph.

2.3 Attention-Based Models

In recent years, the emergence of Transformer [25–28] has made the attention mechanism [29–34] become the mainstream modeling approach for sequence recommendation. The first is the classical SASRec model proposed by Kang W C [25] et al., which introduces Transformer into sequence recommendation.

3 The Proposed Model: DAttRec

3.1 Symbolic Description

A session-based recommendation system is building on the user's historical sessions and predicting based on the user's recent sessions. For each session denoted as $S = [s_1, s_2, \ldots, s_N]$, consists of a sequence of actions, where s_i represents the item clicked by the user at moment i. $S_t = \{s_1, s_2, \ldots, s_t\}$, $1 \leq t \leq N$, denotes the session prefix of session S truncated at moment t. $V = \{v_1, v_2, \ldots v_{|V|}\}$ denotes the set of different items in the sequential recommendation system, called the item dictionary.

Let $X = \{x_1, x_2, \ldots, x_{|V|}\}$ denote the embedding vector of the item dictionary V. The model learns a d-dimensional real-valued embedding vector $x_i \in \mathbb{R}^d$ for each item i in V and represents the last clicked item s_t in the session prefix S_t as a d-dimensional embedding vector $x_t \in \mathbb{R}^d$. The goal of the model is to predict the next clicked item s_{t+1} based on the user's session prefix S_t. To be precise, the model will act as a classifier for the item dictionary V. Each candidate item generates a score such that $\hat{y} = \{\hat{y}_1, \hat{y}_2, \ldots \hat{y}_{|V|}\}$ denotes the output score vector, where $\hat{y}_i$ corresponds to the score of item v_i. After getting the predicted scores, we sort the corresponding items in descending order and use the TOP-K items for recommendation. For the sake of notation, we define the trilinear product of the three vectors as:

$$< a, b, c >= \sum_{i=1}^{d} a_i b_i c_i = a^T (b \odot c) \tag{1}$$

where $a, b, c \in \mathbb{R}^d$ and $\odot$ denotes Hadamard product (the product of elements between the two vectors b and c).

3.2 STAMP Model

Our model is built on top of the STAMP (Short-Term Attention/Memory Priority) model, as shown in Fig. 1.

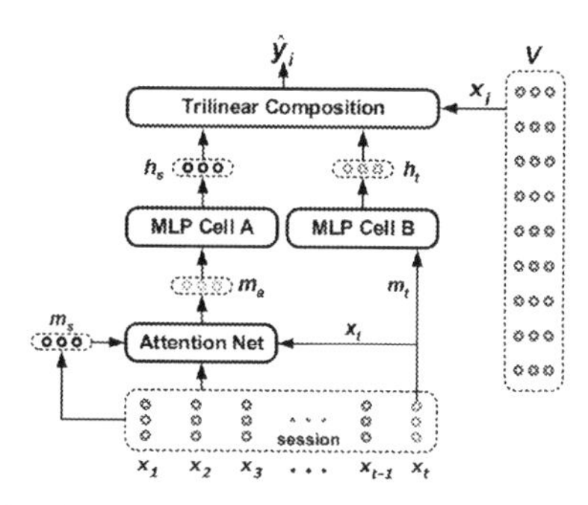

Fig. 1. Schematic diagram of STAMP model

From Fig. 1 we can see that the STAMP model takes as input two embedding vectors (m_s and m_t), where m_s denotes the generic interest of the user's current session, defined as the average of the session's external memory:

$$m_s = \frac{1}{t}\sum_{i=1}^{t} x_i \tag{2}$$

where the term external memory refers to the sequence of commodity embeddings in the current session prefix S_t, and m_t represents the immediate interest of the user in this session. In the STAMP model, the last clicked item represents the immediate interest of the user: $m_t = x_t$. Since x_t is taken from the external memory of the session, we refer to it as the short-term memory of the user's interest, and then the user's generic interest m_s and immediate interest m_t are passed through an attention net to generate a real-valued vector m_a representing the user's true generic interest. The proposed attention net contains two components: (1) a simple forward feedback neural network (FNN) used to generate attention weights for each item in the current session prefix S_t, and (2) an attention composite function is used to compute the user's interest in the general case m_a. The FNN used for attention computation is defined as:

$$\alpha_i = W_0\sigma(W_1x_i + W_2x_t + W_3m_s + b_a) \tag{3}$$

where $x_i \in \mathbb{R}^d$ denotes the i-th item s_i in S_t, $x_t \in \mathbb{R}^d$ denotes the last click, $W_0 \in \mathbb{R}^{1\times d}$ is a weight vector, $W_1, W_2, W_3 \in \mathbb{R}^{d\times d}$ is a weight matrix, b_a is a bias vector, and $\sigma(\cdot)$ denotes the sigmoid function. α_i denotes the current session prefix of item x_i in the attention coefficient. From Eq. (3), we can see that the attention coefficient is calculated based on the embedding of the target item x_i, the last clicked item x_t and the user's generic interest representation m_s, so it can effectively capture the relationship between the long/short term memory of the target item and the user's interest.

After obtaining the attention coefficient $\alpha = (\alpha_1, \alpha_2, \ldots, \alpha_t)$ corresponding to the session prefix S_t, the attention-based general interest m_a about the user of the current session prefix S_t can be calculated as:

$$m_a = \sum_{i=1}^{t} \alpha_i x_i \tag{4}$$

Two Multi-layer Perceptron (MLP) are then used to process the generic interest m_a and the immediate interest m_t for feature abstraction. The two MLPs have the same structure, only with different parameters. A simple MLP without hidden layers is used for feature abstraction, and the operation of m_a is defined as:

$$h_s = f(W_s m_a + b_s) \tag{5}$$

where $h_s \in \mathbb{R}^d$ denotes the output, $W_s \in \mathbb{R}^{d \times d}$ denotes the weight matrix, $b_s \in \mathbb{R}^d$ denotes the bias vector, and $f(\cdot)$ denotes a nonlinear activation function (in this study we used the tanh function). The output vector h_t with respect to m_t is calculated similarly to h_s, and then for a given candidate item x_i, the score function can be defined as:

$$\hat{z}_i = \sigma(< h_s, h_t, x_i >) \tag{6}$$

$\sigma(\cdot)$ denotes the sigmoid function such that $\hat{z} = \{z_1, z_2, \ldots, z_{|V|}\} \in \mathbb{R}^{|V|}$ denotes the unnormalized cosine similarity of the two interests to the candidate item. The probability distribution of each candidate item is then obtained by the SoftMax function represented by:

$$\hat{y} = softmax(\hat{z}) \tag{7}$$

For each of these elements $\hat{y}_i \in \hat{y}$, denotes the probability that the user's next click on the item is v_i in the current session.

For any given session prefix $S_t \in S(t \in [1, 2, \ldots, N])$, the loss function is defined as the cross entropy of the predicted outcome $\hat{y}$:

$$L(\hat{y}) = -\sum_{i=1}^{|V|} y_i \log(\hat{y}_i) + (1 - y_i)\log(1 - \hat{y}_i) \tag{8}$$

where y denotes a one-hot vector activated exclusively by s_{t+1}, e.g., if s_{t+1} denotes the i-th element v_i in the commodity dictionary V, then $y_k = 1$ when $y == k$ and $y_k = 0$ otherwise. An iterative stochastic gradient descent (SGD) optimizer is then executed to optimize the cross-entropy loss.

From Eq. (4), we can see that the STAMP model predicts the next click according to the inner product weighted by candidate projects and users' interests, where the weighted user interest is represented by a bilinear combination of long-term and short-term memory. The effectiveness of this trilinear model is verified in the experiments, and the experimental results show that the short-term memory priority mechanism can effectively capture the user's preferences and facilitate the prediction of the next click, and achieves the best performance on other benchmark datasets.

However, we can see that there are some problems with the model's short-term preferences being captured by the last click. For example, if the last click in the experiment is a wrong click behavior, then it will have a great impact on the experimental data, so we proposed a new model to solve this problem, the proposed model is based on the STAMP model, also giving priority to short-term attention, we call it a Double-layer Attention net Recommendation model (DAttRec).

3.3 DAttRec Model

The DAttRec model is shown in Fig. 2. As we can see in the figure, the difference between the two models is that the DAttRec model extends the last click behavior which captures short-term preferences in the STAMP model to the last L click behaviors, and uses a simple self-attention module for these L behaviors. The purpose of assigning different weights to each click item in the L behaviors is to prevent interest bias in short sequences as well.

Self-attention Module. Self-attention is a special case of the attention mechanism, which refines the representation by matching the sequence with itself, so that context information can be retained and the relationship between each item in the sequence can be captured regardless of the distance between them. We use self-attention to model the last L user behaviors and obtain the user's short-term preferences.

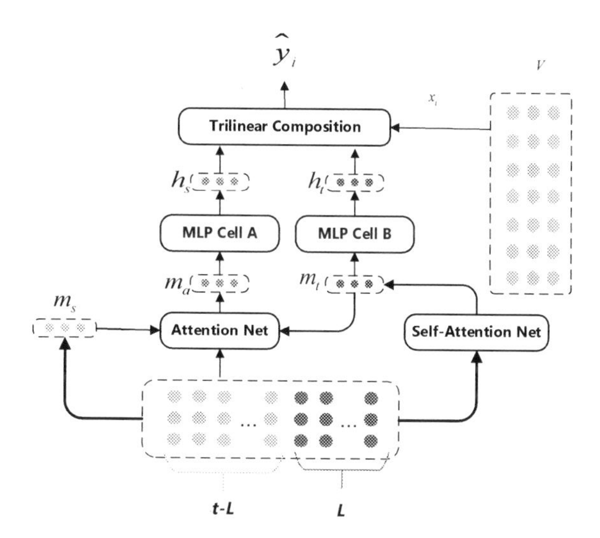

Fig. 2. Schematic diagram of DAttRec model

The input of the attention module contains: *qurey*, *key* and *value*. the output of the attention module is a weighted sum of *value*, where the weight matrix is determined by the *query* and its corresponding *key*. In this paper, these three inputs (*query*, *key*, *value*) are identical, and they are obtained from the embedding of the user's recent L-item click activity (as shown in Fig. 3).

Assuming that the user's short-term preferences can be obtained from the user's last L clicks, each item can be represented as a d-dimensional embedding vector such that $X \in \mathbb{R}^{t \times d}$ as the embedding representation of the entire session prefix, and the most recent L items (from item $t-L+1$ to item t) can be represented as the following matrix:

$$X_t^u = \begin{bmatrix} X_{(t-L+1)1} & \cdots & X_{(t-L+1)d} \\ \vdots & \ddots & \vdots \\ X_{t1} & \cdots & X_{td} \end{bmatrix} \quad (9)$$

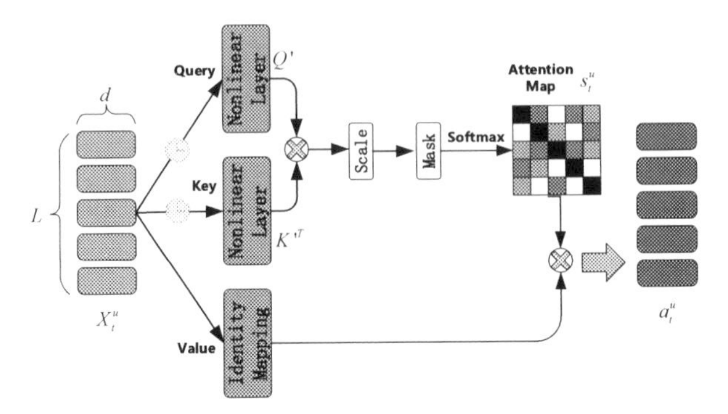

Fig. 3. Schematic diagram of the self-attention module

The *query*, *key*, *value* of the user at time t are equal to X_t^u.

First we project *query* and *key* into the same space by a nonlinear transformation, and they have common parameters.

$$Q^{'} = ReLU\left(X_t^u W_Q\right) \tag{10}$$

$$K^{'} = ReLU\left(X_t^u W_K\right) \tag{11}$$

where $W_Q = W_K \in \mathbb{R}^{d \times d}$, which are the weight matrices of *query* and *key*, respectively, and ReLU introduces nonlinearity into attention as an activation function. Then the attention score mapping matrix can be obtained by computing:

$$S_t^u = softmax\left(\frac{Q^{'} K^{'T}}{\sqrt{d}}\right) \tag{12}$$

The output attention score mapping matrix is an $L \times L$ matrix representing the similarity between the L items. We use a large d value so that the scaling factor can reduce the minimal gradient effect. A mask operation is used before the SoftMax function to mask the diagonal of the attention score mapping matrix to prevent high matching scores due to identical *query* and *key*.

Finally, the attention score mapping matrix is multiplied with *value* to get the final weighted output of the attention module.

$$a_t^u = s_t^u X_t^u \tag{13}$$

Here the attention output $a_t^u \in \mathbb{R}^{L \times d}$ can be regarded as the short-term interest representation of the user, and to obtain the final individual attention representation, the average embedding of the L self-attention representations is taken as the user's final short-term interest representation.

$$m_t^u = \frac{1}{L} \sum_{l=1}^{L} a_{tl}^u \tag{14}$$

The attention model described above loses the temporal relationship of the sequence, so it uses the same approach as Transformer, adding time sequence number before the nonlinear transformation of *query* and *key*, using a time-scaled geometric sequence to add sinusoidal signals of different frequencies to the input signal. The temporal embedding consists of two sinusoidal signals defined as follows:

$$TE(t, 2i) = \sin\left(\frac{t}{10000^{\frac{2i}{d}}}\right) \tag{15}$$

$$TE(t, 2i+1) = \cos\left(\frac{t}{10000^{\frac{2i}{d}}}\right) \tag{16}$$

where t denotes the time step and i denotes the dimension, and the time embedding is simply added to the *query* and *key* before the linear transformation.

4 Experiments

4.1 Datasets and Data Preparation

We evaluated the proposed model on three datasets, the first dataset is the ml-1m dataset in MovieLens, the second dataset is the Video Game dataset in Amazon Mall, and the third dataset is the Digital Music dataset in Amazon Mall. The users, items, and densities of these datasets are shown in Table 1.

Table 1. Experimental dataset statistics

Dataset	Users	Items	Interactions	Density	AVG actions per user
ML-1M	6,040	3,706	1,000,209	4.46%	1655.66
Video Game	7,220	16,334	140,307	0.119%	19.43
Digital Music	2,893	13,183	64,320	0.169%	22.23

4.2 Evaluation Metrics

For each user, we tested using the most recent item and used the second most recent item for hyperparameter tuning. We evaluate the performance of all models using hit ratio(HR) and mean-reverse ranking (MRR). The HR measures the accuracy of the recommendations. We set the value of K in HR@K to 50, defined as follows:

$$HR@50 = \frac{1}{M}\sum_{u\in\mathfrak{U}} 1\left(R_{u,g_u} \leq 50\right) \tag{17}$$

Here g_u is the item that user u interacted with in the most recent time, and R_{u,g_u} is the ranking generated by the model for the real data. The function will return 1 if the model ranks g_u in the top fifty, otherwise it returns 0.

MRR, on the other hand, indicates how the model ranks the item, and intuitively, the higher the real data is ranked in practice, the more desirable it is. MRR is defined as follows:

$$MRR = \frac{1}{M}\sum_{u\in\mathfrak{U}} \frac{1}{R_{u,gu}} \tag{18}$$

where R_{u,g_u} is the ranking of the real data item, MRR focuses on the position of the ranking and calculates the inverse of the ranking where the real data item is located.

4.3 Baselines

We compare the DAttRec model with classical methods and state-of-the-art models to evaluate the following models.

- POP. This method is a primitive recommendation system model that evaluates the popularity of item in the recommendation system and makes recommendations for users accordingly.
- FPMC. This method combines matrix decomposition mechanism with Markov chain for next item recommendation, which can capture both user's interest preference and user's behavior order. It is an advanced hybrid recommendation model.
- Caser. It uses hierarchical and vertical convolutional neural networks to model the user's historical interactions. It also considers skip behavior and optimizes the whole network by minimizing cross entropy.
- GRU4Rec. A deep learning model based on RNN for session-based recommendation, it consists of GRU units that utilize a session-parallel small-batch training process in which a ranking-based loss function is also employed.
- STAMP. This method models the long and short-term interests of users separately through an attention net, using the last click as the short-term preference of the user. It is one of the most advanced models available today.
- AttRec. This method also models users' long- and short-term interests separately and uses metric learning to model users' long-term interests, which is one of the most advanced models nowadays.

4.4 Parameters

We optimize the hyperparameters in our experiments by performing a grid search on the dataset. The hyperparameters selected from the following ranges: embedding dimension d is $\{50, 100, 200, 300\}$, learning rate η is $\{0.001, 0.005, 0.01, 0.05, 0.1, 1\}$, and learning rate decay λ is $\{0.75, 0.8, 0.85, 0.9, 0.95, 1.0\}$. Since we used an adaptive gradient optimizer for DAttRec, the final learning rate for the model was set to 0.05 with no further adjustments. Based on the average performance, we set the potential dimension of DAttRec and all other baseline model potential vectors to 100, the batch size to 512, the epoch to 40, and the weight matrix to be initialized by sampling from a normal distribution $N(0, 0.05^2)$, setting the bias to zero. All item embeddings are randomly initialized with the normal distribution $N(0, 0.002^2)$, the sequence length L is set to 5, and the target length T is set to 3, and then trained jointly with other parameters.

4.5 Performance Comparison

Table 2 shows the experimental results of the six baseline models compared with our model on three datasets, and we can find that DAttRec achieves the best performance on all datasets, which also shows the superiority of our proposed model. Through the data we can see that the traditional POP and FPMC models perform the worst among the various baseline models. In contrast, both Caser and GRU4Rec models have better performance than traditional Markov chain-based models. Our model shows excellent performance on both sparse and dense data sets.

Table 2. Results of model comparison with the baseline model on datasets

Dataset	ML-1M		Video game		Digital music	
Metric	HR@50	MRR	HR@50	MRR	HR@50	MRR
POP	0.144	0.0231	0.0609	0.0126	0.0436	0.0073
FPMC	0.4209	0.1022	0.2226	0.0451	0.158	0.0322
Caser	0.4811	0.0925	0.1438	0.0248	0.1327	0.0228
GRU4Rec	0.5128	0.1049	0.2298	0.0336	0.1647	0.0395
STAMP	0.5309	0.1203	0.2524	0.0518	0.2289	0.0496
SASRec	0.5223	0.1172	0.2414	0.0496	0.2205	0.0467
DAttRec	**0.5551**	**0.1287**	**0.2832**	**0.0568**	**0.2591**	**0.0554**
Improv	3.40%	7.10%	12.20%	9.60%	13.20%	11.60%

4.6 Model Analysis and Discussion

The Effect of the Attention Module. The DAttRec model contains two attention modules, and we want to know if each of them has a positive effect on the model. We remove the attention modules separately to obtain four models, namely the 0–0 model, the 0–1 model, the 1–0 model and the 1–1 model. Where the first position represents the attention module in long-term preferences and the second position represents the attention module in short-term preferences, and when the attention module in that position is removed, the average embedding is used for substitution.

As shown in Table 3, we can find that both attention modules improve the performance of the model to some extent, with the attention module that determines short-term preferences improving the model performance more. It also proves the necessity of our optimization on STAMP.

Table 3. Effects of different attention modules on experimental results in HR@50

Dataset	0–0	1–0	0–1	1–1
ML-1M	0.3409	0.5168	0.5231	0.5551
Video Game	0.0936	0.2708	0.2799	0.2832
Digital Music	0.1068	0.2397	0.2481	0.2591

Effect of Aggregation Methods on the Model. In obtaining the final short-term preferences of users, we used the average user embedding representation. Similarly, we try to change the aggregation methods to maximum, minimum and sum for our experiments. As can be seen from Table 4, the average embedding approach obtains good performance on both dense and sparse datasets, while the other aggregation methods perform slightly worse, especially on sparse datasets. We speculate that this is due to the fact that the average aggregation method can effectively retain more information.

Table 4. Effect of different polymerization methods on experimental results in HR@50

Dataset	avg	min	max
ML-1M	0.5551	0.5483	0.5449
Video Game	0.2832	0.2544	0.2369
Digital Music	0.2591	0.2386	0.2306

Impact of Weights on the Model. Wrences in the model, and we can see in Fig. 4 that the performance of the model is better when the short-term preference is considered but the long-term preference is not considered. Therefore, it is more appropriate to set the weights between 0.2 and 0.4, which also reflects the importance of short-term preferences in the sequential recommendation and proves the feasibility of the short-term. It also reflects the importance of short-term preferences in sequential recommendation and proves the feasibility of the attention mechanism of short-term preferences.

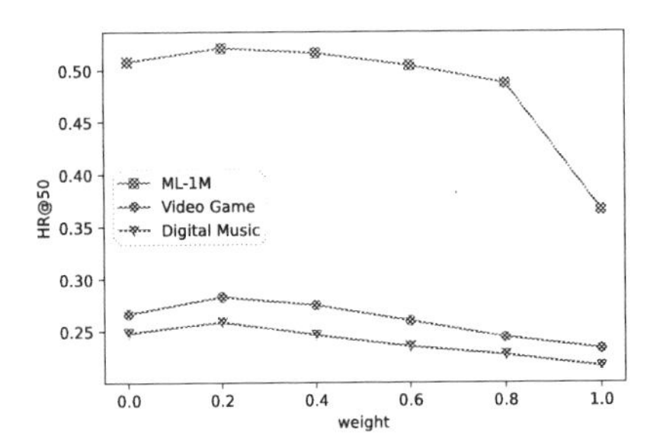

Fig. 4. Effect of different weights on experimental results

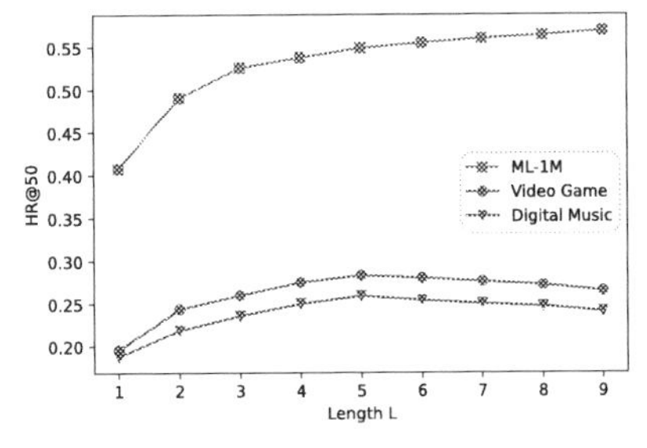

Fig. 5. Effect of different L on experimental results

The Effect of Sequence Length L on the Model. As shown in Fig. 5, since the self-attentive mechanism can capture long-distance relational dependence, it can be applied to long sequences. In dense datasets, the longer the L, the better the performance of the model, and conversely, in sparse datasets, the L is not suitable too long, because the total data is small. Too long L will make the sequence occupied by long-term preferences and reduce the performance of the model. Therefore, the sequence length L for obtaining short-term preferences is set to 5 in this experiment.

5 Conclusion

In this paper, we propose a novel recommendation model based on a double-layer attention mechanism. Our model achieves improvement and optimization of the original algorithm in the extraction and combination of long and short-term preferences. It solves the problem of inaccuracy of recommendation list due to user's accidental clicks. And different weights are assigned to click behaviors to solve the possible interest drift phenomenon in sequential recommendations. Experiments were conducted on three datasets to compare with each model, and the best results were achieved.

Acknowledgements. This work is supported by "Tianjin Project + Team" Key Training Project under Grant No. XC202022.

References

1. Liu, Q., et al.: STAMP: short-term attention/memory priority model for session-based recommendation. In: Proceedings of the 24th ACM SIGKDD International Conference on Knowledge Discovery & Data Mining (2018)
2. Rendle, S., Freudenthaler, C., Schmidt-Thieme, L.: Factorizing personalized markov chains for next-basket recommendation. In: Proceedings of the 19th international conference on World wide web, pp. 811–820 (2010)
3. Hidasi, B., Tikk, D.: General factorization framework for context-aware recommendations. Data Min. Knowl. Disc. **30**(2), 342–371 (2016)
4. He, R., Kang, W.C., McAuley, J.: Translation-based recommendation: a scalable method for modeling sequential behavior. IJCAI, pp. 5264–5268 (2018)
5. Tuan, T.X., Phuong, T.M.: 3d convolutional networks for session-based recommendation with content features. In: RecSys, pp. 138–146 (2017)
6. Tang, J., Wang, K.: Personalized top-n sequential recommendation via convolutional sequence embedding. In: WSDM, pp. 565–573 (2018)
7. Yuan, F., Karatzoglou, A., Arapakis, I., Jose, J.M., He, X.: A simple convolutional generative network for next item recommendation. In: WSDM, pp. 582–590 (2019)
8. Wang, P., Fan, Y., Xia, L., Zhao, W.X., Niu, S., Huang, J.: Kerl: a knowledge-guided reinforcement learning model for sequential recommendation. In: SIGIR, pp. 209–218 (2020)
9. Hidasi, B., Karatzoglou, A., Baltrunas, L., Tikk, D.: Sessionbased recommendations with recurrent neural networks. In: ICLR (2016)
10. Hidasi, B., Quadrana, M., Karatzoglou, A., Tikk, D.: Parallel recurrent neural network architectures for feature-rich sessionbased recommendations. In: RecSys, pp. 241–248 (2016)
11. Wu, C.-Y., Ahmed, A., Beutel, A., Smola, A.J., Jing, H.: Recurrent recommender networks. In: WSDM, pp. 495–503 (2017)
12. Smirnova, E., Vasile, F.: Contextual sequence modeling for recommendation with recurrent neural networks. In: DLRS, pp. 2–9 (2017)
13. Li, J., Ren, P., Chen, Z., Ren, Z., Lian, T., Ma, J.: Neural attentive session-based recommendation. In: CIKM, pp. 1419–1428 (2017)
14. Tan, Y.K., Xu, X., Liu, Y.: Improved recurrent neural networks for session-based recommendations. In: DLRS, pp. 17–22 (2016)
15. Jannach, D., Ludewig, M.: When recurrent neural networks meet the neighborhood for session-based recommendation. In: RecSys, pp. 306–310 (2017)

16. Loyola, P., Liu, C., Hirate, Y.: Modeling user session and intent with an attention-based encoder-decoder architecture. In: RecSys, pp. 147–151 (2017)
17. Li, Z., Zhao, H., Liu, Q., Huang, Z., Mei, T., Chen, E.: Learning from history and present: Next-item recommendation via discriminatively exploiting user behaviors. In: SIGKDD, pp. 1734–1743 (2018)
18. Quadrana, M., Karatzoglou, A., Hidasi, B., Cremonesi, P.: Personalizing session-based recommendations with hierarchical recurrent neural networks. In: RecSys, pp. 130–137 (2017)
19. Wu, S., Tang, Y., Zhu, Y., Wang, L., Xie, X., Tan, T.: Sessionbased recommendation with graph neural networks. In: AAAI, pp. 346–353 (2019)
20. Xu, C., et al: Graph contextualized self-attention network for session-based recommendation. In: IJCAI, pp. 3940–3946 (2019)
21. Qiu, R., Yin, H., Huang, Z., Chen, T.: Gag: Global attributed graph neural network for streaming session-based recommendation. In: SIGIR, pp. 669–678 (2020)
22. Wang, Z., Wei, W., Cong, G., Li, X.-L., Mao, X.-L., Qiu, M.: Global context enhanced graph neural networks for session based recommendation. In: SIGIR, pp. 169–178 (2020)
23. Pan, Z., Cai, F., Chen, W., Chen, H., de Rijke, M.: Star graph neural networks for session-based recommendation. In: CIKM, pp. 1195–1204 (2020)
24. Chen, T., Wong, R.C.-W.: Handling information loss of graph neural networks for session-based recommendation. In: SIGKDD, pp. 1172–1180 (2020)
25. Kang, W.-C., McAuley, J.: Self-attentive sequential recommendation. In: ICDM, pp. 197–206 (2018)
26. Ren, R., et al.: Sequential recommendation with self-attentive multiadversarial network. In: SIGIR, pp. 89–98 (2020)
27. Bai, T., et al.: Ctrec: a long-short demands evolution model for continuoustime recommendation. In: SIGIR, pp. 675–684 (2019)
28. Tang, J., et al.: Towards neural mixture recommender for long range dependent user sequences. In: WWW, pp. 1782–1793 (2019)
29. Zhou, K., et al.: S3-rec: self-supervised learning for sequential recommendation with mutual information maximization. In: CIKM, pp. 1893–1902 (2020)
30. Wu, J., Cai, R., Wang, H.: Déjà vu: a contextualized temporal attention mechanism for sequential recommendation. In: WWW, pp. 2199–2209 (2020)
31. Ji, W., Wang, K., Wang, X., Chen, T., Cristea, A.: Sequential recommender via time-aware attentive memory network. In: CIKM, pp. 565–574 (2020)
32. Ye, W., et al.: Time matters: Sequential recommendation with complex temporal information. In: SIGIR, pp. 1459–1468 (2020)
33. Wang, C., Zhang, M., Ma, W., Liu, Y., Ma, S.: Make it a chorus: knowledge-and time-aware item modeling for sequential recommendation. In: SIGIR, pp. 109–118 (2020)
34. Brauwers G., Frasincar, F.: A general survey on attention mechanisms in deep learning., IEEE Transactions on Knowledge and Data Engineering (TKDE) (2022)

Fault Detection in Seismic Data Using Graph Attention Network

Patitapaban Palo[1(✉)], Aurobinda Routray[1], and Sanjai Kumar Singh[2]

[1] Department of Electrical Engineering, Indian Institute of Technology, Kharagpur, Kharagpur 721302, West Bengal, India
patitapabanpalo@iitkgp.ac.in

[2] Oil and Natural Gas Corporation Ltd., Dehradun 248003, Uttarakhand, India

Abstract. This paper presents a natural attention-based approach for automated fault detection in seismic data. Fault analysis in seismic data is important for drilling and exploration in the oil and natural gas industries. The seismic fault is a perceptual phenomenon, and manual fault detection is still practiced in various industries. The convolutional neural network (CNN) is the most commonly used method in the newly conducted research for automated fault detection. However, our paper uses a graph attention network (GAT) based approach. We first extract 2D patches centered around the points of concern. Next, we present these extracted patches in the graph domain using the k-nearest neighbor graph. The graph representation of patches is connectional in the graph domain based on seismic amplitude similarity. Then, we apply GAT to classify the faults. Both the training and testing sets contain both synthetic and real data. The proposed methodology gives good accuracy when applied to field data.

Keywords: Graph attention network (GAT) · Graph neural network (GNN) · Seismic fault detection

1 Introduction

The seismic data acquisition takes place over an area by sending acoustic signals below ground and receiving the reflected signals. The acquired seismic data is then processed to remove noise and enhance the resolution. The subsurface image is generated in the next step. After analyzing these pre-processed seismic data, further subsurface geological information is extracted. In the oil and natural gas industries, detecting faults in seismic data is a prerequisite for identifying hydrocarbon sources. Generally, fault in seismic data occurs due to displacement between blocks of rocks that can span from few millimetres to a few kilometres. If the fault spans over a large area, then the area becomes potential to drill and produce oil and natural gas. However, the non-linear and irregular structure of

We are grateful to the Oil and Natural Gas Corporation (ONGC), India, for supporting the work and providing real seismic data used in this paper.

C. Strauss et al. (Eds.): DEXA 2022, LNCS 13427, pp. 97–109, 2022.
https://doi.org/10.1007/978-3-031-12426-6_8

seismic data often requires interpretation from an expert with high graphics and computational tools to locate the faults. The challenge is to automate the fault detection process.

Some methods like model-based tracking [2], the ant tracking algorithm [22], and a few more partially automate fault detection. One more commonly used approach to automate fault detection is Hough transform method [3]. Pre-processing of data before applying the Hough transform can improve the results as well. To name a few of the pre-processing steps are highlighting the likely fault points [27], semblance based matching attribute [12], and enhancement of data using dip attribute [18]. On the other hand, seismic amplitude values are the input to a few of the convolutional neural network (CNN) based approaches to fault detection. Authors in [9] divide a given seismic cube into two sections, train the neural network using the data of one section and test it on the other section. The FaultSeg3D method [30] treats the fault detection problem as an image segmentation problem and the FaultNet3D method [29] as an image classification problem. Transfer learning can also be used by training a neural network on synthetic data and using the trained parameters to detect faults in real data [7].

Using seismic features/attributes as input to the neural networks is a familiar way to fault detection. Seismic attribute values can help improve the network's performance on real seismic data. In the scalable deep learning method [15], seismic attributes are extracted from synthetic data and used in the neural network. One supervised deep learning approach [33] is to implement a network that is designed to use a combination of Dip and Azimuth. The multi-attribute support vector machine (SVM) method combines 14 attributes to detect the fault [10]. However, seismic attributes can be computationally expensive; that is why we do not use any attribute.

Like an image, a graph is also a way to present data that contains a set of nodes (or objects or signals), and the edges define the relationships between them. Presenting seismic data in the graph domain can help inspect the connections between data. Data belonging to a similar class will be connected when we present data in the nearest neighbor graph. We can detect faults by analyzing these connections. Graph representation learning has the main objectives of node classification, link prediction, and clustering. As we present the extracted patches as nodes in the graph domain, our application is node classification. Node classification using Graph neural network (GNN) is introduced in [4]. Basically, GNN applies deep learning methods to data in the graph domain.

Graph Laplacian regularization and graph embedding are the two approaches to learning about graph representation. Two examples of Laplacian regularization are manifold regularization [5] and deep semi-supervised embedding [28]. Two such examples for embedding based approaches are DeepWalk [23], and node2vec [13]. On the other hand, graph convolutional propagation is introduced in [11] for the task of graph-level classification. However, for graph convolution, a single weight matrix per layer can also be used [16]. Authors in [19] define a new CNN architecture, specifically applicable for graph space using a graph matching procedure between the input signal and filters. Spectral graph convolution is a way to implement graph convolutional networks with fast localized

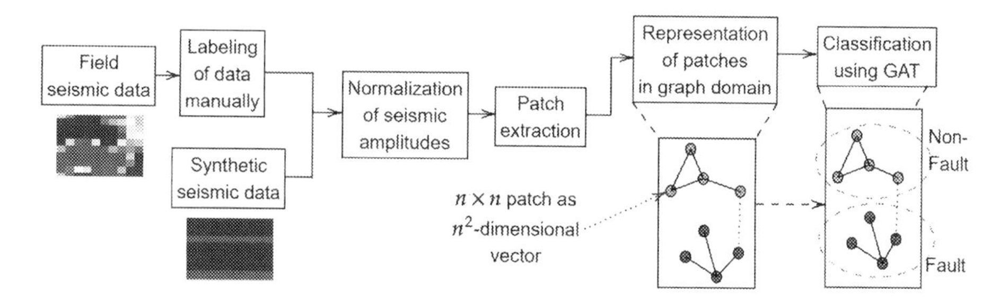

Fig. 1. Block diagram of the proposed work. The training dataset is prepared by combining synthetic and real data. Extracted patches contain seismic amplitudes. In a kNN graph, patches are represented as nodes, and GAT is used to determine their classification and fault location.

convolutions [8]. GNN has a wide range of applications in various fields such as citation networks [16], natural language processing [32], hyperspectral image processing [14], social networks [1] and others.

The graph attention network (GAT) leverages the masked self-attention layers in structured graph data [26]. Different weights are applied to different nodes in a neighborhood by stacking layers in which nodes have control over their neighborhood's features. An advantage of attention mechanisms is that they are capable of dealing with variable-sized inputs. Self-attention is the process of computing a representation of a single sequence with the aid of an attention mechanism. Machine reading [6] and learning sentence representations [17], among other tasks, have been shown to benefit from self-attention.

A few CNN-based methods involve the extraction of patches and then inputting the patches to neural networks [20,25,31]. The method proposed in [25] is: first, extract patches and classify them as fault or non-fault patches using CNN; next, identify the fault line using the Hough transform. In contrast, authors in [31] extract 3D patches centering each point before applying CNN. Similarly, this paper extracts 2D patches of different dimensions centering on a single seismic data point (pixel). In order to implement GAT, we represent these patches in the graph domain. Figure 1 shows the block diagram of our work. The rest of the paper is organized as follows: Sect. 2 describes the data used in our experiments. In Sect. 3, we discuss the method for extracting patches, and in Sect. 4, we present the graph representation of the patches. The model and its application are described in Sect. 5. Sections 6 and 7 describe the different results obtained. We conclude the paper in Sect. 8 of the paper.

2 Description of Data

After acquiring seismic data, first, data are pre-processed, and the next task is interpreting the data. Figure 2 shows an example of the 3D structure of seismic

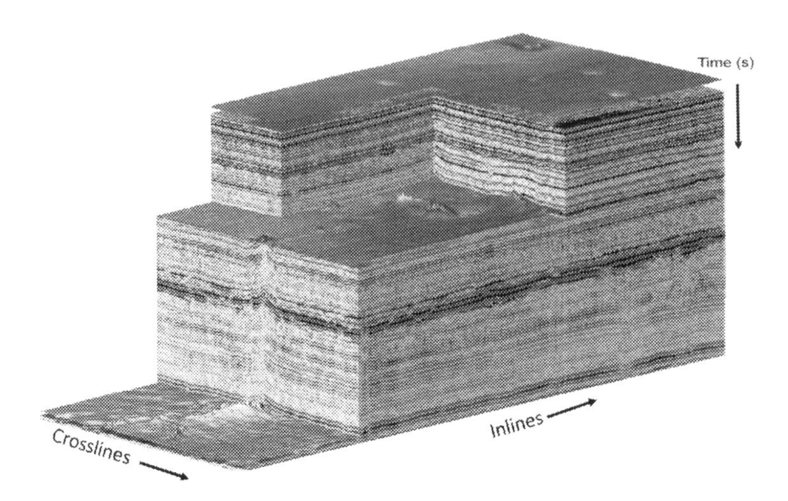

Fig. 2. An example of a seismic data (image source: https://www.boem.gov/sites/default/files/environmental-stewardship/Environmental-Studies/Gulf-of-Mexico-Region/Kramer-Shedd.pdf).

data. The inline and crossline directions are perpendicular to each other and are along the earth surface. The time samples of data are into the depth of the earth surface. The seismic data contain seismic amplitude values that measure the relative change in rock property impedance between two distinct seismic layers. The seismic amplitude can take positive, negative or zero values depending on the contrast between two layers. The real data we use in our experiments are 3D processed post-stack time migrated volumes. These data are collected from the KG (Krishna-Godavari) Basin area from the Bay of Bengal, India. KG Basin data consist of 2536 inlines, 2601 crosslines with a step size of 2, and the data is

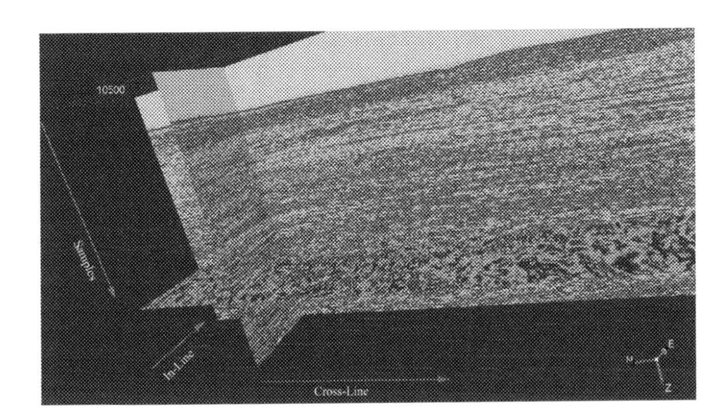

Fig. 3. Demonstration of a section from 3D KG Basin data. This section of the KG Basin data has inline ranges from 1982 to 2100, crossline ranges from 101 to 1020, and 1001 samples.

recorded for 4 s, with a sampling interval of 4 ms. Figure 3 shows a small section of KG Basin data. Under expert geophysicists, we manually label some parts of data to be included in the training set. Every pixel in the data is manually examined to create the labels. Due to the human inference, there may be some label noise, causing some mislabeled or unlabeled faults. One single trace in seismic data can be presented as a simple convolution operation.

$$s(t) = w(t) * r(t) + n(t) \tag{1}$$

where $s(t)$ is the seismic trace, $w(t)$ presents a wavelet, $r(t)$ is the reflectivity series, and $n(t)$ is noise. We can also use this equation to produce synthetic seismic data. We assume zero noise and create a synthetic reflectivity series. Then we convolve this reflectivity series with a Ricker wavelet to generate a synthetic trace. The next task is to create faults in these synthetic data. In seismology, the major fault categories are dip-slip, strike-slip, and oblique-slip based on the slip type. We create three fault types inferring these fault categories: straight fault, slant fault, and curve fault. Figure 4a shows an example of ideal synthetic data with three faults. This synthetic data has linear layers, but in reality, the seismic layers will be very complicated. Figure 4b shows a real data with faults. As shown in Fig. 4b, the structure of real data will be highly irregular. The idea behind using synthetic data is to create a suitable size of training data for the GAT. Because the labelled seismic data are not readily available, and for any network to perform efficiently, it should also have enough training data. Using a mixture and synthetic data and real data helps to create a large training dataset for GAT.

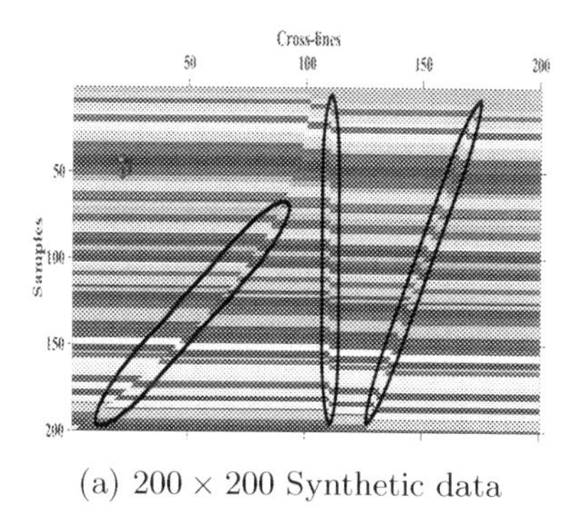

(a) 200 × 200 Synthetic data

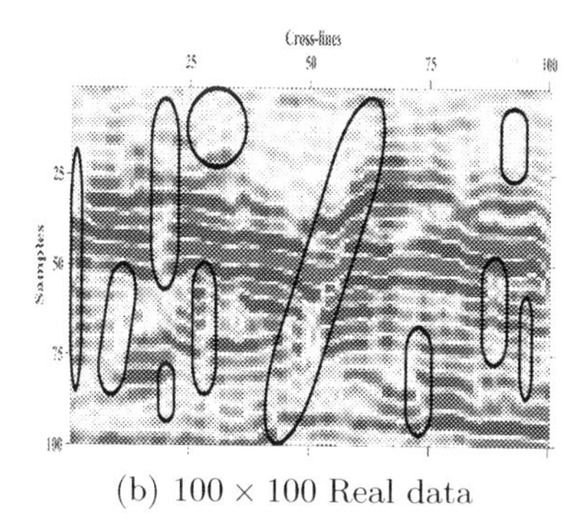

(b) 100 × 100 Real data

Fig. 4. Data are of 1 inline section. (a) is 200 × 200 synthetic data with faults, and (b) is 100 × 100 real data, from KG basin data, with faults.

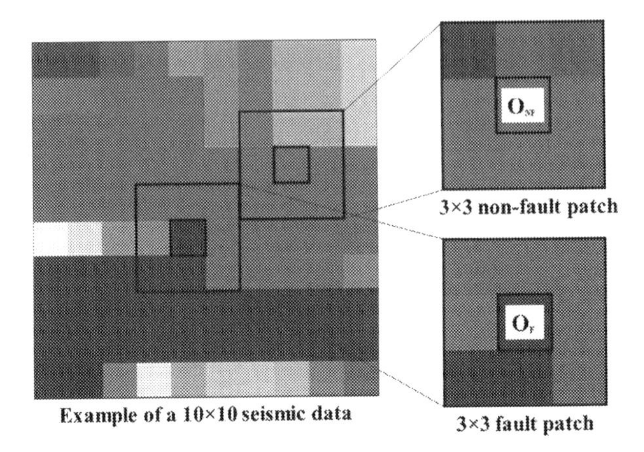

Fig. 5. Two central points O that has to be analyzed, and illustrating extraction of patches of dimension 3×3.

3 Extraction of Patches

Fault in seismic data is a kind of discontinuity among the seismic layers. Further, the faults or fractures appear vertically; horizontal faults are unlikely to be of concern. We train the GAT to detect these faults from the neighborhood of the point of concern by using the patches. We extract 2D patches by considering 1 inline, n crosslines and n samples centred around a point (pixel). After analyzing the patches, we interpret if the point is a fault point or not. Likewise, we analyze all the points in a given data by extracting patches for all points. Figure 5 shows a simple example of patch extraction for two central points O_F and O_{NF} in a 10×10 data. The dimension of patches will always be odd, and Fig. 5 shows patches of dimension 3×3. However, we vary the dimension of patches for our experiments, and we extract patches up to 25×25. For an input patch of dimension $n \times n$, the label will be that of the central point O, denoted by fault (1) or non-fault (0). Zero paddings are done for the patch extraction of points situated at the edges of data.

4 Representation of Patches in the Graph Domain

A 2D-grid graph is one option for representing the patches in the graph domain. In this graph, each point in the data will be seen as a single node. However, for extensive data, this graph representation has an issue of the high number of nodes. For extracting patches of dimension $n \times n$ from seismic data of I inlines J crosslines and K samples, we will have $I \times J \times K$ graphs, each with $n \times n$ nodes. However, in a k-nearest neigh- bor (kNN) graph, the whole patch will be seen as a single node in the n^2-dimensional vector space. So, for extracting patches of dimension $n \times n$ from $I \times J \times K$ seismic data, we will have one graph with $I \times J \times K$ nodes. Hence we consider the kNN graph for our experiments.

The kNN graph computes the distance between the nodes, and the edges are connected between the k nearest neighbors. The kNN graph is very popular for the clustering application of graph nodes. Because in a kNN graph, nodes belonging to one class will be connected, and at the same time, a connection between nodes belonging to different classes is improbable.

We can express a graph as $G = (V, \mathbf{A})$, where $V = \{v_0, ..., v_{N-1}\}$ is a set of N nodes, and $\mathbf{A}$ is the $N \times N$ adjacency matrix. In our case, the N nodes are the N patches of dimension $n \times n$ in the n^2-dimensional vector space. Further, we consider the value of k in the kNN graph to be six. Because the value six will be optimal for having data of two classes in terms of efficiency and time complexity. The graph construction is done by connecting a node with its six nearest neighbors. The nearest neighbors are chosen by calculating the distance between the nodes. The patches contain the amplitude values. Hence the amplitude distance between two nodes is calculated by considering the Euclidean distance between patchs' amplitude values. We take undirected edges, and the weight of each edge is $A_{i,j} = exp(-d_{i,j}^2)$, where $d_{i,j}$ is the amplitude distance between ith and jth vertex.

5 Application of GAT

Authors in [21] apply a graph signal regularization method to classify the patches after presenting the patches using the kNN graph. However, we apply GAT after presenting the patches in the graph domain. For an input graph with N nodes, the node features $h = \{\overrightarrow{h_1}, \overrightarrow{h_2}, ..., \overrightarrow{h_N}\}, \overrightarrow{h_i} \in \mathbb{R}^F$, where F is the number of features. In our case, we use $n \times n$ patches so the number of features per node $F = n^2$. The output layer $h' = \{\overrightarrow{h'_1}, \overrightarrow{h'_2}, ..., \overrightarrow{h'_N}\}, \overrightarrow{h'_i} \in \mathbb{R}^{F'}$.

Performing the self-attention with a weight matrix $\mathbf{W} \in \mathbb{R}^{F' \times F}$ we get the attention coefficient

$$e_{i,j} = a(\mathbf{W}\overrightarrow{h_i}, \mathbf{W}\overrightarrow{h_j}) \tag{2}$$

The masked attention is performed by computing $e_{i,j}$ for the nodes $j \in N_i$, where N_i is the neighbourhood of the node i. Further, the coefficients are normalized using softmax function.

$$a_{i,j} = softmax_j(e_{i,j}) \tag{3}$$

In GAT, the attention mechanism a is a single-layer feed-forward network. With LeakyReLU, the attention coefficient can be expressed as

$$a_{i.j} = \frac{exp(LeakyReLU(\overrightarrow{a}^T[\mathbf{W}\overrightarrow{h_i}||\mathbf{W}\overrightarrow{h_j}]))}{\sum_{k \in N_i} exp(LeakyReLU(\overrightarrow{a}^T[\mathbf{W}\overrightarrow{h_i}||\mathbf{W}\overrightarrow{h_k}]))} \tag{4}$$

where $.^T$ is the transposition and $||$ is the concatenation operation. Hence the final output feature for every node can be expressed as

$$\overrightarrow{h'_i} = \sigma(\sum_{j \in N_i} a_{i,j}\mathbf{W}\overrightarrow{h_j}) \tag{5}$$

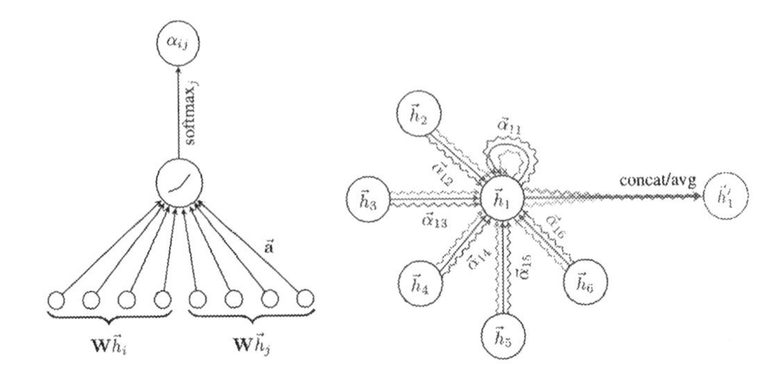

Fig. 6. Left: The attention mechanism a Right: An multihead attention example. Separately colored arrows indicate independent attention computations [26]

where σ is the non-linear activation unit (Fig. 6).

Instead of concatenation, using averaging in multihead attention, the final output layer can be written as

$$\overrightarrow{h'_i} = \sigma(\frac{1}{K}\sum_{k=1}^{K}\sum_{j\in N_i} a_{i,j}^k \mathbf{W}^k \overrightarrow{h_j}) \tag{6}$$

Table 1 provides the number of data used for designing the GAT model, and this pattern is followed for all dimensions of patches.

Table 1. Numbers of data used in GAT

Data	Fault data			Non-fault data		
	Synth	Real	Total	Synth	Real	Total
Training	5000	5000	10000	2000	8000	10000
Validation	1000	1000	2000	500	1500	2000
Testing	1000	2000	3000	500	2500	3000

6 Experimental Results

We have done the experiments with a system of 128 GB RAM, 4 GB GPU and a processor of Intel(R) Xeon(R) Gold 5115 CPU @ 2.40 GHz 2.39 GHz. For constructing the graphs, we use the graph signal processing toolbox [24], and we use PyTorch to design the model. The GAT architecture we use has 16 hidden layers and 16 number of head attentions. We also use 0.6 drop out, 0.005 learning rate and α for LeakyReLU is 0.2. We run one training model for 200 epochs and then test to get the accuracy; we repeat this experiment 100 times and record the average accuracy. Table 2 shows the performance of GAT on different sizes of patches. The accuracy increases with the increased dimension of patches up to 13 × 13, but after that, it almost saturates. The highest accuracy is 90.54% with patches of dimension 13 × 13.

Table 2. Performance of GAT on various dimensions of patches

Patch size	Accuracy	Patch size	Accuracy
3 × 3	78.25%	15 × 15	90.16%
5 × 5	81.26%	17 × 17	89.67%
7 × 7	86.89%	19 × 19	89.39%
9 × 9	88.64%	21 × 21	88.72%
11 × 11	88.97%	23 × 23	89.06%
13 × 13	90.54%	25 × 25	88.05%

7 Application on Field Seismic Data

The actual evaluation of any method lies in its application to real field data. We have included some real data in the testing set, and in this section, we apply our method directly to real field seismic data. We take the highest accuracy of 90.54% for patch dimensions of 13 × 13.

First, we take 200 × 200 synthetic data with three fault types of different lengths. Figure 7a1 shows the synthetic data and Fig. 7b1 shows the original label of the data. Figure 7c1 shows the result obtained using GAT and 13 × 13 patches. For testing the application to field data, we take 100 different 100 × 100 data, which are not a part of training data (described in the Table 1). Figure 7a2-a3 show two examples of such 100 × 100 real data and Fig. 7b2-b3 show the target labels for these data. These labels are created manually and can have some label noise ($<5\%$) due to some unlabeled or mislabeled faults. We calculate the accuracies of the obtained results from these manually created labels. Figure 7c2-c3 demonstrate results on real data obtained using GAT. Next, we compare the performance of our method with some of the existing standard methods on the 100 × 100 × 100 data from the KG Basin. We take one Hough transform based method and two CNN based methods to compare with our GAT based method. The first method enhances the resolution of seismic data by using a multi-scale fusion of the dip attribute followed by the Hough transform to detect the faults [18]. The faultseg3D method is a simple binary image segmentation of 3D seismic volumes using CNN [30]. Authors in [31] directly apply CNN to the extracted 3D patches, but we implement the same using 2D patches. Table 3 gives the accuracies obtained from different methods. GAT method detects both the major faults, and the minor faults. Other methods fail at detecting all the minor faults, and false fault detection also affects the accuracy.

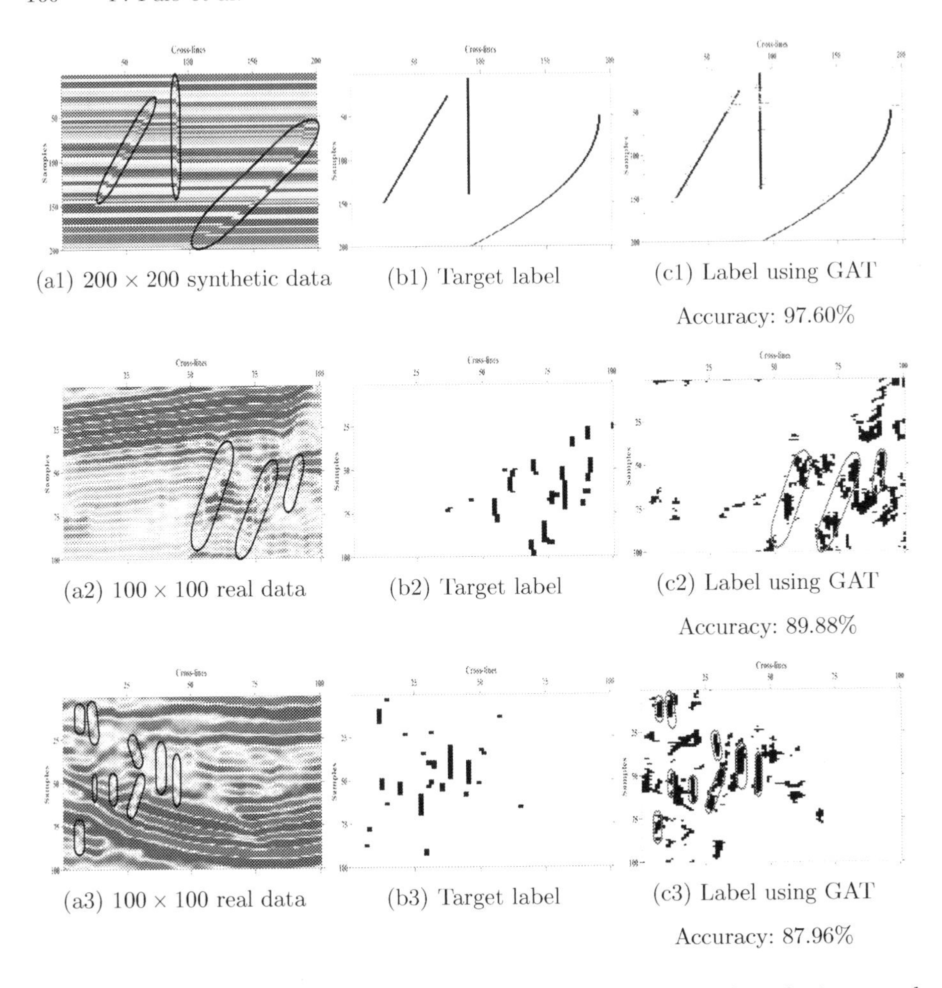

(a1) 200 × 200 synthetic data (b1) Target label (c1) Label using GAT Accuracy: 97.60%

(a2) 100 × 100 real data (b2) Target label (c2) Label using GAT Accuracy: 89.88%

(a3) 100 × 100 real data (b3) Target label (c3) Label using GAT Accuracy: 87.96%

Fig. 7. Examples of obtained results (with fault regions highlighted in the input and the outputs).

Table 3. Accuracy of different methods on 100 × 100 × 100 data from KG Basin

Method	Accuracy
Resolution enhanced Hough transform	74.00%
FaultSeg3D	83.35%
CNN on 2D patches	84.79%
GAT based method	89.51%

8 Conclusion

We propose a methodology that uses seismic amplitude values as input to GAT that can successfully detect a fault in a 2D seismic section. The proposed method extracts patches from data using a patch extraction method and then represents the patches using the kNN graph in a graph domain. And then, in the next step, we classify the patches using GAT. The classified patches detect the fault location. The proposed method outperforms some existing fault detection methods in seismic data. The proposed method in this paper is for 2D data, and in future work, we will try to extend it for 3D data.

References

1. Abu-El-Haija, S., Kapoor, A., Perozzi, B., Lee, J.: N-GCN: multi-scale graph convolution for semi-supervised node classification. CoRR arXiv:abs/1802.08888 (2018). http://arxiv.org/abs/1802.08888
2. Admasu, F., Back, S., Toennies, K.: Autotracking of faults on 3D seismic data. Geophysics **71**, A49–A53 (2006). https://doi.org/10.1190/1.2358399
3. AlBinHassan, N.M., Marfurt, K.: Fault detection using hough transforms. In: SEG Technical Program Expanded Abstracts, pp. 1719–1721 (2005). https://doi.org/10.1190/1.1817639
4. Atwood, J., Towsley, D.: Diffusion-convolutional neural networks. In: Advances in Neural Information Processing Systems, vol. 29, pp. 2001–2009. Curran Associates, Inc. (2016). https://doi.org/10.5555/3157096.3157320
5. Belkin, M., Niyogi, P., Sindhwani, V.: Manifold regularization: a geometric framework for learning from labeled and unlabeled examples. J. Mach. Learn. Res. **7**, 2399–2434 (2006). JMLR:v7:belkin06a
6. Cheng, J., Dong, L., Lapata, M.: Long short-term memory-networks for machine reading. CoRR abs/1601.06733 (2016). http://arxiv.org/abs/1601.06733
7. Cunha, A., Pochet, A., Lopes, H., Gattass, M.: Seismic fault detection in real data using transfer learning from a convolutional neural network pre-trained with synthetic seismic data. Comput. Geosci. **135**, 104344 (2020). https://doi.org/10.1016/j.cageo.2019.104344
8. Defferrard, M., Bresson, X., Vandergheynst, P.: Convolutional neural networks on graphs with fast localized spectral filtering. In: Proceedings of the 30th International Conference on Neural Information Processing Systems, NIPS 2016, pp. 3844–3852. Curran Associates Inc. (2016). https://doi.org/10.5555/3157382.3157527
9. Di, H., Wang, Z., AlRegib, G.: Seismic fault detection from post-stack amplitude by convolutional neural networks. In: 80th EAGE Conference and Exhibition, pp. 1–5 (2018). https://doi.org/10.3997/2214-4609.201800733
10. Di, H., Shafiq, A., Alregib, G.: Seismic-fault detection based on multiattribute support vector machine analysis. In: SEG Technical Program Expanded Abstracts, pp. 2039–2044 (2017). https://doi.org/10.1190/segam2017-17748277.1
11. Duvenaud, D.K., et al.: Convolutional networks on graphs for learning molecular fingerprints. In: Advances in Neural Information Processing Systems, vol. 28, pp. 2224–2232. Curran Associates, Inc. (2015). https://doi.org/10.5555/2969442.2969488
12. Gibson, D., Spann, M., Turner, J., Wright, T.: Fault surface detection in 3-D seismic data. IEEE Trans. Geosci. Remote Sens. **43**, 2094–2102 (2005). https://doi.org/10.1109/TGRS.2005.852769
13. Grover, A., Leskovec, J.: Node2vec: scalable feature learning for networks. In: Proceedings of the 22nd ACM SIGKDD International Conference on Knowledge Discovery and Data Mining, KDD 2016, pp. 855–864. Association for Computing Machinery (2016). https://doi.org/10.1145/2939672.2939754

14. Hong, D., Gao, L., Yao, J., Zhang, B., Plaza, A., Chanussot, J.: Graph convolutional networks for hyperspectral image classification. IEEE Trans. Geosci. Remote Sens. **59**, 1–13 (2020). https://doi.org/10.1109/TGRS.2020.3015157
15. Huang, L., Dong, X., Clee, T.E.: A scalable deep learning platform for identifying geologic features from seismic attributes. Lead. Edge **36**(3), 249–256 (2017). https://doi.org/10.1190/tle36030249.1
16. Kipf, T.N., Welling, M.: Semi-supervised classification with graph convolutional networks. CoRR arXiv:abs/1609.02907 (2016). http://arxiv.org/abs/1609.02907
17. Lin, Z., et al.: A structured self-attentive sentence embedding. CoRR abs/1703.03130 (2017). http://arxiv.org/abs/1703.03130
18. Mahadik, R., Routray, A.: Fault detection and optimization in seismic dataset using multiscale fusion of a geometric attribute. In: IECON 2019–45th Annual Conference of the IEEE Industrial Electronics Society, vol. 1, pp. 107–112 (2019). https://doi.org/10.1109/IECON.2019.8927569
19. Martineau, M., Raveaux, R., Conte, D., Venturini, G.: Graph matching as a graph convolution operator for graph neural networks. Pattern Recogn. Lett. **149**, 59–66 (2021). https://doi.org/10.1016/j.patrec.2021.06.008
20. Palo, P., Routray, A., Mahadik, R., Singh, S., Tandon, R., Bahuguna, C.S.: Fault interpretation using neural networks. In: 13th Biennial International Conference and Exhibition. SPG, February 2020. https://www.spgindia.org/Kochi2020-expanded-abstracts/id-389-fault-interpretation-using-neural-networks.pdf
21. Palo, P., Routray, A., Singh, S.: Seismic fault analysis using graph signal regularization. In: 2021 29th European Signal Processing Conference (EUSIPCO), pp. 1835–1839 (2021, forthcoming)
22. Pedersen, S.I., Randen, T., Sonneland, L., Steen, Ø.: Automatic fault extraction using artificial ants. In: SEG Technical Program Expanded Abstracts, pp. 512–515 (2005). https://doi.org/10.1190/1.1817297
23. Perozzi, B., Al-Rfou, R., Skiena, S.: Deepwalk: online learning of social representations. In: Proceedings of the 20th ACM SIGKDD International Conference on Knowledge Discovery and Data Mining, KDD 2014, pp. 701–710. Association for Computing Machinery (2014). https://doi.org/10.1145/2623330.2623732
24. Perraudin, N., et al.: GSPBOX: a toolbox for signal processing on graphs. arXiv e-prints, August 2014
25. Pochet, A., Diniz, P.H.B., Lopes, H., Gattass, M.: Seismic fault detection using convolutional neural networks trained on synthetic poststacked amplitude maps. IEEE Geosci. Remote Sens. Lett. **16**(3), 352–356 (2019). https://doi.org/10.1109/LGRS.2018.2875836
26. Veličković, P., Cucurull, G., Casanova, A., Romero, A., Liò, P., Bengio, Y.: Graph attention networks. In: International Conference on Learning Representations (2018)
27. Wang, Z., AlRegib, G.: Fault detection in seismic datasets using hough transform. In: IEEE International Conference on Acoustic, Speech and Signal Processing, pp. 2372–2376, May 2014. https://doi.org/10.1109/ICASSP.2014.6854024
28. Weston, J., Ratle, F., Collobert, R.: Deep learning via semi-supervised embedding. In: Proceedings of the 25th International Conference on Machine Learning, ICML 2008, pp. 1168–1175. Association for Computing Machinery (2008). https://doi.org/10.1145/1390156.1390303
29. Wu, X., Shi, Y., Fomel, S., Liang, L., Zhang, Q., Yusifov, A.Z.: FaultNet3D: predicting fault probabilities, strikes, and dips with a single convolutional neural network. IEEE Trans. Geosci. Remote Sens. **57**(11), 9138–9155 (2019). https://doi.org/10.1109/TGRS.2019.2925003

30. Wu, X., Liang, L., Shi, Y., Fomel, S.: FaultSeg3D: using synthetic data sets to train an end-to-end convolutional neural network for 3D seismic fault segmentation. Geophysics **84**, IM35–IM45 (2019). https://doi.org/10.1190/geo2018-0646.1
31. Xiong, W., et al.: Seismic fault detection with convolutional neural network. Geophysics **83**(5), O97–O103 (2018). https://doi.org/10.1190/geo2017-0666.1
32. Yao, L., Mao, C., Luo, Y.: Graph convolutional networks for text classification. In: Proceedings of the AAAI Conference on Artificial Intelligence, vol. 33, no. 01, pp. 7370–7377 (2019). https://doi.org/10.1609/aaai.v33i01.33017370
33. Zheng, Y., Zhang, Q., Yusifov, A., Shi, Y.: Applications of supervised deep learning for seismic interpretation and inversion. Lead. Edge **38**(7), 526–533 (2019). https://doi.org/10.1190/tle38070526.1

Skeleton-Based Mutual Action Recognition Using Interactive Skeleton Graph and Joint Attention

Xiangze Jia[1], Ji Zhang[2(✉)], Zhen Wang[3], Yonglong Luo[4], Fulong Chen[4], and Gaoming Yang[5]

[1] Nanjing University of Aeronautics and Astronautics, Nanjing, China
jiaxiangze@nuaa.edu.cn
[2] University of Southern Queensland, Toowoomba, Australia
Ji.Zhang@usq.edu.au
[3] Zhejiang Lab, Hangzhou, China
wangzhen@zhejianglab.com
[4] Anhui Normal University, Wuhu, China
ylluo@ustc.edu.cn, long005@mail.ahnu.edu.cn
[5] Anhui University of Science and Technology, Huainan, China

Abstract. Skeleton-based action recognition relies on skeleton sequences to detect certain predetermined types of human actions. The existing related works are inadequate in mutual action recognition. We thus propose an innovative interactive skeleton graph to represent the skeleton data. In addition, because the GCN pays attention to the information about the edges in the skeleton graph which represent the interaction between joints, we propose a joint attention module that assists the model in paying attention to the pattern of vertices which represent the joints in the skeleton graph. We validate our model on the NTU RGB-D datasets, and the experimental results demonstrate the superiority of our model against other baseline methods in terms of recognition effectiveness in understanding mutual actions.

Keywords: Interactive skeleton graph · Joint attention · Action recognition

1 Introduction

Technically, various types of data can be leveraged for action recognition, such as images, depth maps, and skeleton data [1,2]. Conventional methods rely on the handcrafted feature representation and models to handle skeleton sequences. Inspired by research on image feature extraction, models based on CNN extract features from skeleton sequences projected into a pseudo image. Recently, GCN has been introduced and has been successfully applied in many fields, such as document classification, semi-supervised learning, and point cloud analysis.

C. Strauss et al. (Eds.): DEXA 2022, LNCS 13427, pp. 110–116, 2022.
https://doi.org/10.1007/978-3-031-12426-6_9

Each vertex learns a more advanced feature representation from its surrounding vertices through adjacent edges using GCN.

It is observed that many scenes in the study are mutual actions. In previous related works [3–6], solutions for mutual action recognition can be divided into two categories.: 1) only the skeleton sequence of the main subject is used for action recognition; 2) the skeleton sequences of all subjects are used for motion recognition, but each subject is processed independently.All these methods cannot obtain accurate feature representations of mutual action because 1)there are two key parties in mutual action in which only using one subject for action recognition will inevitably lead to feature loss; 2) processing each subject separately cannot achieve the aggregation of features.

We use the **G**CN-based **L**STM model as a backbone and propose two key innovative components: **I**nteractive Skeleton Graph and Joint **A**ttention Module, and our model is named **GLIA**. With the interactive skeleton graph, GCN fuses features of multiple subjects involved in the action with the interactive skeleton graph. The joint attention module we proposed assigns attention weights to the corresponding vertices representing joints,

Our contributions can be summarized as follows:

- We propose an innovative interactive skeleton graph that contributes to a better feature fusion of multiple subjects involved in the mutual actions;
- We introduce a joint attention module that focuses on the vertices in the skeleton graph, i.e., joints, which make the model make full use of the skeleton graph;
- Our interactive skeleton graph and joint attention module are plug and play, and can be used in other networks, such as ST-GCN;
- We conduct experiments on two popular benchmark datasets for mutual action recognition, i.e., the NTU60 and NTU120 datasets, and the experimental results show GLIA achieves SOTA performance for mutual action recognition.

2 Our Model

2.1 GCN with Distance Grouping Strategy

We use graph structure to represent skeleton data. Let $G = (V, A)$ represents the skeleton graph, where $V = \{v_i\}_{i=1}^{N}$ represents the set of N body joints, and A is the adjacency matrix. If there is a bone between joint i and joint j, we have $A_{i,j} = 1$; otherwise, $A_{i,j} = 0$. The input features are expressed as $X = \{x_1, x_2, \ldots, x_d\} \in \mathbb{R}^{N \times d}$. We define $\Phi_k(A)$ as the polynomial of the adjacency matrix A; that is, only when the shortest distance between v_i and v_j is equal to k then we have $[\Phi_k(A)]_{i,j} = 1$. We define $\Phi_0(A) = I$ as the self-connection of the vertices, and $\Phi_k(A)$ can be obtained from A^k and $\Phi_j(A), j < k$. Different from the regular GCN, we have:

$$Y_{gc}(X) = \frac{1}{K_1} \sum_{k=0}^{K_1-1} D_k^{-1} \Phi_k(A) X W_k \tag{1}$$

where K_1 is a hyper-parameter used to determine the receptive field of the vertex. D_k is the degree matrix of $\Phi_k(A)$, and its role is to normalize the features. $W_k \in \mathbb{R}^{d \times d'}$ is the learnable weight.

We use the above improved GCN for spatial feature representation and LSTM for temporal information modeling. For spatiotemporal feature fusion, We use GCN-based LSTM [7] as the backbone.

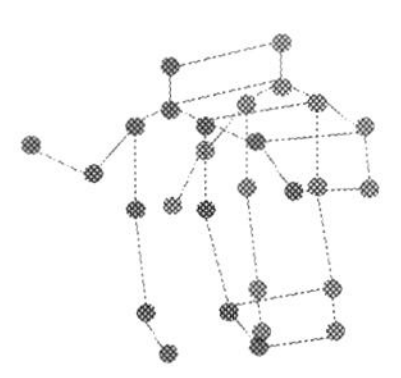

Fig. 1. The interactive skeleton graph involves two subjects. The joint edge connects the joints in the same subject, and the subject edge (red) binds the same joints between different subjects. Only part of the subject edges is given in the figure for brevity. (Color figure online)

2.2 Interactive Skeleton Graph

Given the limitations of existing works in dealing with mutual actions, we propose an interactive skeleton graph for skeleton data. As shown in Fig. 1, there are two types of edges in our interactive skeleton graph: the joint edges and subject edges. The joint edge indicates that two joints in the same subject are adjacent. The subject edge connects the same joints of different subjects. We introduce $\mathcal{A} \in \mathbb{R}^{MN \times MN}$ to represent the joint edges, where M is the number of subjects, MN is the number of joints, and $\mathcal{A} = diag(A, \ldots, A)$ is a diagonal matrix composed of N independent A. We use the adjacency matrix $\mathcal{B} \in \mathbb{R}^{MN \times MN}$ to represent the subject edges. We refine the Eq. 1 about GCN as

$$Y_{gc}(X) = \frac{1}{K_1} \sum_{k=0}^{K_1-1} D_k^{-1} \Phi_k(\mathcal{A}) X W_k + \frac{1}{K_2} \sum_{k'=1}^{K_2} D'_{k'} \Phi'_{k'}(\mathcal{B}) X W_{k'} \tag{2}$$

where K_2 is a hyper-parameter used to determine the receptive field of the vertices on other subjects. Y_{gc} consists of two parts: the first part is the feature in each subject, and the second part is the interactive feature between subjects. We define $\Phi'_{k'}(\mathcal{B})$ as

$$\Phi'_{k'}(\mathcal{B}) = \mathcal{B}\Phi_{k'-1}(\mathcal{A}) \quad k' > 0 \tag{3}$$

2.3 Joint Attention Module

We introduce a joint attention module to learn the importance of the vertices representing joints. The joint attention module learns from the feature map and assigns weights to different joints. We squeeze the feature map as an $(M \times N, c)$

dimensional tensor by pooling through temporal dimension and feed the tensor to Multilayer perceptron (MLP) to generate attention weights. Finally, we multiply (by element) the activation value with the original feature map, and each joint is given different attention weights.

Table 1. The recognition accuracy of the SBU dataset.

Methods	Acc
ST-LSTM [1]	93.3%
2sGCA-LSTM [8]	94.9%
VA-LSTM [9]	97.2%
LSTM-*IRN* [10]	**98.2%**
SGCConv [11]	94.0%
GLIA (ours)	**98.2%**

3 Experimental Evaluation

3.1 Datasets

SBU Dataset. SBU dataset is constructed by Kinect and contains the RGB-D sequence information of mutual actions. Seven participants in an experimental environment complete it. There are 282 sequence data, which are divided into eight categories. The skeleton sequences contain 15 joints, and three-dimensional coordinates represent each joint.

NTU RGB-D. NTU60 has 56,880 action samples, divided into 60 classes and 11 mutual action categories. The advantages of this data set lie in the rich action categories, the vast number of video clips, and the diversity of camera perspectives. NTU120 extends NTU60, adding 60 classes and another 57,600 video samples. There are two ways to evaluate the recognition accuracy in NTU60 and NTU120, Cross-Subject (CS) and Cross-View (CV) accuracy evaluation.

3.2 Results

SBU. As shown in Table 1, we compare our GLIA to several major existing models, including LSTM-IRN, VA-LSTM, and GCA-LSTM. We achieve similar accuracy as LSTM-IRM. However, unlike LSTM-IRN using all inter subjects and intrasubject joint pairs as input, our GLIA doesn't need to pair joints. The fusion of joint features depends on the interactive skeleton graph, which is a more straightforward feature fusion method.

Table 2. The recognition accuracy of mutual actions of NTU60 and NTU120.

Methods	NTU60		NTU120	
	CS acc (%)	CV acc (%)	CS acc (%)	CV acc (%)
ST-LSTM [1]	83.0	87.3	63.0	66.6
LSTM-IRN [10]	90.5	93.5	77.7	79.6
AGC-LSTM [7]	89.2	95.0	73.0	73.3
SAN [4]	88.2	93.5	-	-
VACNN [12]	88.9	94.7	-	-
ST-GCN [3]	83.3	88.7	-	-
AS-GCN [13]	87.6	95.2	82.9	83.7
ST-TR [5]	90.8	**96.5**	85.7	87.1
2sshift-GCN [14]	90.3	96.0	86.1	86.7
MS-G3D [15]	91.7	96.1	-	-
2sKA-AGTN [6]	90.4	96.1	86.7	88.2
GLIA	**93.7**	**96.5**	**88.2**	**88.9**

NTU60 and NTU120. We report the recognition accuracy of mutual actions in Table 2. Among the 11 categories involving mutual action, our GLIA achieves 93.7% and 96.5% accuracy in Cross-Subject and Cross-View, respectively. We classify these works as LSTM-based, CNN-based, GCN-based. Compared to GCN-based models, e.g., AS-GCN, AAM-GCN, and newer self-attention models, e.g., ST-TR, KA-AGTN, our GLIA is still the best in mutual action recognition. Our GLIA is more advantageous in multi-subject feature fusion. As an extension of NTU60, the data form and distribution of NTU120 and NTU60 are similar. With the increase of classes and samples, the recognition accuracy of all models decreased compared with those in NTU60. However, our GLIA is still the best in the comparative models in terms of the Cross-Subject and Cross-View.

4 Conclusion

We propose the interactive skeleton graph that helps realize a better feature fusion between multiple subjects. In addition, we propose a joint attention module to further improve the skeleton graph utilization. We conduct the experimental evaluations on two popular datasets, and the results demonstrate the effectiveness of our GLIA in mutual action recognition. Furthermore, we believe that our GLIA can still be improved. For example, Transformer is more expressive than GCN and can be used in Spatio-temporal dimensions; Pre-training can improve the representation and transfer ability of the models.

Acknowledgment. The authors would like to thank the support from Natural Science Foundation of China (No. 62172372), Zhejiang Provincial Natural Science Foundation (No. LZ21F030001) and Henan Center for Outstanding Overseas Scientists (GZS2022011).

References

1. Liu, J., Shahroudy, A., Xu, D., Wang, G.: Spatio-temporal LSTM with trust gates for 3D human action recognition. In: Leibe, B., Matas, J., Sebe, N., Welling, M. (eds.) ECCV 2016. LNCS, vol. 9907, pp. 816–833. Springer, Cham (2016). https://doi.org/10.1007/978-3-319-46487-9_50
2. Ke, Q., Bennamoun, M., An, S., Sohel, F., Boussaid, F.: A new representation of skeleton sequences for 3D action recognition. In: Proceedings of The IEEE Conference on Computer vision and Pattern Recognition, Honolulu, pp. 3288–3297. IEEE (2017)
3. Yan, S., Xiong, Y., Lin, D.: Spatial temporal graph convolutional networks for skeleton-based action recognition. In: Thirty-Second AAAI Conference on Artificial Intelligence, New Orleans. AAAI (2018)
4. Cho, S., Maqbool, M., Liu, F., Foroosh, H.: Self-attention network for skeleton-based human action recognition. In: Proceedings of the IEEE/CVF Winter Conference on Applications of Computer Vision, Snowmass Village, pp. 635–644. IEEE (2020)
5. Plizzari, C., Cannici, M., Matteucci, M.: Spatial temporal transformer network for skeleton-based action recognition. In: Del Bimbo, A., et al. (eds.) ICPR 2021. LNCS, vol. 12663, pp. 694–701. Springer, Cham (2021). https://doi.org/10.1007/978-3-030-68796-0_50
6. Liu, Y., Zhang, H., Xu, D., He, K.: Graph transformer network with temporal kernel attention for skeleton-based action recognition. Knowl.-Based Syst. **240**, 108146 (2022)
7. Si, C., Chen, W., Wang, W., Wang, L., Tan, T.: An attention enhanced graph convolutional LSTM network for skeleton-based action recognition. In: Proceedings of the IEEE/CVF Conference on Computer Vision and Pattern Recognition, Long Beach, pp. 1227–1236. IEEE (2019)
8. Liu, J., Wang, G., Duan, L.Y., Abdiyeva, K., Kot, A.C.: Skeleton-based human action recognition with global context-aware attention LSTM networks. IEEE Trans. Image Process. **27**(4), 1586–1599 (2017)
9. Zhang, P., Lan, C., Xing, J., Zeng, W., Xue, J., Zheng, N.: View adaptive recurrent neural networks for high performance human action recognition from skeleton data. In: Proceedings of the IEEE International Conference on Computer Vision, Honolulu, pp. 2117–2126. IEEE (2017)
10. Perez, M., Liu, J., Kot, A.C.: Interaction relational network for mutual action recognition. IEEE Trans. Multimedia **24**, 366–376 (2021)
11. Wu, F., Souza, A., Zhang, T., Fifty, C., Yu, T., Weinberger, K.: Simplifying graph convolutional networks. In: International Conference on Machine Learning, Long Beach, pp. 6861–6871. PMLR (2019)
12. Zhang, P., Lan, C., Xing, J., Zeng, W., et al.: View adaptive neural networks for high performance skeleton-based human action recognition. IEEE Trans. Pattern Anal. Mach. Intell. **41**(8), 1963–1978 (2019)

13. Li, M., Chen, S., Chen, X., Zhang, Y., et al.: Actional-structural graph convolutional networks for skeleton-based action recognition. In: Proceedings of the IEEE/CVF Conference on Computer Vision and Pattern Recognition, Long Beach, pp. 3595–3603. IEEE (2019)
14. Cheng, K., Zhang, Y., He, X., Chen, W., et al.: Skeleton-based action recognition with shift graph convolutional network. In: Proceedings of the IEEE/CVF Conference on Computer Vision and Pattern Recognition, pp. 183–192. IEEE (2020)
15. Liu, Z., Zhang, H., Chen, Z., Wang, Z., Ouyang, W.: Disentangling and unifying graph convolutions for skeleton-based action recognition. In: Proceedings of the IEEE/CVF Conference on Computer Vision and Pattern Recognition, pp. 143–152. IEEE (2020)

Comparison of Sequence Variants and the Application in Electronic Medical Records

Yuqing Li[1(✉)], Hieu Hanh Le[1], Ryosuke Matsuo[2], Tomoyoshi Yamazaki[3], Kenji Araki[3], and Haruo Yokota[1]

[1] Tokyo Institute of Technology, Tokyo, Japan
{li,hanhlh}@de.cs.titech.ac.jp, yokota@cs.titech.ac.jp
[2] Life Data Initiative, Kyoto, Japan
matsuo@ldi.or.jp
[3] University of Miyazaki Hospital, Miyazaki, Japan
{yama-cp,taichan}@med.miyazaki-u.ac.jp

Abstract. A sequence variant is a data structure that represents elements with partial order, and it can be regarded as a sequence with branches. The comparison of sequence variants is an important problem in practical applications. To conduct comparisons, it is necessary to detect the common and uncommon parts of sequence variants, but a suitable method is not available. In this study, we developed a method for comparing sequence variants by providing appropriate definitions and algorithms. The longest common subsequence variant is defined based on the longest common subsequence and a merged sequence variant is proposed for comparison. We also propose algorithms for calculating these sequence variants. As an example, we applied the methods to real data from electronic medical records (EMRs) to determine the diversity of the treatment patterns in different hospitals, before presenting the results to medical workers to help them recognize the differences and improve their medical actions. Frequent treatment patterns were extracted from a real treatment order database of EMRs by using sequential pattern mining, and the differences in the treatment patterns were calculated and visualized as longest common subsequence variants and merged sequence variants.

Keywords: Sequence variant · Electronic medical records · Medical support · Comparison of medical institutes · Sequential pattern mining

1 Introduction

A sequence is a very common structure that appears almost everywhere. An ordered list can be represented as a sequence. The comparison of two or more sequences is an important problem. Identifying the longest common subsequence (LCS) that represents the common part of several sequences is a well-known method for solving this problem. However, the sequence structure and LCS cannot be identified in all situations in practical applications. For various reasons,

C. Strauss et al. (Eds.): DEXA 2022, LNCS 13427, pp. 117–130, 2022.
https://doi.org/10.1007/978-3-031-12426-6_10

elements may have partial order and they cannot be represented as a sequence, and thus the LCS cannot be used for comparing the elements. Thus, it is necessary to expand the sequence and LCS concepts. A sequence variant (SV) with branches has been proposed as an expansion of a sequence. If we consider an SV as a directed graph, then to compare SVs we can try to use methods used to compare graphs, such as graph edit distance [1]. However, SVs have some characteristics such as unique supremum and infimum, and such graph methods could not treat such points appropriately. Therefore we need to consider methods more specifically. To conduct comparisons of SVs, it is necessary to detect the common and uncommon parts of the SVs, but a suitable method is not available.

A clinical pathway is a typical example of a practical problem that involves the comparison of SVs. A clinical pathway is the typical treatment process for a given disease and it is usually generated by medical workers. In recent years, electronic medical records (EMRs) have been rapidly replacing traditional medical records written on paper and the reuse of data in EMRs is expected. As an example of the reuse of EMRs, the automatic extraction of frequent treatment patterns for a given disease has been investigated widely. In 2015, the Government of Japan initiated the Millennial Medical Record Project [2] to manage and reuse medical records on a nationwide scale. Due to the increasing number of medical institutes that are participating in the Millennial Medical Record Project, it is expected that research will be conducted to analyze the data from multiple hospitals. This research may allow the identification of the diversity of treatment procedures, and the medical workers in one hospital will be able to compare the differences in their treatment patterns with those in other medical institutes.

In this study, we developed a method for comparing SVs by providing appropriate definitions and algorithms. We then applied these methods to analyze the EMRs from multiple medical institutes and to help improve their medical actions. To address the research aims, we focused on medical treatment order data in EMRs. In the proposed method, the longest common subsequence variant (LCSV) is defined based on the LCS, and the merged sequence variant (MSV) is used for comparison. Frequent treatment patterns were then extracted from real treatment order databases of EMRs by using sequential pattern mining (SPM), and the differences in the treatment patterns were calculated and visualized as LCSVs and MSVs.

The remainder of this paper is organized as follows. Background knowledge and related work are summarized in Sect. 2. The proposed methods and experimental evaluation are described in Sect. 3 and Sect. 4, respectively. We give our conclusions and suggestions for future work in Sect. 5.

2 Related Work

In this section, we introduce related works about SPM and SVs.

2.1 SPM

SPM is based on using important data-mining algorithms for discovering frequent patterns in a sequence database and it has various application domains, such as in medicine, e-commerce, and the World Wide Web [3]. Given a database based on a set of sequences, the problem involves extracting frequent patterns where the percentage of sequences containing them is greater than the predefined minimum support value, i.e., minsup.

The Apriori-based frequent pattern mining algorithms are well-known SPM algorithms [4], but they are very time-consuming with large data sets and many irrelevant patterns are generated. To exclude irrelevant patterns, PrefixSpan [5] was proposed to mine the complete set of patterns while also reducing the candidate pattern generation effort by exploring prefix projection. To further improve the efficiency, CSpan [6] was proposed for mining closed sequential patterns. This algorithm uses a pruning method called occurrence checking, which allows the early detection of closed sequential patterns during mining.

2.2 Analysis of Medical Data by SPM

Wright et al. [7] used an SPM-based algorithm called SPADE [8] to extract the frequent patterns in the medical orders of diabetic patients and recommended the dosages of medications according to the rules detected in the frequent patterns. Uragaki et al. [9] constructed an item as a group comprising the type, explanation, code, and name, and then conducted extraction considering the information for medicines. Their results showed that more information could be extracted by mining based on the code representing the effect of a medicine rather than mining based on the name of the medicine. Moreover, the time interval was calculated statistically so it was not necessary to set the time interval in advance. Le et al. [10] proposed methods by considering privacy.

2.3 Analysis of SVs

Honda et al. [11] detected SVs by detecting the common part of a closed frequent pattern with the same number of items for each relevant treatment day. Le et al. [12] analyzed the reasons why variants might appear by using multivariate analysis.

3 Proposed Methods

To our knowledge, no previous studies have proposed methods for comparing SVs, and thus we introduce the LCSV and MSV as new concepts.

In the following, we provide definitions of the Sequence, SV, LCS, LCSV and MSV, and introduce the methods used to calculate LCSV and MSV. Moreover, we introduce a method for comparing frequent treatment patterns. The proposed methods are based on the premise that frequent sequences are extracted by SPM and the sequences are combined as SVs.

We refer to items as nodes. A node has four attributes comprising *name* as the explanation, *nextList* as the list of adjacent nodes, *id* as its unique number, and *label* as the SV to which it belongs. In particular, sv.nextList $= \{sv' \in SV | (sv, sv') \in E\}$.

3.1 Sequence

Definition 1 (Itemset). *An itemset I is defined as follows:*

$$I = \{i_1, i_2, \ldots, i_n\},$$

where $i \in I$ is an item.

Definition 2 (Sequence). *For an itemset I, a sequence S is defined as follows:*

$$S = (\{s_1, s_2, \ldots, s_m\}, \prec_S),$$

where $s_j = (id, i), i \in I$ and id is a unique index in S and $\prec_S$ is the total order on S, i.e., $\forall s_1, s_2 \in S, s_1 \prec_S s_2 \vee s_2 \prec_S s_1$.

3.2 SV

Definition 3 (SV). *For an itemset I, a lattice SV is defined as follows:*

$$SV = (\{sv_1, sv_2, \ldots, sv_l\}, \prec_{SV}),$$

where $sv_j = (id, i), i \in I$ and id is a unique index in SV and $\prec_{SV}$ is the partial order on SV. Moreover, SV has a unique supremum and unique infimum.

When capital letters (e.g., A, B, C) are used to refer to an item in the following, the letter refers to the name of the item with ids. In addition, $A \prec B, B \prec C, B \prec D, C \prec E, D \prec E$ is expressed as $<A, B, (C, D), E>$.

We introduce a sequence variant graph (SVG) as a method for visualizing SVs.

Definition 4 (SVG). *For an SV, a directed acyclic graph is defined as an SVG, as follows:*

$$G = (V, E),$$

where the vertex set V is the itemset of SV, and $\forall e = (sv_{out}, sv_{in}) \in E, sv_{out} \prec_{SV} sv_{in}$.

It is easy to show the corresponding SV by visualizing an SVG. For example, if A, B, C, D, E are items, then the SVG of $SV = <A, B, (C, D), E>$ is shown in Fig. 1.

SVs and SVGs can be transformed from each other, so we treat them as the same in the following.

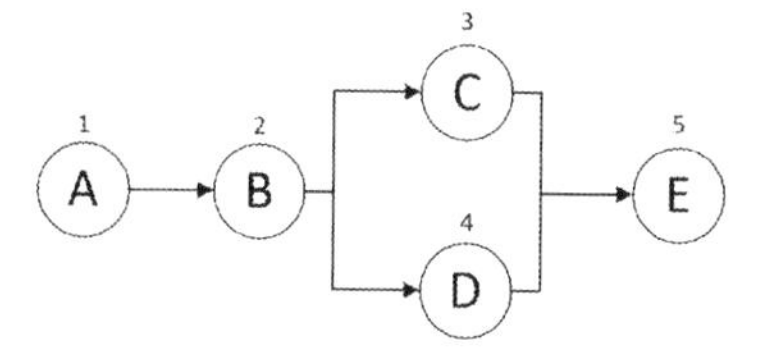

Fig. 1. SVG of SV.

3.3 LCS

Definition 5 (Common subsequence (CS)). *For the sequences S_α and S_β, the CS of S_α and S_β is defined as a sequence that satisfies the following conditions:*

$$\forall cs_i, cs_j \in CS, cs_i, cs_j \in S_\alpha \cap S_\beta, \text{ and } cs_i \prec_{CS} cs_j \iff cs_i \prec_{S_\alpha} cs_j \wedge cs_i \prec_{S_\beta} cs_j.$$

A subsequence of a CS is also a CS, so there may be multiple CSs. Thus, we can define the LCS as follows.

Definition 6 (LCS). *For the sequences S_α and S_β, the LCS of S_α and S_β is defined as a CS that satisfies the following condition:*

$$\text{For any common subsequence } CS, |CS| \leq |LCS|,$$

where $|CS|$ refers to the number of nodes in CS.

This problem is NP-hard for the general case of an arbitrary number of input sequences [13]. However, the problem is solvable in polynomial time by dynamic programming when the number of sequences is constant.

3.4 LCSV

In practice, a frequent pattern extracted from a database is not unique, and thus it cannot be described as a sequence and the LCS cannot be used for comparing patterns directly. Therefore, analogous to SVs, we introduce the common subsequence variant (CSV) and LCSV concepts.

Definition 7 (CSV). *For SV_α and SV_β, the CSV of SV_α and SV_β is defined as an SV that satisfies the following conditions:*

$$\forall csv_i, csv_j \in CSV, csv_i, csv_j \in SV_\alpha \cap SV_\beta \text{ and}$$

$$csv_i \prec_{CSV} csv_j \iff csv_i \prec_{SV_\alpha} csv_j \wedge csv_i \prec_{SV_\beta} csv_j.$$

The subsequence variant of a CSV is also a CSV, so there may be multiple CSVs. Thus, we can define the LCSV.

Algorithm 1. LCSV

1: input: SV_1 and SV_2
2: output: $LCSV$
3: number SV_1 from 1 to s
4: number SV_2 from $s+1$ to t
5: $r_1 \leftarrow$ the number of paths of SV_1
6: $r_2 \leftarrow$ the number of paths of SV_2
7: **for** $i = 1, 2, \ldots, r_1$ **do**
8: **for** $j = 1, 2, \ldots, r_2$ **do**
9: $LCS(i, j) \leftarrow$ the LCS of path i of SV_1 and path j of SV_2
10: **end for**
11: $LCS(i) \leftarrow$ the longest in

$$LCS(i,1), LCS(i,2), \ldots, LCS(i, r_2)$$

12: **end for**
13: $LCSV \leftarrow$ combination of $LCS(1), LCS(2), \ldots, LCS(r_1)$
14: renumber nodes in $LCSV$ from $t+1$ and also renumber those in SV_1 and SV_2

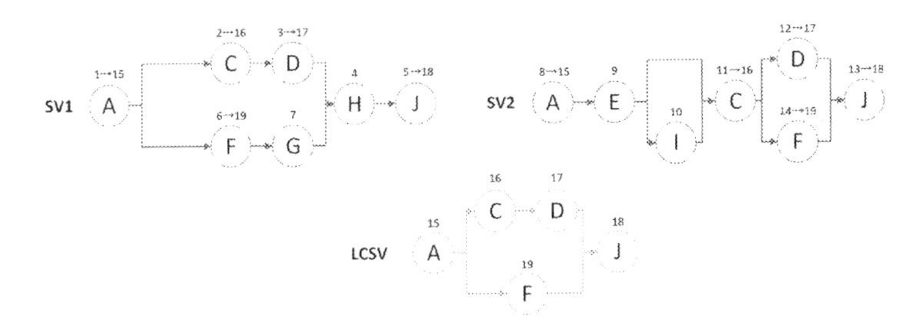

Fig. 2. Finding the LCSV of SV_1 and SV_2.

Definition 8 (LCSV). *For SV_α and SV_β, the LCSV of SV_α and SV_β is defined as a CSV that satisfies the following condition:*

$$\textit{For any common subsequence variant } CSV, |CSV| \leq |LCSV|,$$

where $|CSV|$ refers to the number of nodes in CSV.

The LCSV is calculated using Algorithm 1, as illustrated in Fig. 2. We use dynamic programming to calculate the LCS of paths.

3.5 MSV

To compare the frequent treatment patterns, we introduce the MSV as follows.

Definition 9 (MSV). *For SV_α, SV_β, and their $LCSV$, the MSV is defined as an SV that satisfies the following conditions.*
$\forall sv_i, sv_j \in SV_\alpha \cup SV_\beta, sv_i, sv_j \in MSV$, *and* $sv_i \prec_{MSV} sv_j \iff sv_i \prec_{SV_\alpha} sv_j \wedge sv_i \prec_{SV_\beta} sv_j . \forall msv_i, msv_j \in MSV, msv_i \prec_{MSV} msv_j \Longrightarrow (msv_i, msv_j \in$

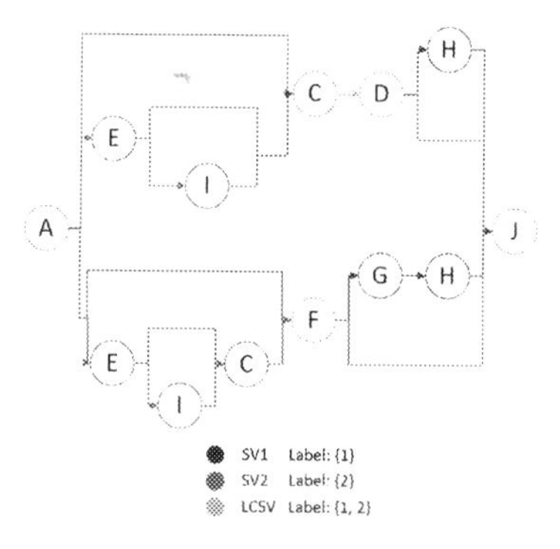

Fig. 3. MSV of SV_1 and SV_2.

Algorithm 2. MSV

1: input: SV_1, SV_2
2: output: MSV
3: $LCSV \leftarrow LCSV(SV_1, SV_2)$
4: $map \leftarrow NodeMappingGeneration(SV_1, SV_2, LCSV)$
5: restore the MSV from *map* (*map* contains label and order information)

$SV_\alpha \wedge msv_i \prec_{SV_\alpha} msv_j) \vee (msv_i, msv_j \in SV_\beta \wedge msv_i \prec_{SV_\beta} msv_j)$. *Moreover, the nodes in LCSV are labeled as* $\{\alpha, \beta\}$ *and the other nodes are labeled as* α *or* β *according to the SV to which they belong.*

Thus, we propose the method for calculating the MSV of SV_1 and SV_2 in Algorithms 2, 3, and 4.

The mapping *map* is a key–value structure where the key is the id of the node and the value is the node. In the last step of Algorithm 3, when we combine the two mappings, if the node is in map_1 and map_2, then we combine the nextList of them and place it in *map*.

To make it easier to understand the labels of the nodes, we use blue for SV_1, red for SV_2, and orange for the common part, i.e., the $LCSV$. We then apply the algorithm in Fig. 2 and the result is shown in Fig. 3.

In fact, constructing the MSV involves several problems. First, if several nodes have the same name in the SV, they need to be distinguished. This problem is solved by numbering the nodes when calculating the LCSV. Another problem occurs when some nodes have order relations with several patterns. In the MSV, the nodes must be multiplied and some of them belong to the LCSV whereas others only belong to some SVs. For example, as shown in Fig. 2, node C and node F both occur in parallel and in sequence, so in the MSV shown in Fig. 3, there are two node Cs with different label as {2} (red) and {1,2} (yellow) to ensure that the positional relationship is retained in MSV.

Algorithm 3. NodeMappingGeneration

```
1: input: SV_1, SV_2, LCSV
2: output: a mapping map
3: map_1 ← ∅
4: map_2 ← ∅
5: list ← node(LCSV)
6: n ← maximum number of nodes in LCSV
7: while list ≠ ∅ do
8:     CN_c ← node with the smallest number in list
9:     CN_1 ← node in SV_1 with the same number as CN_c
10:    CN_2 ← node in SV_2 with the same number as CN_c
11:    newCN_c ← CN_c
12:    newCN_c.next ← CN_1.next ∪ CN_2.next
13:    put {newCN_c.id : newCN_c} into map_1 and map_2
14:    nextList ← CN_c.next
15:    map_1 ← SearchNext(CN_1, nextList, CN_c, n, map_1)
16:    nextList ← CN_c.next
17:    map_2 ← SearchNext(CN_2, nextList, CN_c, n, map_2)
18:    list.remove(CN_c)
19: end while
20: combine map_1 and map_2 as map
```

4 Experiments

In experiments, we applied the proposed methods to real treatment order data from the University of Miyazaki Hospital (UMH) and Miyazaki Medical Association Hospital (MMAH) to verify their effectiveness.

4.1 Experimental Method and Environment

In the medical order database, orders without medical meanings such as "Fees" and "Dietary therapy" were deleted, and SPM was then applied to obtain frequent patterns in the data from each hospital. We combined the sequences from one hospital as a SV. Next, the LCSV algorithm was applied to calculate the LCSV of the SVs from the two hospitals. Finally, the MSV was constructed and visualized in a graph. The experiment used the original patterns and only the longest were considered because they contained the maximum number of medically meaningful medical orders.

The execution environment is described in Table 1.

4.2 Data Sets

The target data comprised medical treatment data coded as clinical pathways and recorded from April 2015 to March 2020 in the EMR system called the Clinical Research Information Infrastructure at the Faculty of Medicine, UMH. The data did not include information that could be used to uniquely identify a patient to ensure patient privacy. When we extracted medical treatment data from patient

Algorithm 4. SearchNext

```
1: input: CN, nextList, CN_c, n, map
2: output: map
3: if nextList = ∅ or CN.next=∅ then
4:     return
5: end if
6: for node in CN.next do
7:     if map.keySet().contains(node.id) and CN_c.next.contains(node) then
8:         newNode ← node
9:         newNode.id ← n + 1
10:        n + +
11:        put {newNode.id : newNode} into map
12:    else
13:        put {node.id : node} into map
14:    end if
15:    if node ∈ nextList then
16:        map.get(node.id).label ← {1, 2}
17:        nextList.remove(node)
18:    else
19:        SearchNext(node, nextList, CN_c, n, map)
20:    end if
21: end for
```

Table 1. Execution environment

CPU	AMD Ryzen Threadripper 3960X 24-Core Processor
Memory	128 GB
OS	Ubuntu 18.04.1 LTS
Java version	Java 11.0.11
R version	R 3.6.3

records, we used anonymous patient IDs to prevent the recovery of patient identification. The use of data from this EMR system to support medical treatments is described in [14], which is the website of the University of Miyazaki. Our study was approved by the Ethics Review Board of the University of Miyazaki and the Research Ethics Review Committee of the Tokyo Institute of Technology.

The EMR system used to develop our methods included time data with accuracy to the minute and the SPM algorithm was based on time. Thus, we aimed to apply the methods and conduct the analysis based on time information. However, we wanted to compare multiple hospitals, so we needed to use the data recorded in the Clinical Research Information Infrastructure database. This treatment order information only included date information and no actual times unlike the data we used before. Due to the lack of time information, we used the data in the order recorded in the database and actual time information was not used in the experiments. Thus, we could not guarantee the correctness of the order relation during one day. Therefore, the experiments should be conducted again based on date information in the future.

Table 2. Characteristics of the PCI data set.

	UMH	MMAH
Number of patients	438	4,265
Maximum clinical order for each patient	816	1,100
Average clinical order for each patient	85.61	69.51
Maximum length of stay (days)	48	50
Average length of stay (days)	11.52	6.87

Table 3. Information for the orders

Node name	Receipt code	Point	Explanation
CBC	160008010	21	Complete blood count
HR and RR	160102510	150	Heart rate and respiratory rate
B-CK-MB	160114710	90	Biochemical test
BUN	160019010	11	Biochemical test
T-BIL	160017010	11	Biochemical test
B-Cr	160019210	11	Biochemical test
B-CRP	160054710	16	Immunological test
ECG12	160068410	136	Electrocardiogram
B-BNP	160162350	130	Endocrinological test
PCI	150374910	36,000	Percutaneous coronary intervention
	150375010	22,000	
	150375110	19,300	
	150260350	28,280	
	150284310	24,720	
	150359310	24,720	
	150375210	34,380	
	150375310	24,380	
	150375410	21,680	
	160107550	17,720	
	150318310	19,640	

The target data for the experiments comprised medical treatment pathways for *percutaneous coronary intervention* (PCI). PCI was selected because it is a common surgical intervention in both hospitals. Table 2 shows the characteristics of the data set.

4.3 Experimental Results

Extraction of SVs. The information for the orders is shown in Table 3. In particular, PCI had several types and we unified them as PCI. Each order had a point referring to the cost. As there were several PCI types, we used the average point of 24,802. To acquire more information, the order name or receipt code

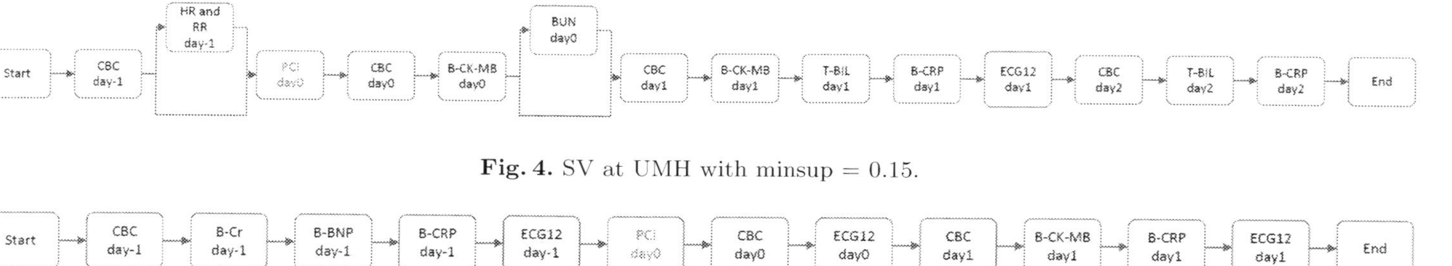

Fig. 4. SV at UMH with minsup = 0.15.

Fig. 5. SV at MMAH with minsup = 0.25.

Fig. 6. LCSVs at UMH (0.15) and MMAH (0.25).

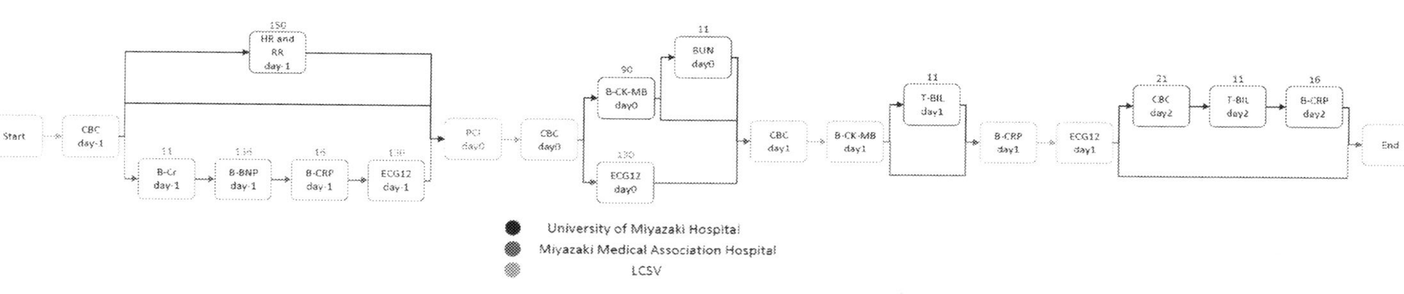

Fig. 7. MSVs at UMH (0.15) and MMAH (0.25).

could be searched in [15] to obtain the kubun code (e.g., A000-00), and a detailed explanation could then be found in [16] using the kubun code.

We tested several minsup values but due to the page count limitation, we used 0.15 for UMH and 0.25 for MMAH. The SVs are shown in Figs. 4 and 5.

The total points were calculated for each hospital. Four paths were present in the SV for UMH. The total points in the four paths were 25,411, 25,400, 25,261, and 25,250, and the average was 25,330.5. The total number of points in the SV for MMAH was 25,524. The difference between the SVs was under 0.5%, and thus it was difficult to compare the results.

Only one path was present in the SV for MMAH, possibly because the number of patients was large and it was difficult to find a sufficiently frequent pattern.

Calculation of LCSVs. The LCSVs in the SVs for UMH and MMAH are shown in Fig. 6.

We considered the order explanation and relative date as a whole, e.g., "B-CK-MB day0" and "B-CK-MB day1" are different orders.

Construction of MSVs. The MSVs for the two hospitals are shown in Fig. 7, where the points in the medical orders in the different parts are shown at the top.

By comparing the blue and red parts in the figure, we identified the differences in the treatment patterns in the two hospitals. For example, T-BIL occurred several times in the patterns in UMH but it did not occur in those in MMAH. In addition, ECG12 was conducted every day in MMAH but only once in UMH.

We discussed the differences with medical workers and the effectiveness of the comparison was confirmed. In particular, we discussed why the difference in the use of ECG12 was likely to have occurred. A three-lead electrocardiogram was applied continuously after surgery. It was considered that it may be sufficient to apply a 12-lead electrocardiogram only once in UMH. By contrast, the doctors at MMAH considered that it was necessary every day. We searched the original database and found that T-BIL was sometimes conducted in UMH but the frequency was too low to be extracted.

5 Conclusion and Future Work

In this study, we focused on comparing medical treatment order data in EMRs. Given that no methods have been developed for making such comparisons, we proposed definitions of the LCSV and MSV. We extracted frequent treatment patterns for a specific disease from a real treatment order database of EMRs by using SPM and applied our proposed methods for calculating the LCSV and MSV for the SVs from two hospitals. We also proposed a method for visualizing the differences in the SVs. We applied our proposed methods to real treatment order data from the two hospitals in experiments, and the differences between the two hospitals were visualized.

As explained above, the EMR system used in the present study could not guarantee the order relations in one day, and thus extraction based on date information should be considered in future research. Therefore, during the extraction process, the orders that occurred in one day were treated as occurring at the same time in the present study. The treatment patterns for other diseases could also be analyzed in future. In addition, a visualization tool for automatically representing the MSV could be developed to make it easier to understand the differences in SVs. As many medical institutes have joined the Millennial Medical Record Project, it would be useful to extend the method to comparing more than two hospitals.

Acknowledgement. This research is supported by Grants-in-Aid for Scientific Research (B) (#20H04192) and Grants-in-Aid for Early-Career Scientists (#21K17746) from the Japan Society for the Promotion of Science.

References

1. Gao, X., Xiao, B., Tao, D., Li, X.: A survey of graph edit distance. Pattern Anal. Appl. **13**, 113–129 (2010)
2. Yoshihara, H.: Millennial medical record project: toward establishment of authentic Japanese version EHR and secondary use of medical data. J. Inf. Process. Manag. **60**, 767–778 (2018)
3. Fournier-Viger, P., Lin, J.C.-W., Kiran, R.U., Koh, Y.S., Thomas, R.: A survey of sequential pattern mining. Data Sci. Pattern Recogn. **1**(1), 54–77 (2017)
4. Agrawal, R., Srikant, R.: Fast algorithms for mining association rules in large databases. In: Proceedings of the 20th International Conference on Very Large Data Bases, pp. 487–499 (1994)
5. Pei, J., Han, J., Mortazavi-Asl, B., Pinto, H., Chen, Q., Dayal, U., Hsu, M.: PrefixSpan: mining sequential patterns efficiently by prefix-projected pattern growth. In: Proceedings of the 2001 International Conference on Data Engineering, pp. 215–224 (2001)
6. Raju, V.P., Varma, G.S.: Mining closed sequential patterns in large sequence databases. Int. J. Database Manag. Syst. **7**(1), 29–39 (2015)
7. Wright, A.P., Wright, A.T., McCoy, A.B., Sittig, D.F.: The use of sequential pattern mining to predict next prescribed medications. J. Biomed. Inform. **53**, 73–80 (2015)
8. Zaki, M.J.: SPADE: An efficient algorithm for mining frequent sequences. Mach. Learn. **42**, 31–60 (2001)
9. Uragaki, K., et al.: Sequential pattern mining on electronic medical records with handling time intervals and the efficacy of medicines. In: Proceedings of the 21st IEEE International Symposium on Computers and Communications, pp. 20–25 (2016)
10. Le, H.H., Kushima, M., Araki, K., Yokota, H.: Differentially private sequential pattern mining considering time interval for electronic medical record systems. In: Proceedings of the 23rd International Database Engineering and Applications Symposium, pp. 95–103 (2019)

11. Honda, Y., Kushima, M., Yamazaki, T., Araki, K., Yokot, Y.: Detection and visualization of variants in typical medical treatment sequences. In: Proceedings of the 3rd VLDB Workshop on Data Management and Analytics for Medicine and Healthcare. Springer, pp. 88–101 (2017)
12. Le, H.H., Yamada, T., Honda, Y., Kayahara, M., Kushima, M., Araki, K., Yokota, H.: Analyzing sequence pattern variants in sequential pattern mining and its application to electronic medical record systems. In: Hartmann, S., Küng, J., Chakravarthy, S., Anderst-Kotsis, G., Tjoa, A.M., Khalil, I. (eds.) DEXA 2019. LNCS, vol. 11707, pp. 393–408. Springer, Cham (2019). https://doi.org/10.1007/978-3-030-27618-8_29
13. Maier, D.: The complexity of some problems on subsequences and supersequence. J. ACM **25**(2), 322–336 (1978)
14. The Section of Medical Information at the Faculty of Medicine, University of Miyazaki Hospital. http://www.med.miyazakiu.ac.jp/home/jyoho/
15. Various Information on Medical Fees. https://shinryohoshu.mhlw.go.jp/shinryohoshu/paMenu/doPaDetailSpNext&100
16. Index of Treatment Orders. https://www.ichikawa568.com/ika-sinryouhousyu-tensuuhyo.html

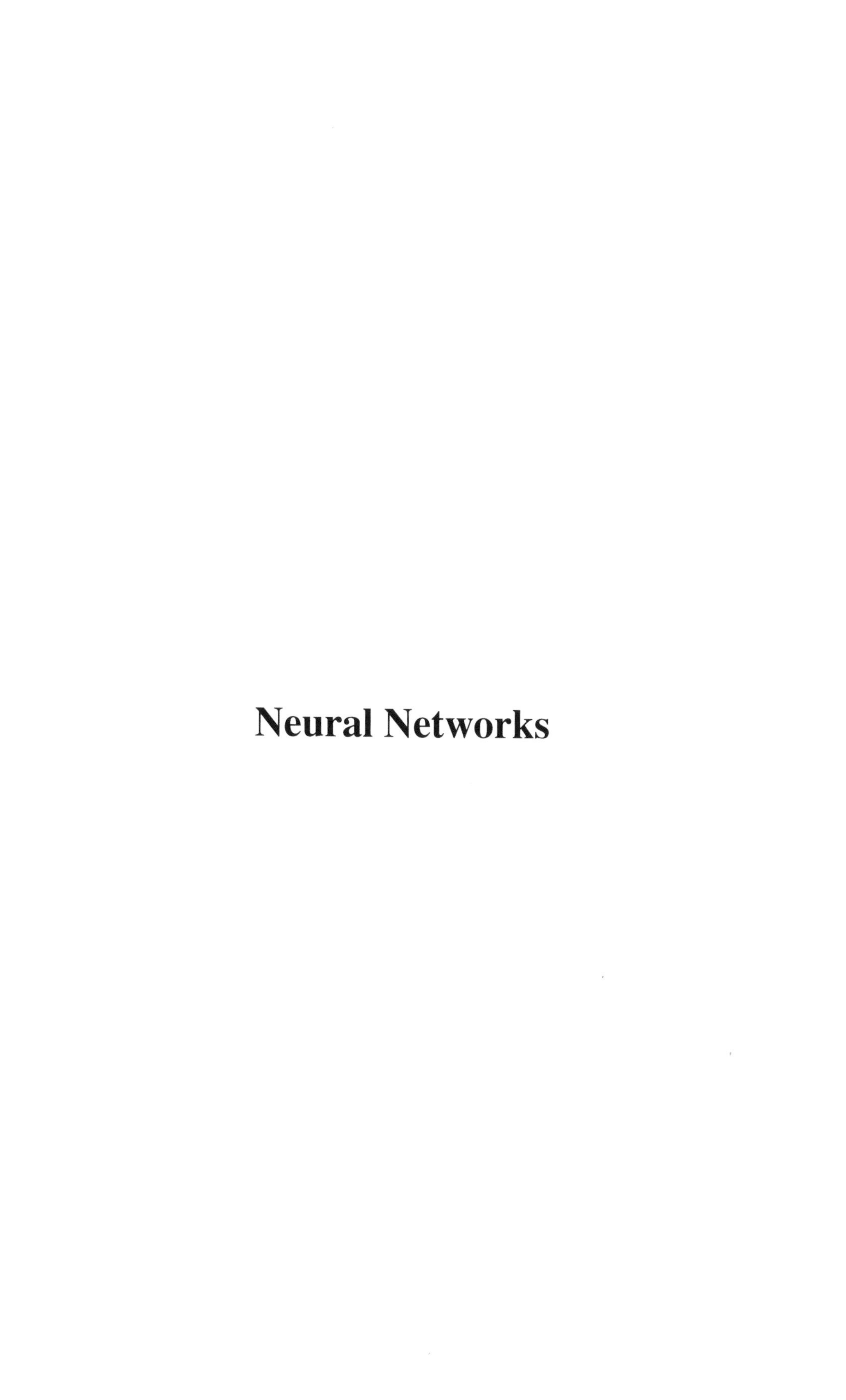

Neural Networks

Reconciliation of Mental Concepts with Graph Neural Networks

Lorenz Wendlinger[1(✉)], Gerd Hübscher[2], Andreas Ekelhart[3], and Michael Granitzer[1]

[1] Universität Passau, Passau, Germany
{lorenz.wendlinger,michael.granitzer}@uni-passau.de
[2] Hübscher & Partner Patentanwälte GmbH, Linz, Austria
gerd@huebscher.at
[3] SBA Research, Floragasse 7, Vienna, Austria
aekelhart@sba-research.org

Abstract. In the digital age, knowledge processes can be formalized and simplified using task management systems. As they evolve, so must the underlying schemata to retain harmony and concurrency with the real world. In this work we present a graph neural network model that can help in reconciling these data. It can do so by leveraging a novel propagation rule that does not presume reciprocal dependency but is able to represent it still. Thereby it can predict structures in the form of usage links with high accuracy and assist in the reconstruction of missing information. We evaluate this model on a new knowledge management dataset and show that it is superior to traditional embedding methods. Further, we show that it outperforms related work in an established general link prediction task.

Keywords: Knowledge graph · Link prediction · Graph neural networks

1 Introduction

For today's knowledge and information society [6,10,28], one of the greatest challenges is to increase the efficiency of knowledge work [7,8] through successful digitalisation. Knowledge work is based to a large extent on communication between knowledge workers, such as researchers, developers, consultants or attorneys. However, the observable data and information occurring in this communication is largely unstructured and requires a-priori knowledge to understand, extract and apply semantic concepts.

Externalising this required knowledge is a demanding process, which requires the repetitive articulation and internalisation [22,23]. Despite the increasing support of knowledge management systems, this process of knowledge transfer often fails due to the necessary generalisation on the part of knowledge workers or due to the impossibility of integrating and applying this knowledge appropriately on new tasks.

C. Strauss et al. (Eds.): DEXA 2022, LNCS 13427, pp. 133–146, 2022.
https://doi.org/10.1007/978-3-031-12426-6_11

One form of externalising and structuring knowledge that has gained a lot of interest in the last decade is knowledge graphs. They can broadly be described as "*a representation of knowledge (however defined) in the structural form of a directed (mostly acyclic) graph*" [3].

More precisely, knowledge graphs can be seen as knowledge bases of types, instances and their attributes that are created user-driven in a dynamic context, guarded by constraints, and organised in a logical and computable graph [16]. Types and constraints are important for the stability and the expressiveness of a knowledge graph used by several users, because they support a uniform representation of information artefacts.

However, the dynamic development of knowledge graphs, especially in the context of knowledge work with unknown processes and information artefacts, raises the problem that the quality of the assignment of instances to types depends strongly on the respective user and thus not only the stability and expressiveness of a knowledge graph, but also that the linking or integration of several knowledge graphs into each other becomes difficult or impossible [14].

The support of users through automated or semi-automated classification or typification of information artefacts, as well as the prediction of links between these information artefacts, is therefore an essential prerequisite for the applicability of knowledge graphs in the context of knowledge work.

We propose a Graph Neural Network model that tackles this task via linkage reconciliation. It can assist in discovering missing links between existing data, which is a frequent problem when adding new data or incorporating it into new tasks. Importantly, the presented approach works on an anonymized knowledge graph that only contains data objects, but not their values, making this a suitable method even for sensitive data and remote processing. We also compare this method against a strong baseline and established graph embedding methods, where it shows superior performance. Additionally, we evaluate it in a general link prediction task on a citation graph and find performance to be satisfactory. Further, the effect of model components is investigated and found to be positive and essential to high performance in all cases. Lastly, we construct a purely structural prediction scenario without any node information and demonstrate that even then the model is able to perform adequately.

2 Related Work

There is a large catalogue of work on the analysis of and learning on networks. We focus here on introducing methods and previous results that concern link prediction and graph neural networks.

Topology methods, such as the Adamic-Adar index [1] or the Resource Allocation index [33], are an established class of algorithms that exploit simple structural patterns based on local node structure. They are generally not suited to directed graphs and incur a high degree of information loss therefore.

There are multiple methods that can learn embeddings, a dense representation that preserves some structural information and other graph characteristics in an unsupervised manner. The influential DeepWalk algorithm by Perozzi et al. [24] is based on random walks through the graph. These sequences can then be factorized with the skip-gram technique of Mikolov et al. [21]. The weights of this factorization matrix are the learned representations for each node. They can be used as features for a simple classifier and thereby outperform traditional methods in multi-label node classification tasks.

NetMF [25] is a closed-form solution that approximates the weight matrix generated in a deep walk. It uses sparse truncated SVD to learn embeddings for the pointwise mutual information index matrix and achieves node classification performance similar or superior to other embedding methods in established prediction tasks. The node representations generated by such an embedding method can be used as features to predict links between nodes by comparing their concordance as a basis for attachment likelihood. A popular choice for such a method of comparison is the dot product between two embedding vectors. In recent years, graph neural networks have been established as a suitable candidate to build end-to-end models that can learn supervised tasks directly. Most of them are built around the graph convolution propagation rule developed by Kipf et al. in [18] and also known as Graph Convolutional Networks (**GCN**s). In the context of link prediction, the GCN is designated as the encoder that produces embeddings and used in conjunction with a decoder, that transforms pairs of node embeddings into link likelihoods. There are multiple flavours of graph convolution that e.g. perform neighbourhood sampling, [11], or employ attention mechanisms, c.f. [29]. Recent work by Wang et al. [30] has shown that in established link prediction benchmarks, these adaptations essentially perform equivalently. Salha et al. [27] compare symmetric and asymmetric decoding methods to their own gravity-inspired decoding in three link prediction tasks and different scenarios. They find that, depending on the reciprocity in the prediction scenario, asymmetric decoding can be vital to making usable predictions.

3 Asymmetric Bidirectional Residual GNN

In this section we define the **A**symmetric **B**idirectional **Res**idual **G**raph **C**onvolutional **N**etwork (**ABRes-GCN**), which is the core of our work. This includes the novel bidirectional adaptation that was designed specifically for application in knowledge graphs. We also cover the basic propagation rule and functional basics of auto-encoders that it makes use of.

Graph Convolutional Networks, introduced by [18], are an instance of Graph Neural Networks that compute node representations from neighbouring node features. Similar to the neighbourhood concepts that enable convolutions on image data, these features are processed via linear combination in multiple parallel filters learned via gradient descent with back-propagation, c.f. [26]. They natively process arbitrary directed graph data and can be extended to cover heterogeneous nodes and edges. Through the nesting of multiple propagation passes, meaningful dense representations can be extracted from structured but unordered data. This makes them a suitable candidate for graph auto-encoding.

In such a setting, they are trained to recover the structure of a graph, i.e. pairwise links between an intact node set.

Formally, we represent a knowledge graph as $\mathcal{G} = (\mathcal{V}, \mathcal{E}, \lambda_l)$ with nodes (or vertices) $\mathcal{V}$ and edges $\mathcal{E} \subseteq \{(u, v) : \ u, v \in \mathcal{V} \wedge u \neq v\}$ between them. In most knowledge graphs, nodes are heterogeneous and their one-hot-encoded type is given by $\lambda_l : \mathcal{V} \rightarrow \{0, 1\}^{n_l}$. For simplicity, here we consider only graphs with homogeneous edges of a singular type, however methods can easily be adapted to heterogeneous links. In layer ℓ of L layers with $F^{(\ell)}$ filters, the convolution of node v in $\mathcal{G}$ with in-neighbourhood $\Gamma(\mathcal{G}, v) = \{u : (u, v) \in \mathcal{E}\}$ is

$$f^{(\ell+1)}(\mathcal{G}, v) = \sum_{u \in \Gamma(\mathcal{G}, v)} \Theta^{(\ell+1)} f^{(\ell)}(\mathcal{G}, u) z(v) \tag{1}$$

with a learned weight matrix $\Theta^{(\ell+1)} \in \mathbb{R}^{F^{(\ell)} \times F^{(\ell+1)}}$ for each layer. To avoid the accumulation of large feature values in high in-degree nodes, they are scaled with the inverse square root of the node degree in normalization $z(v)$. To ensure that the representation $f^{(\ell+1)}(\mathcal{G}, v)$ also contains the representation of v itself, $\mathcal{G}$ is extended with self-loops $\mathcal{E} = \mathcal{E} \cup \{(v, v) : \ v \in \mathcal{V}\}$. Input features are the initial node representations in layer 0, i.e. $f^{(0)}(\mathcal{G}, v) = \lambda_l(v)$ and $F^{(0)} = n_l$.

Bidirectional Graph Convolution. In a knowledge graph, there are links that can introduce reciprocal dependencies, such as *part of* relationships. This information can be made available to the model by including the reverse direction, expressed as the reverse graph $G^{-1} = (\mathcal{V}, \mathcal{E}^{(-1)}, \lambda_l) = (\mathcal{V}, \{(v, u) : (u, v) \in \mathcal{E}\}, \lambda_l)$, into the propagation rule. Both modalities are integrated into a common vertex representation by linear combination. This novel propagation rule for layer ℓ is:

$$f^{(\ell+1)}(\mathcal{G}, v) = \Theta_c^{(\ell+1)} \left(f^{(\ell+1)}(\mathcal{G}, v), f^{(\ell+1)}(\mathcal{G}^{-1}, v) \right) \tag{2}$$

$\Theta_c^{(\ell+1)} \in \mathbb{R}^{F^{(\ell+1)} \times 2F^{(\ell+1)}}$ is a matrix of learned weights for each layer. This improves the modelling capabilities to cover reciprocal relationships with minimal additional complexity, preserving adequate convergence.

Residual Computation. To facilitate the training of ABRes-GCN, skip connections [12] are added between same size layers for explicit residuals. They allow for the stacking of multiple GCN layers without the problematic vanishing gradients that plague back-propagation in deep architectures. This can help deeper and more expressive models' convergence and thereby improve performance [5].

The rectification $\sigma : \mathbb{R} \rightarrow \mathbb{R}, x \mapsto \max(0, x)$ introduces non-linearity to enhance model capabilities beyond linear functions. The last of L layers is not activated in this manner, i.e. $f_{\text{res}}^{(L)} = f^{(L)}$.

Asymmetric Decoding. To obtain the likelihood P of a link forming between two nodes u and v, their embeddings $f(u)$ and $f(v)$ are compared. This coefficient is scaled through the sigmoid function σ to ensure $0 < P(u,v) < 1$. For the inner product, we can see how this leads to a symmetric distance metric:

$$P(u,v) = \sigma(f(u)^T f(v)) = \sigma(f(v)^T f(u)) = P(v,u) \tag{3}$$

This is unsuitable for any directed graph where relationships between nodes are not guaranteed to be mutual. Therefore substitution with an asymmetric decoder is prudent for modelling undirected linkage. It is easily derived along the lines of [31] from any combination of decoder and encoder by specifying a separate source and target embedding $f_s(u)$ and $f_t(u)$ for each node u. We then arrive at an asymmetric inner product of the form:

$$P(u,v) = \sigma(f_{(s)}(u)^T f_{(t)}(v)) \neq \sigma(f_{(s)}(v)^T f_{(t)}(u)) = P(v,u) \tag{4}$$

These source and target vectors can be obtained from any embedding by assigning a slice to each role, as such: $f_{(s)}(u) = f(u)_{[1:\frac{F(L)}{2}]}$ and $f_{(t)}(u) = f(u)_{[\frac{F(L)}{2}+1:F(L)]}$ This decoder does not require any other adaptations to the encoder and is natively suited to undirected link prediction. Reducing the embedding in this way requires encoding of sufficient information in both slices, which is not guaranteed for arbitrary representations. In ABRes-GCN it is however learned jointly through back-propagation and does not induce a penalty on available decoded information beyond the halving of dimensionality.

Training Procedure. During model training, the cross entropy reconstruction loss is minimized through gradient descent with back-propagation, as in [19]. ABRes-GCN is trained with a negative sampling procedure that produces random pairs of nodes that are not connected in the training set for each epoch. This provides a degree of stochasticity that can introduce some helpful noise during training, similar to the shuffling of mini batches. For the convenience of a balanced setting, the number of negative samples is set to equal the number of positive samples.

The adaptive momentum gradient descent method of [17] is used to train ABRes-GCN in mini-batches. We use an exponential learning rate decay with a factor γ at each epoch to encourage exploitation in the later stages of training. For improved convergence behaviour and added regularization, batch normalization is also applied to each graph convolution layer's output before rectification. We use the dropout method of [13] after the last feature computation to encourage the learning of more robust representations and decoupled weight decay [20] of 0.02 for regularization.

4 Dataset

The TEAM-IP-1[1] dataset is a special instance of a knowledge graph based on the in**T**egrated knowl**E**dge and t**A**sk **M**anagement (TEAM) model and system presented in [15,16]. The TEAM model combines several perspectives of knowledge

[1] Publicly available at https://github.com/wendli01/abres_gcn/blob/master/team_ip_1.zip.

work in a multi-dimensional graph that includes both an instance model and a domain model (type level). The data perspective of this graph is described at instance level by data objects and data object relations of different kinds. Each data object corresponds to a mental concept that can be either value-bearing or value-unbearing. For example, the data object of the pure *String* "*John Doe*" is value-bearing, while the data object of the *Name* "*John Doe*" based on this text is value-unbearing (see Fig. 1). The TEAM model ensures that all data objects within the graph are unique. The data object relations connecting the data objects are directed edges pointing from the data object with higher represented information content towards the data object with lower represented information content. For example, the *Natural Person* with the *Name* "*John Doe*" contains more information than the *Name* alone. Several data object relations exist, such as "has", "roleOf" and "hasValue".

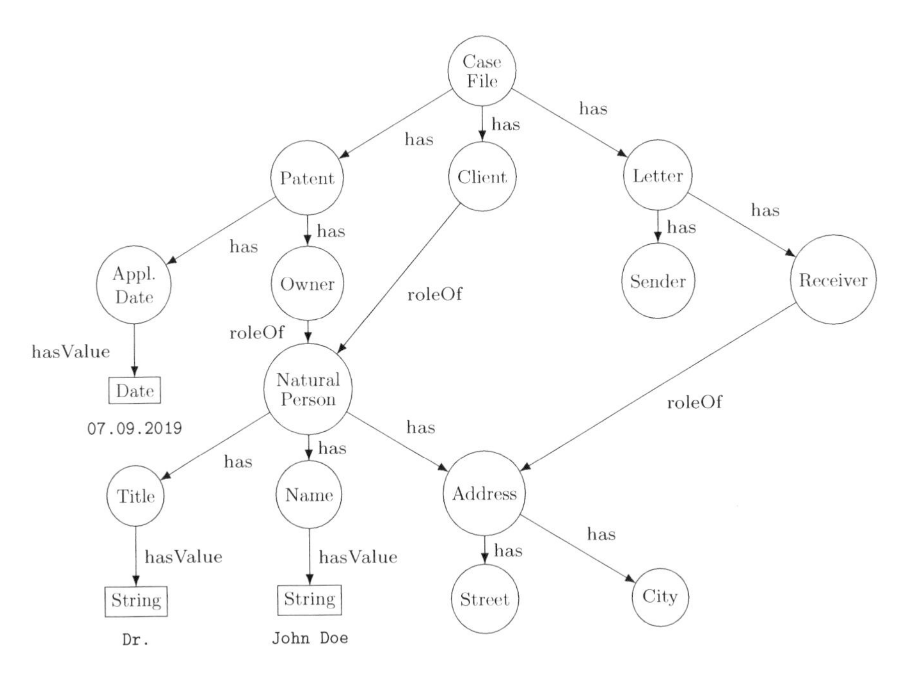

Fig. 1. Exemplary representation of the data perspective of the TEAM-IP-1 dataset, showing data objects as nodes and data object relations as edges directed from nodes with higher information content to nodes with lower information content.

Each data object and each data object relation is assigned to a data object type or a data object relation type in the domain model.

The TEAM-IP-1 dataset includes randomly selected real-world transaction data of an IP management system, which were imported into a TEAM system. The resulting TEAM model was then reduced to the data perspective and subsequently anonymised by removing all node and edge attributes. The resulting dataset encompasses 68 846 nodes of 124 different node types and 160 452 relations. The support of the individual data object types and data object type relations in the instance model can be seen in Fig. 2. Despite the high number of data object relations, the density of the data set is comparatively low, which is partly due to the random selection of the imported data. The resulting density is at 3.4×10^{-5}, far lower than in social networks, such as karateclub [32] (0.14), and slightly lower than in citation networks like Citeseer [9] (0.00083).

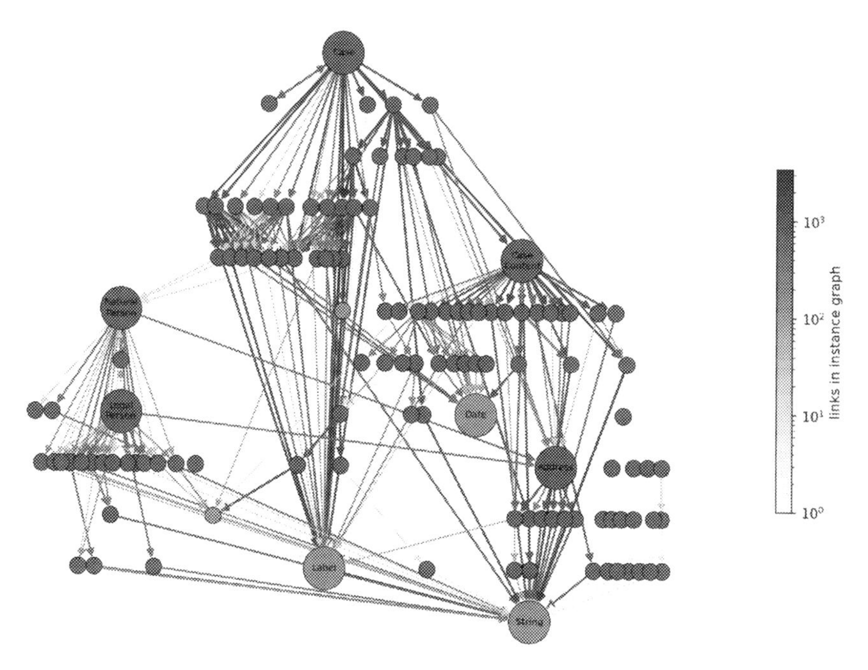

Fig. 2. Node type graph with the number of links expressed by edge width and a colormap. Central nodes (degree centrality ≥ 0.11) are named. Value unbearing node types are colored blue, value bearing types orange. (Color figure online)

5 Experiments

Here we describe the reconciliation experiment that forms the core of our work. This includes a deeper look into the recall-precision trade-off as well as prediction difficulty on a node type basis. In this section the evaluation methodology for this link prediction task is laid out in detail. We also study how network depth affects ABRes-GCN and seek to gauge the importance of node information relative to structural information.

All code, experiment setups, final and intermediate results are available at https://github.com/wendli01/abres_gcn.

5.1 Methodology

We compare the predictive performance of the ABRes-GCN to a selection of tailored and established baselines. These include a Bag-of-Words inspired approach dubbed Bag-of-Types that represents a node as the type counts of its undirected neighbours. We also compare the performance of various established graph representation learning methods. Chief among them is the traditional DeepWalk [24] with 10 walks of length 80. NetMF [25] is run in 10 iterations with 32 dimensions and PMI matrix powers of order 2. The parameter setting of ABRes-GCN is listed in Table 1, with invariable hyper-parameters except for 0 dropout for Citeseer like in [27]. For all baseline encoding methods, the same asymmetric decoding schema is used, as it showed improved performance across all of them. This is despite the methods not being designed to encode information uniformly in all dimensions. Because these baseline models are not learned end-to-end, the concordance between node embeddings, e.g. in the form of the dot product, is not normalized or meaningful. Because the decoder is separate from the encoder and cannot adjust the embeddings to produce link likelihoods. Hence, a mechanism that can learn to map decoded edge representations to likelihoods is necessary. For this purpose, we use a random forest [4] with 100 trees split via the Gini impurity for all unsupervised algorithms ant denote it as *+RF*.

When evaluating an estimator, producing predictions for all 4.7 billion pairs of nodes in TEAM-IP-1 is intractable. We therefore perform negative sampling to create a set of n_{test} negatives to complete the n_{test} positives. To gain an estimate of repeatability and robustness, 10 trials are performed. In each fold, $n_{test} = \lceil 0.2 \times 160\,452 \rceil$ relations are reserved for testing, with the remaining $n_{train} = \lfloor 0.8 \times 160\,452 \rfloor$ designated for estimator fitting. For Citeseer, we use a test set of proportion 0.1, in concordance with other work.

Predictions are evaluated using the average precision score (AP), which corresponds to the area under the precision-recall curve. We chose this metric as it gives insight into model performance across the whole range of decision thresholds, and thereby indicates whether it produces usable prediction probabilities. For comparison with related work, we also list the area under the receiver-operator-characteristic curve (ROC-AUC) where possible.

5.2 Predictive Performance

The reconciliation task on TEAM-IP-1 factorizes into a network link prediction. We set the Bag-of-Types model as a strong baseline, against which other predictors are measured.

The result of all methods are listed in Table 2. Traditional walk-based node embedding methods cannot be used to accurately predict links in this task. This deficit can be attributed to the structure of TEAM-IP-1, especially short path lengths and disconnected components, that is not well suited to random traversal. Even so, the performance of NetMF shows a large increase over DeepWalk, mirroring some of the findings of Qiu in [25]. The GCN without the asymmetric decoding performs significantly worse than the baseline, while adopting the

Table 1. Parameter setting for the ABRes-GCN for link prediction.

Parameter name	Default value
GCN layer sizes	(32, 32, 32, 32, 32)
Epochs	200
Dropout probability	0.1 for TEAM-IP-1, 0 otherwise
Initial learning rate	0.01
Learning rate decay	0.98

Table 2. Link prediction performance on TEAM-IP-1 averaged over 10 folds with standard deviation indicated. Statistically significant (according to a paired t-test with $\alpha = 0.05$) improvement or degradation against the *Bag-of-Types + RF* baseline is indicated by (+) or (−), respectively.

Method	AP	ROC AUC
Random Guessing	0.325 ± 0.0028 (−)	0.501 ± 0.003 (−)
Bag-of-Types + RF	0.801 ± 0.002	0.859 ± 0.002
DeepWalk [24] + RF	0.399 ± 0.025 (−)	0.598 ± 0.028 (−)
NetMF [25] + RF	0.738 ± 0.0177 (−)	0.809 ± 0.004 (−)
GCN	0.711 ± 0.0205 (−)	0.785 ± 0.012 (−)
A-GCN	0.775 ± 0.0064 (−)	0.802 ± 0.007 (−)
ARes-GCN	0.885 ± 0.0035 (+)	0.907 ± 0.007 (+)
ABRes-GCN	**0.901** ± 0.0025 (+)	**0.926** ± 0.0025 (+)

asymmetric decoder results in a score that is still lacking. Only with explicit residuals can the model surpass baseline performance, then by a large margin.

A look at the precision recall curve in Fig. 3 confirms that the ABRes-GCN produces usable link likelihoods. Recall stays high even for configurations focused on precision and tapers off at a consistent rate. This model behaviour makes tuning to scenarios with different type 1 and type 2 error requirements easy and predictable. The precision curves are similar across multiple trials, which shows model consistency beyond only low variance in average precision.

The effect of our GCN adaptations is of special interest and their score distributions are shown in Fig. 4. There is a substantial increase in consistency and overall performance effected by the asymmetric decoding, which confirms the notion that it is obligatory for directed link prediction. ABRes-GCN performs significantly better than a model without the bidirectional propagation rule, according to a paired t-test with $p = 3.19 \times 10^{-9}$.

Due to the underlying type hierarchy, node type plays a large role in to which degree attachment can be predicted. Moreover, there are types for which a target role is easier to identify than a source role and vice versa. The score distribution

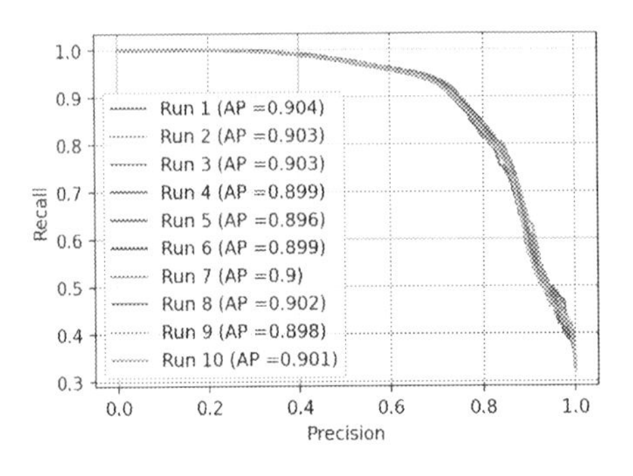

Fig. 3. Precision vs. Recall for ABRes-GCN predictions in 10 trials. The average precision, i.e. area under the curve, is annotated for each fold.

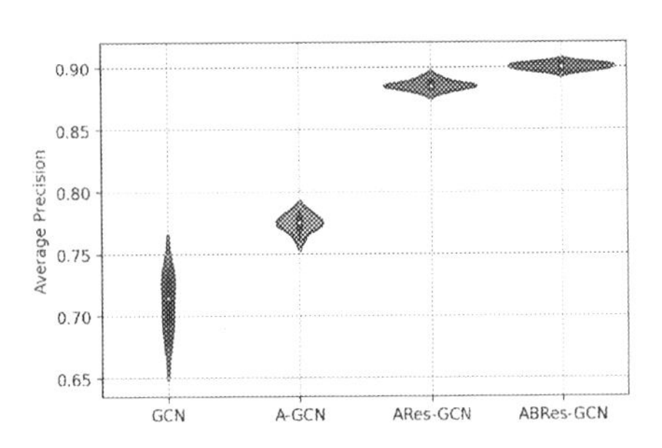

Fig. 4. TEAM-IP-1 link prediction performance of GCN model variants. Distributions of average precision in 10 folds.

over the 10 folds per node type is depicted in Fig. 5, split by node role. There is a cluster of well connected node types that all show high source scores but inconsistent target scores. This is a pattern induced by the type hierarchy in which predicting links in lower levels is easier, as the more specific value bearing nodes get more common.

5.3 General Link Prediction on Citeseer

Table 3. Link prediction performance on Citeseer averaged over 10 folds with standard deviation indicated.

Method	AP	ROC AUC
Source/Target Graph AE [27]	$0.75_{\pm 0.034}$	$0.74_{\pm 0.031}$
Gravity Graph VAE [27]	$\mathbf{0.898}_{\pm 0.01}$	$0.877_{\pm 0.011}$
Gravity Graph AE [27]	$0.848_{\pm 0.011}$	$0.784_{\pm 0.016}$
ABRes-GCN	$0.836_{\pm 0.015}$	$\mathbf{0.913}_{\pm 0.011}$

For a better insight into the performance of ABRes-GCN, we also evaluate it in an established link prediction task on the Citeseer citation dataset [9]. It achieves appreciably higher AUC scores than the gravity based GNNs of [27], that have been tuned for high AUC in this general link prediction task. It also performs markedly better than the Source/Target Graph AE of [27], which is functionally similar to our A-GCN model (Table 3).

5.4 Ablation Studies

Layer Configuration. The depth of neural networks is a contentious issue, as it epitomizes the trade-off between complexity and usability. We investigate how it affects the predictive performance of our ABRes-GCN model, which was designed with residuals to tackle this exact issue. Like the other eponymous extension to the original GCN propagation rule, skip connections are imminently necessary for high performance in the link prediction task.

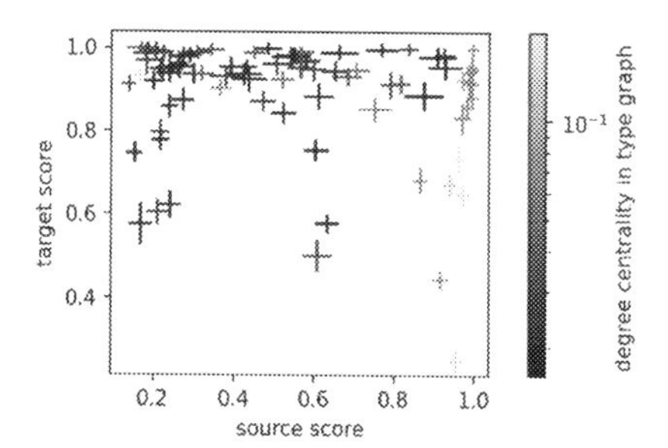

Fig. 5. Mean average precision per node type, split by their role as source or target nodes. Types with fewer than 10 links in all 10 folds are excluded. Marker colour denotes the degree centrality in the type graph (c.f. Fig. 2), lobe size is the standard deviation.

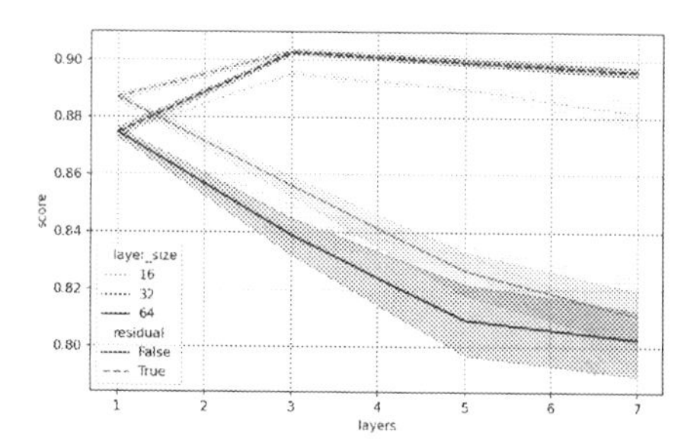

Fig. 6. Effect of network depth for different widths with and without explicit residuals on TEAM-IP-1 link prediction performance. Mean over all folds with standard deviation shaded.

Their positive effect on any model that consists of more than one layer is clearly visible in Fig. 6. Without this adaption, the average precision of the model degrades rapidly with increasing depth, especially for configurations with larger widths, i.e. filter sizes. This demonstrates that even for models of medium depth, convergence and resulting performance can be improved significantly through explicit residual computation. In the TEAM-IP-1 link prediction scenario, they are required for any configuration with hidden layers.

Node Information and Structure in the Context of User Privacy. To gauge the importance of structure, we perform an ablation on the available information in the relational model. If the node features, comprising only of their respective type, is obscured, the ABRes-GCN model still achieves high predictive performance. This corresponds to an average precision of $0.807_{\pm 0.0071}$ relative to the full models 0.901 $_{\pm 0.0025}$. This is equal to the baseline score of the *Bag-of-Types + RF* model. Here it is important to visualize the difference between the two models: the Bag-of-Types model is functionally equivalent to a latent single filter in the first layer of a normal, typed, GCN constructed over the undirected graph. Essentially, one method makes use of structure only, while the other uses no structural information beyond the simple neighbourhood but they achieve very similar performance. We can infer from this result that ABRes-GCN can predict relations in knowledge graphs based on structure alone, even if there exists an underlying type hierarchy that is ignored. Because it does not rely on this data, it lends itself to application in scenarios which need to protect sensitive or confidential data by privacy preserving methods. What constitutes confidential information depends on the underlying business and may include e.g., trade secrets and market strategies. Personally identifiable information (PII) on the other hand is well-defined and comprises e.g., a full name, a social security number, document numbers, email addresses or telephone numbers.

However, it has been shown that removing the identity of each node in a graph before publishing does not always guarantee privacy, as the structure of the graph, combined with prior knowledge of an attacker, could still leave the graph vulnerable to re-identification of individuals [2]. Thus, in future work we want to explore how anonymization approaches on (heterogeneous) graphs would impact the performance of the presented approach and study how the optimal trade-off between privacy vs. loss of utility can be reached.

We can also surmise that accurate prediction in the absence of node typing might be an effect of a type definition that is too general and cannot adequately represent the role of a node in relation to its descendants. This information is evidently present in the embeddings of ABRes-GCN, as it can recover it to a high degree. A schema to incorporate this information into a refined type model can be devised easily and is left as future work.

6 Conclusion

In this work, we identify incomplete linkage as a problem in knowledge graphs used for knowledge management. We create a data-driven graph dataset derived from a real-world system that consists of an anonymized data model, dubbed TEAM-IP-1. We present ABRes-GCN, an extension to an established Graph Neural Network model that improves the representation of data flow in such graphs by a mechanism that respects reciprocity. We show that ABRes-GCN outperforms traditional and graph embedding methods in a link prediction task on TEAM-IP-1, that presents an analogous challenge. In this capacity, ABRes-GCN can assist users in the management of knowledge by reconciling existing and new data. We also demonstrate that our model performs adequately in other link prediction challenges. Because of its high performance on prediction with anonymized and purely structural data, further work into model applications in the context of user privacy is planned.

Acknowledgements. The research reported in this paper has been supported by the FFG BRIDGE project KnoP-2D (grant no. 871299).

References

1. Adamic, L.A., Adar, E.: Friends and neighbors on the web. Soc. Netw. **25**(3), 211–230 (2003)
2. Backstrom, L., Dwork, C., Kleinberg, J.: Wherefore art thou R3579X? Anonymized social networks, hidden patterns, and structural steganography. In: Proceedings of the 16th International Conference on World Wide Web, WWW 2007, pp. 181–190. Association for Computing Machinery (2007)
3. Bergman, M.K.: Common sense view of knowledge graphs (2019)
4. Breiman, L.: Random forests. Mach. Learn. **45**(1), 5–32 (2001). https://doi.org/10.1023/A:1010933404324
5. Bresson, X., Laurent, T.: Residual gated graph convnets. arXiv preprint arXiv:1711.07553 (2017)
6. Bürstenbinder, J., et al.: Auf dem Weg in die Wissens- und Informationsgesellschaft. In: Jung, V., Warnecke, H.J. (eds.) Handbuch für die Telekommunikation, pp. 1273–1410. Springer, Heidelberg (2002). https://doi.org/10.1007/978-3-642-55450-6_6
7. Drucker, P.F.: Landmarks of Tomorrow: A Report on the New "Post-Modern" World. Harper & Brothers, New York (1959)
8. Drucker, P.F.: Knowledge-worker productivity: the biggest challenge. Calif. Manag. Rev. **41**(2), 79–94 (1999). https://doi.org/10.2307/41165987
9. Giles, C.L., Bollacker, K.D., Lawrence, S.: CiteSeer: an automatic citation indexing system. In: Proceedings of the Third ACM Conference on Digital Libraries, pp. 89–98 (1998)
10. Greco, P.: The knowledge society. J. Sci. Commun. **06**(04), C01 (2007). https://doi.org/10.22323/2.06040301. https://jcom.sissa.it/archive/06/04/Jcom0604(2007)C01
11. Hamilton, W., Ying, Z., Leskovec, J.: Inductive representation learning on large graphs. In: Advances in Neural Information Processing Systems, vol. 30 (2017)
12. He, K., Zhang, X., Ren, S., Sun, J.: Deep residual learning for image recognition. In: Proceedings of the IEEE Conference on Computer Vision and Pattern Recognition, pp. 770–778 (2016)
13. Hinton, G.E., Srivastava, N., Krizhevsky, A., Sutskever, I., Salakhutdinov, R.R.: Improving neural networks by preventing co-adaptation of feature detectors. arXiv preprint arXiv:1207.0580 (2012)
14. Hübscher, G., et al.: Graph-based managing and mining of processes and data in the domain of intellectual property. Inf. Syst. **106**, 101844 (2022). https://doi.org/10.1016/j.is.2021.101844
15. Hübscher, G., Geist, V., Auer, D., Hübscher, N., Küng, J.: Integration of knowledge and task management in an evolving, communication-intensive environment. In: ACM (ed.) The 22nd International Conference on Information Integration and Web-Based Applications & Services (iiWAS 2020), pp. 407–416. ACM (2020). https://doi.org/10.1145/3428757.3429260
16. Hübscher, G., Geist, V., Auer, D., Hübscher, N., Küng, J.: Representation and presentation of knowledge and processes - an integrated approach for a dynamic communication-intensive environment. Int. J. Web Inf. Syst. **17**(6), 669–697 (2021). https://doi.org/10.1108/IJWIS-03-2021-0031
17. Kingma, D.P., Ba, J.: Adam: a method for stochastic optimization. arXiv preprint arXiv:1412.6980 (2014)

18. Kipf, T.N., Welling, M.: Semi-supervised classification with graph convolutional networks. arXiv preprint arXiv:1609.02907 (2016)
19. Kipf, T.N., Welling, M.: Variational graph auto-encoders. arXiv preprint arXiv:1611.07308 (2016)
20. Loshchilov, I., Hutter, F.: Decoupled weight decay regularization. arXiv preprint arXiv:1711.05101 (2017)
21. Mikolov, T., Sutskever, I., Chen, K., Corrado, G.S., Dean, J.: Distributed representations of words and phrases and their compositionality. In: Advances in Neural Information Processing Systems, vol. 26, pp. 3111–3119 (2013)
22. Nonaka, I.: The knowledge-creating company: reprint of the 1991 article, managing for the long term, best of HBR, Nov.–Dec. 1991. Harv. Bus. Rev. 162–171 (2007)
23. Nonaka, I., Takeuchi, H.: The Knowledge-Creating Company: How Japanese Companies Create the Dynamics of Innovation. Oxford University Press, Oxford (1995)
24. Perozzi, B., Al-Rfou, R., Skiena, S.: DeepWalk: online learning of social representations. In: Proceedings of the 20th ACM SIGKDD International Conference on Knowledge Discovery and Data Mining, pp. 701–710 (2014)
25. Qiu, J., Dong, Y., Ma, H., Li, J., Wang, K., Tang, J.: Network embedding as matrix factorization: unifying DeepWalk, LINE, PTE, and node2vec. In: Proceedings of the Eleventh ACM International Conference on Web Search and Data Mining, pp. 459–467 (2018)
26. Rumelhart, D.E., Hinton, G.E., Williams, R.J.: Learning representations by back-propagating errors. Nature **323**(6088), 533–536 (1986)
27. Salha, G., Limnios, S., Hennequin, R., Tran, V.A., Vazirgiannis, M.: Gravity-inspired graph autoencoders for directed link prediction. In: Proceedings of the 28th ACM International Conference on Information and Knowledge Management, pp. 589–598 (2019)
28. Steinbicker, J.: Zur Theorie der Informationsgesellschaft: Ein Vergleich der Ansätze von Peter Drucker, Daniel Bell und Manuel Castells. Lehrtexte Soziologie, Leske + Budrich, Opladen (2001)
29. Veličković, P., Cucurull, G., Casanova, A., Romero, A., Lio, P., Bengio, Y.: Graph attention networks. arXiv preprint arXiv:1710.10903 (2017)
30. Wang, X., Vinel, A.: Benchmarking graph neural networks on link prediction. arXiv preprint arXiv:2102.12557 (2021)
31. Yu, Y., Wang, X.: Link prediction in directed network and its application in microblog. Math. Probl. Eng. **2014** (2014)
32. Zachary, W.W.: An information flow model for conflict and fission in small groups. J. Anthropol. Res. **33**(4), 452–473 (1977)
33. Zhou, T., Lü, L., Zhang, Y.C.: Predicting missing links via local information. Eur. Phys. J. B **71**(4), 623–630 (2009). https://doi.org/10.1140/epjb/e2009-00335-8

I-PNN: An Improved Probabilistic Neural Network for Binary Classification of Imbalanced Medical Data

Ivan Izonin[1](✉), Roman Tkachenko[1], and Michal Greguš[2]

[1] Lviv Polytechnic National University, Lviv, Ukraine
ivanizonin@gmail.com
[2] Comenius University in Bratislava, Bratislava, Slovakia
Michal.Gregus@fm.uniba.sk

Abstract. The modern development of Medicine relies heavily on effective data mining. However, many practical diagnostic tasks operate on small data samples with an asymmetric number of instances of different classes. This paper considers the binary classification task in the case of a short unbalanced set of medical data. The authors improved the implementation of the Probabilistic Neural Network (I-PNN). It is based on a new method of forming the outputs of the PNN's summation layer, which, as in the analog, retains the condition of ensuring a complete system of events (formation of a set of probabilities of belonging to each class that in the sum equal 1). However, in contrast to analogs, this method takes into account the uneven representation of all classes in a stated data set, which provides the ability to effectively solve classification tasks in the case of an unbalanced dataset. The authors substantiate the proposed approach and describe all the steps of algorithmic implementation of the proposed I-PNN. Modeling of I-PNN operation using a well-known unbalanced medical dataset was performed. The optimal parameters of I-PNN operation are selected. An experimental increase in the accuracy of the proposed I-PNN (up to 5% based on F1-score) compared to the existing PNN was found. All these advantages create many prerequisites for the practical use of I-PNN in the case of processing a short set of unbalanced data in various areas of medical diagnostics.

Keywords: Medical data · Imbalanced classification task · Small data approach · Probabilistic Neural Network

1 Introduction

The modern development of Medicine is accompanied by the digitization of all information in the form of structured and unstructured data. Many information technologies for various purposes are developed for their processing [1]. Intelligent components of such systems, in particular, based on Artificial Neural Networks (ANN) [2], play an essential role in the processing of both large and small datasets [3]. Effective solution to basic Data Mining tasks is increasingly becoming additional information for the doctor during the diagnosis, study of the effectiveness of ongoing therapy, etc.

C. Strauss et al. (Eds.): DEXA 2022, LNCS 13427, pp. 147–157, 2022.
https://doi.org/10.1007/978-3-031-12426-6_12

Based on the analysis of known medical histories, physicians are increasingly using decision support systems that provide access to the experience of thousands of colleagues across the country and create the necessary datasets for analysis. In medical practice, such datasets can be both large and small [4, 5]. In the first case, a vast arsenal of ML-based analysis methods for such data is developed. However, in the case of small amounts of information, intellectual analysis is complicated [6]. The problem is exacerbated in the case of asymmetry of the representation of classes in the dataset intended for diagnosis [7, 8]. Therefore, it is necessary to improve existing or develop new ones, particularly neural network methods, which will ensure the required accuracy.

A Probabilistic Neural Network (PNN) was developed quite a long time ago, but it is very widely used today in the intellectual analysis of medical data. It is due to a number of its advantages, in particular:

- does not require a training procedure;
- has only one parameter to be configured;
- provides high accuracy;
- demonstrates satisfactory results when processing short datasets;
- provides the ability to obtain output signals both in the form of belonging to a particular class and in the form of a set of probabilities of belonging current point to all classes of the task.

The last advantage of this ANN is that it plays an essential role in medical diagnosis [9, 10]. It is because the set of probabilities of belonging of the current point to all classes of the task, especially when it describes the full system of events [7], can serve as a very informative additional indicator when making a decision a doctor. This approach allows the specialist to more deeply evaluate the results of the analysis of a particular patient with an intelligent subsystem [11] based on PNN before making a diagnosis.

Despite many advantages, PNN has some disadvantages. In particular, it becomes enormous and slow when processing large data sets. In addition, like most classifiers, it reduces its effectiveness when analyzing unbalanced datasets.

Therefore, this work aims to develop an approach for implementing PNN, which should increase the accuracy of its work in the case of processing short unbalanced datasets. We called it Imbalanced PNN (I-PNN).

The main contribution of this paper can be summarized as follows:

- we proposed an improved version of the Probabilistic Neural Network (I-PNN) based on a new method of calculating the set of probabilities belonging to each class of the task, which meets the condition of a complete system of events and takes into account the uneven number of vectors of different classes in the available short medical dataset;
- we substantiated the proposed approach and experimentally confirmed a significant increase in the classification accuracy of I-PNN when solving the binary classification task on three short sets of medical data with different levels of Imbalanced Ratio.

The practical value of this study lies in the possibility of significantly improving the accuracy of solving medical diagnostics tasks with intelligent services based on I-PNN in the case of processing short sets of unbalanced medical data.

2 State-of-the-Arts

According to the following request to the Scopus scientometric database (executed on April 1, 2022):

TITLE-ABS-KEY (pnn OR probabilistic AND neural AND network AND classification) AND (LIMIT-TO (PUBYEAR, 2022) OR LIMIT-TO (PUBYEAR, 2021) OR LIMIT-TO (PUBYEAR, 2020) OR LIMIT-TO (PUBYEAR, 2019) OR LIMIT-TO (PUBYEAR, 2018)) over the last five years, more than 1,400 published works on this topic have been indexed.

This number of articles confirms the relevance of the use of PNN in solving various applied medical and technical diagnostics tasks.

Most of these articles deal with classification in the case of a balanced dataset. Still, some approaches have been developed to improve the accuracy of PNN in the case of unbalanced data sets [6]. Among the best-known methods are undersampling, oversampling, and ensemble learning. Let us look at some of them in more detail.

The authors in [12] researched the selection of optimal features from a dataset that will reduce the sensitivity of PNN to omissions, anomalies, and imbalances in the dataset. It is experimentally established that this approach will improve the accuracy of PNN. However, the authors used the classical data pre-processing methods, and they did not perform any changes in the topology or algorithm of PNN operation.

The authors in [13] researched predicting the development of diabetes using an unbalanced data set. The proposed approach uses the SMOTE method for the oversampled minority class. This approach demonstrated an increase in the accuracy, where PNN demonstrated the highest results among the five methods considered. However, an artificial increase in the number of instances of the minor class can provoke overfitting [14].

In [15] the authors investigated the problems of constructing the most accurate model to predict drug resistance. The applied value of such a model lies in helping to choose HIV + therapy. Performing feature selection procedures and balancing four data sets using undersampling has avoided some problems in solving the classification problem. PNN here has demonstrated the highest accuracy of work compared to other researched methods that once again confirm the urgency of its use in practical research in Medicine.

The paper [16] deals with the classification task in the case of an unbalanced dataset based on PNN. The authors have developed an ensemble model based on Neural Networks of this type that demonstrates significant accuracy advantages. However, extensive time, energy, and memory resources are required to implement this approach to accompany the implementation of the ensemble.

In [17], the authors compared the effectiveness of the classical PNN and Naïve Bayesian Classifier for solving the classification task on unbalanced datasets. The basis of this study was the search for skewed features and evaluation of the two above classifiers with it presence and without it. According to practical experiments on six unbalanced datasets, Naïve Bayesian Classifier showed higher accuracy than the classical PNN in both cases. It is because the implementation of classical PNN does not meet the condition of output signals formation according to the complete system of events. Moreover, it does not consider the asymmetry of classes and therefore demonstrates significantly lower accuracy. The authors of [18] eliminated the first of these two shortcomings. They have replaced the procedure for generating PNN outputs in the summation layer to form a complete system of events. However, the second disadvantage of the above is that it does not provide sufficient accuracy of PNN in the case of data processing with an asymmetric number of classes.

Therefore, in this paper, we propose an improved approach to implementing PNN, which should increase the accuracy of its work in the case of processing short unbalanced data sets. We called it Imbalanced PNN (I-PNN).

3 I-PNN: Improved PNN for Imbalanced Classification Tasks in Medicine

The PNN has four layers: input, pattern, summation, and output. The PNN topology is shown in Fig. 1.

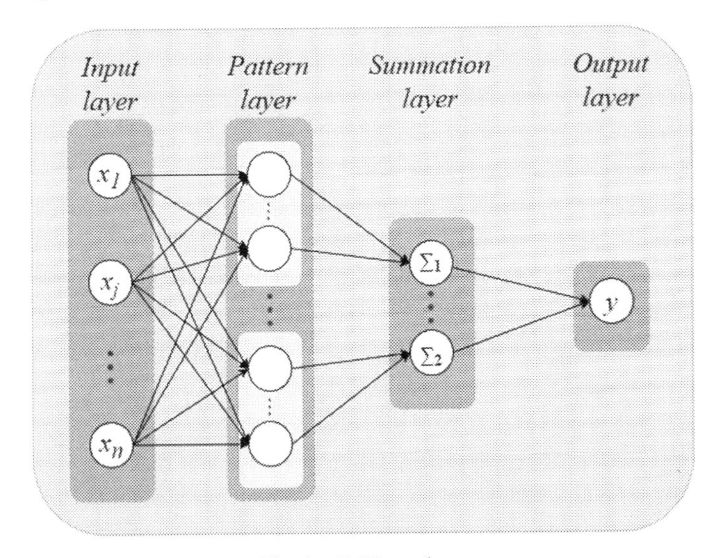

Fig. 1. PNN topology.

Each layer of the topology from Fig. 1 is responsible for performing specific actions of the general PNN's operation algorithm. In particular, in the initial layer, the input parameters of the Neural Network are initialized. The pattern layer is designed to calculate Gaussian functions from Euclidean distances between the current point and all points from the dataset. In the summation layer, a set of probabilities of belonging of the current point to each class is calculated. In particular, in [18], the formula for calculating the outputs of the summation layer (Fig. 1) forms a complete system of events. Although this method has many advantages, including the increased accuracy of the PNN, it is accompanied by some limitations in the case of unbalanced data sets. That is why this level of the PNN topology is of the most significant interest in this work. The output layer of the PNN topology generates an output signal of belonging of a given observation to a specific class of the task based on the use of the largest value from the set of probabilities obtained in the summation layer.

To describe the improved PNN version proposed in this paper to increase classification accuracy in case of unbalanced datasets lets us consider the mathematical basis of the Probabilistic Neural Network.

Suppose we have a tabular dataset, which is represented as a set of vectors in the total number of N in the form: $x_{i,1}, ..., x_{i,j}, ..., x_{i,n} \rightarrow y_i, i, \left(i = \overline{1, N}\right)$. Accordingly, the number of vectors N_k in the form $x_{i_k,1}, ..., x_{i_k,j}, ..., x_{i_k,n} \rightarrow y_i$ which belong to one of $k, \left(k = \overline{1, K_{\max}}\right)$ classes under the condition $N = \sum_{k=1}^{K_{\max}} N_k$ is also known. In this case, $k, \left(k = \overline{1, K_{\max}}\right)$ is the number of the class to which the corresponding vector belongs $i = \overline{1, N}$; $i_k, \left(i_k = \overline{1, N_k}\right)$ is the number of the vector from $K_{\max}$ classes; $j, \left(j = \overline{1, n}\right)$ is the attribute number of the corresponding vector.

A set of vectors $x_{m,1}, ..., x_{m,j}, ..., x_{m,n}$, is also specified, for which it is unknown to belong to certain classes of the task, i.e., there is no output element y_m, $m = \overline{1, M}$.

The PNN implementation algorithm involves performing the following steps:

1. Determine the number of input attributes and specified task classes.
2. Construct the PNN topology according to step 1.
3. Perform normalization of all vectors $\overline{x}_{i,j}$ in columns according to the expression:

$$x_{i,j} = \frac{x_{i,j}}{\max\limits_{1 \le j \le n} \left|x_{i,j}\right|}, \quad i = \overline{1, N}, \quad j = \overline{1, n}. \tag{1}$$

4. Perform normalization of all vectors $\overline{x}_{m,j}$ based on the maximum absolute value of the element obtained in item 3 in each column.
5. Calculate the Euclidean distances between $\overline{x}_{m,j}$ and all $\overline{x}_{i_k,j}$ for each of the presented classes of the task, according to the expression:

$$E_{m,i_k} = \sqrt{\sum_{j=1}^{n} (x_{m,j} - x_{i_k,j})^2}. \tag{2}$$

6. Calculate the Gaussian functions from (2) according to the expression:

$$G_{m,i_k} = \exp\left(-\frac{\left(E_{m,i_k}\right)^2}{\sigma^2}\right). \tag{3}$$

   where σ is a smooth factor (the only PNN parameter to be configured).
7. Calculate the probabilities $P(m, k)$ of belonging of the current vector $\overline{x}_{m,j}$ to each of the given classes of the problem, provided that $\sum_{k=1}^{K_{\max}} P(m, k) = 1$ [18]:

$$P(m, k) = \frac{\sum\limits_{i_k=1}^{N_k} G_{m,i_k}}{\sum\limits_{i_1=1}^{N_1} G_{m,i_k} + ... + \sum\limits_{i_k=1}^{N_k} G_{m,i_k} + ... + \sum\limits_{i_{K_{\max}}=1}^{N_{K_{\max}}} G_{m,i_{K_{\max}}}}, \quad k = \overline{1, K_{\max}} \tag{4}$$

Expression (4) describes the complete system of events, which is a tremendous advantage in Medicine. It is because it is crucial to know the magnitude of the probability of each of the events when $\sum_{k=1}^{K_{\max}} P(m,k) = 1$ [18]. However, this expression does not consider the uneven representation of the number of vectors from different classes. Thus minority classes are suppressed by the majority ones. In the case of the need to solve the binary classification task, which often arises in medical practice when it is necessary to process small data and even their imbalance - this is a significant disadvantage. Therefore, in this paper, we propose a new approach to the formation of the output signals of the PNN summation layer to determine the probability of belonging of the current vector to one of k, $\left(k = \overline{1, K_{\max}}\right)$ classes:

$$P(m,k) = \frac{\frac{\sum_{i_k=1}^{N_k} G_{m,i_k}}{N_k}}{\frac{\sum_{i_1=1}^{N_1} G_{m,i_k}}{N_1} + \ldots + \frac{\sum_{i_k=1}^{N_k} G_{m,i_k}}{N_k} + \ldots + \frac{\sum_{i_{K_{\max}}=1}^{N_{K_{\max}}} G_{m,i_{K_{\max}}}}{N_{K_{\max}}}}, \quad k = \overline{1, K_{\max}} \tag{5}$$

This relatively simple but effective solution satisfies both requirements: it preserves the complete events system. Also, it takes into account the uneven presentation of all classes of the task in a given dataset.

8. For each vector with unknown output $\max\{P(m,k)\}$ should be find. It should be noted that $\{P(m,k)\}$ are calculated based on (5). Expression (4) in the proposed I-PNN implementation algorithm is ignored. The maximum value $y_m = \max\{P(m,k)\}$ is the output signal of the I-PNN, which will determine whether the current vector belongs to one of the specified classes of the task.

4 Modeling and Results

The authors developed software in Python, which was used for practical experiments. We have worked on binary classification tasks in the field of Medicine.

Modeling I-PNN was performed using three well-known unbalanced short medical datasets with different amounts of instanced, attributes, and representation of classes (minority - min., and majority - maj. classes). The primary characteristics of the datasets and the Imbalanced Ratio for each of them are presented in Table 1. In addition, this table contains links to open repositories to get acquainted with them in more detail.

The proposed I-PNN and the base PNN are Neural Networks without training. Therefore, the three data sets in Table 2 were randomly divided into preparatory and test sets at 80% to 20%. In addition, the dataset [21] was pre-processed because it contained many gaps in the data. All vectors with omissions were removed from this dataset. I-PNN and base PNN contain only one parameter to be configured. It is a smooth factor - σ. The optimal value of this parameter for I-PNN during the analysis of each of the three medical datasets was found by the brute force method in the interval [0.01: 10] with a step of 0.01.

Table 1. Short descriptions of the binary imbalanced datasets used for modeling.

Dataset	The number of instances	The number of attributes	The number of maj.: min instances	IR**
Haberman's Survival Data Set [19]	306	3	225:81	2.78
Pima Indians Diabetes [20]	768	8	500:268	1.87
Breast Cancer Wisconsin [21]	682*	9	443:239	1.85

* *number of instances after data clearing (deleted vectors with missing data)*
** *IR (Imbalanced Ratio) = number of majority class (maj.) / number of minority class (min.)*

Table 2. I-PNN results on three different medical datasets.

Dataset	The optimal value of the smooth factor σ	Precision	Recall	F1-score
Haberman's Survival Data Set [19]	0.09	0,9104	0,9016	0,9059
Pima Indians Diabetes [20]	0.53	0,7675	0,7742	0,7708
Breast Cancer Wisconsin [21]	0.12	0,9796	0,9782	0,9789

Among the performance indicators used for I-PNN's modeling, the authors used Precision, Recall, and F1-score. Total accuracy was not considered because, as is known [22, 23], this indicator should not be used when evaluating the performance of artificial intelligence tools in the case of processing unbalanced data sets.

The results based on Precision, Recall, and F1-score for the three data sets and the selected optimal value σ that provided the highest accuracy of I-PNN are summarized in Table 2.

5 Comparison and Discusion

To compare the effectiveness of the proposed I-PNN in the case of solving binary classification tasks on short, unbalanced medical datasets with different Imbalanced Ratios, authors used a basic version of PNN from [18]. The optimal value σ was determined according to the same optimization method used for I-PNN.

The comparison results of both studied Artificial Neural Networks based on three different performance indicators are shown in Fig. 2. Conventionally, Fig. 2 is divided into three areas - each of which shows the results of experiments for three different data sets.

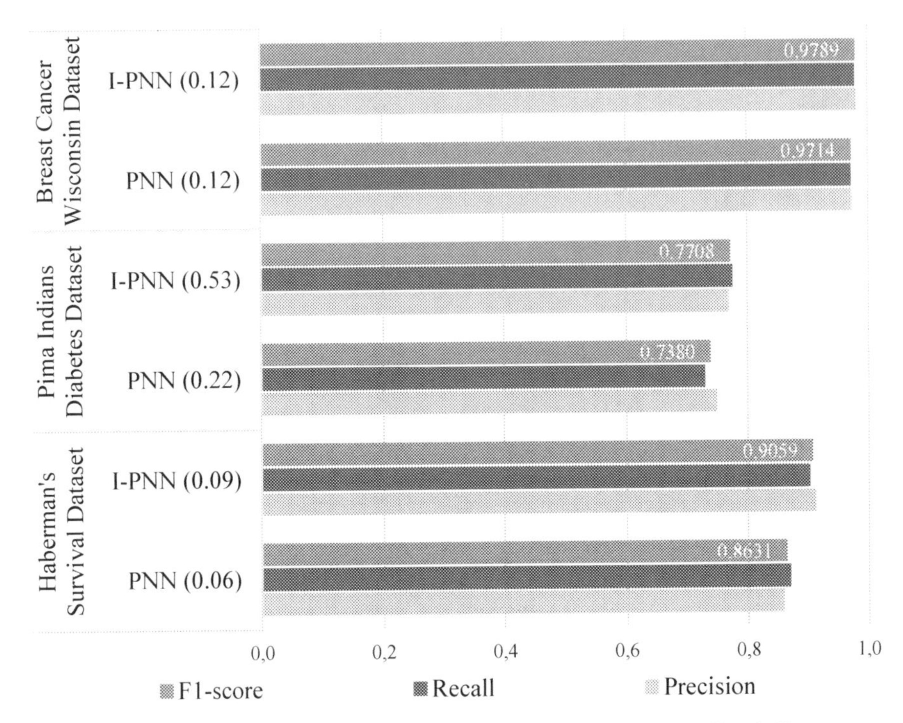

Fig. 2. Comparisons results of I-PNN and PNN using Precision, Recall and F1-score.

It should be noted that in parentheses next to each name of the studied ANN indicates the optimal value σ that ensured obtaining such results.

From Fig. 2 it can be stated:

- I-PNN provides an increase in accuracy compared to PNN [18] based on the F1-score for the three studied data sets;
- a significant increase in accuracy (4–5%) according to F1-score was obtained for data sets with low imbalance (according to [24], datasets [20] and [21] are low imbalance);
- very small accuracy increases are received for a dataset from [19]. Both can explain this: the small amount of data (twice lower as the other two datasets) and the significantly higher Imbalance Ration (more than 1.5 times higher than the other two datasets);
- if we consider the values of Precision, Recall indicators for both studied ANN for all three datasets, they are higher (to a greater or lesser extent) after applying the proposed I-PNN.

All the above advantages allow the use I-PNN: in its current form for further assessing the patient's condition by a doctor (as a set of probabilities); or by integrating it into automated hybrid diagnostic systems, such as [25], to improve the accuracy of their work in case of processing small unbalanced datasets.

6 Conclusions and Future Work

The paper proposes an improved scheme of Probabilistic Neural Network. We called it I-PNN. The authors have developed a new method for forming the outputs of the PNN summation layer, which, in contrast to the existing one, considers the uneven representation of all classes in a current dataset. This problem is quite common in Medicine, significantly when solving the binary classification tasks in various intelligent diagnostic systems. The authors substantiate the proposed approach and present a complete algorithm for implementing I-PNN.

Experimental modeling conducted by the authors on three data sets showed an increase in accuracy for all cases using three performance indicators - Precision, Recall, and F1-score. In particular, the F1-score showed an increase in accuracy from 0.5% to 5%, depending on the data set's characteristics, volume, and Imbalanced Ratio.

Further research will be conducted to evaluate the effectiveness of I-PNN in solving multi-class imbalanced classification tasks. In addition, it is planned to develop a homogeneous I-PNN's ensemble and its integration into hybrid diagnostic systems to improve the accuracy of their work in case of the need to process unbalanced data sets.

Acknowledgment. The authors would like to thank the reviewers for the correct and concise recommendations that helped present the materials better. We would also like to thank the Armed Forces of Ukraine for providing security to perform this work. This work has become possible only because of the resilience and courage of the Ukrainian Army. The National Research Foundation of Ukraine funded this research (grant number 2021.01/0103).

References

1. Chumachenko, D., Chumachenko, T., Meniailov, I., Pyrohov, P., Kuzin, I., Rodyna, R.: On-line data processing, simulation and forecasting of the coronavirus disease (COVID-19) propagation in ukraine based on machine learning approach. In: Babichev, S., Peleshko, D., Vynokurova, O. (eds.) DSMP 2020. CCIS, vol. 1158, pp. 372–382. Springer, Cham (2020). https://doi.org/10.1007/978-3-030-61656-4_25
2. Tsmots, I., Teslyuk, V., Vavruk, I.: Hardware and software tools for motion control of mobile robotic system. In: 2013 12th International Conference on the Experience of Designing and Application of CAD Systems in Microelectronics (CADSM), p. 368 (2013)
3. Shakhovska, N., Fedushko, S., Greguš ml., M., Melnykova, N., Shvorob, I., Syerov, Y.: Big Data analysis in development of personalized medical system. Procedia Comput Sci. **160**, 229–234 (2019). https://doi.org/10.1016/j.procs.2019.09.461
4. Sokolovskyy, Y., Levkovych, M., Mokrytska, O., Yatsyshyn, S., Kaspryshyn, Y., Strauss, C.: Mathematical Models and Analysis of Deformation Processes in Biomaterials with Fractal Structur, vol. 2488, pp. 133–144. Lviv, Ukraine (2019)
5. Khavalko, V., Tsmots, I., Kostyniuk, A., Strauss, C.: Classification and recognition of medical images based on the SGTM neuroparadigm. In: Proceedings of the 2nd International Workshop on Informatics & Data-Driven Medicine (IDDM 2019), 11–13 Nov 2019, vol. 2488, pp. 234–245. Lviv, Ukraine (2019)
6. Salazar, A., Vergara, L., Safont, G.: Generative adversarial networks and markov random fields for oversampling very small training sets. Expert Syst. Appl. **163**, 113819 (2021). https://doi.org/10.1016/j.eswa.2020.113819

7. Shakhovska, N., Yakovyna, V.: Feature selection and software defect prediction by different ensemble classifiers. In: Strauss, C., Kotsis, G., Tjoa, A.M., Khalil, I. (eds.) DEXA 2021. LNCS, vol. 12923, pp. 307–313. Springer, Cham (2021). https://doi.org/10.1007/978-3-030-86472-9_28
8. Shakhovska, N., Yakovyna, V., Kryvinska, N.: An improved software defect prediction algorithm using self-organizing maps combined with hierarchical clustering and data preprocessing. In: Hartmann, S., Küng, J., Kotsis, G., Tjoa, A.M., Khalil, I. (eds.) DEXA 2020. LNCS, vol. 12391, pp. 414–424. Springer, Cham (2020). https://doi.org/10.1007/978-3-030-59003-1_27
9. Kotsovsky, V., Geche, F., Batyuk, A.: On the computational complexity of learning bithreshold neural units and networks. In: Lytvynenko, V., Babichev, S., Wójcik, W., Vynokurova, O., Vyshemyrskaya, S., Radetskaya, S. (eds.) ISDMCI 2019. AISC, vol. 1020, pp. 189–202. Springer, Cham (2020). https://doi.org/10.1007/978-3-030-26474-1_14
10. Kotsovsky, V., Batyuk, A., Yurchenko, M.: New approaches in the learning of complex-valued neural networks. In: 2020 IEEE Third International Conference on Data Stream Mining Processing (DSMP), pp. 50–54 (2020)
11. Gromaszek, K., Bykov, M.M., Kovtun, V.V., Raimy, A., Smailova, S.: Neural network modelling by rank configurations. Photonics Applications in Astronomy, Communications, Industry, and High-Energy Physics Experiments 2018, p. 93. SPIE, Wilga, Poland (2018)
12. Shahadat, N., Rahman, B., Ahmed, F., Anwar, F.: Dropout effect on probabilistic neural network. In: 2017 International Conference on Electrical, Computer and Communication Engineering (ECCE), pp. 217–222. IEEE, Cox's Bazar, Bangladesh (2017)
13. Ramezankhani, et al.: The impact of oversampling with SMOTE on the performance of 3 classifiers in prediction of type 2 diabetes. Med. Decis. Making **36**, 137–144 (2016)
14. Guan, H., et. al.: SMOTE-WENN: Solving class imbalance and small sample problems by oversampling and distance scaling. Appl Intell. **51**, 1394–1409 (2021)
15. Raposo, L.M., Arruda, M.B., de Brindeiro, R.M., Nobre, F.F.: Lopinavir resistance classification with imbalanced data using probabilistic neural networks. J. Med. Syst. **40**(3), 1–7 (2016). https://doi.org/10.1007/s10916-015-0428-7
16. Chandrasekara, V., Tilakaratne, C., Mammadov, M.: an improved probabilistic neural network model for directional prediction of a stock market index. Appl. Sci. **9**, 5334 (2019). https://doi.org/10.3390/app9245334
17. Shahadat, N., Pal, B.: An empirical analysis of attribute skewness over class imbalance on Probabilistic Neural Network and Naïve Bayes classifier. In: 2015 International Conference on Computer and Information Engineering (ICCIE), pp. 150–153. IEEE, Rajshahi, Bangladesh (2015). https://doi.org/10.1109/CCIE.2015.7399301
18. Izonin, I., Tkachenko, R., Ryvak, L., Zub, K., Rashkevych, M., Pavliuk, O.: Addressing medical diagnostics issues: essential aspects of the PNN-based approach. CEUR-WS: Proceedings of the 3rd International Conference on Informatics & Data-Driven Medicine, 19–21 Nov 2020, vol. 2753, pp. 209–218. Växjö, Sweden (2020)
19. UCI Machine Learning Repository: Haberman's Survival Data Set. https://archive.ics.uci.edu/ml/datasets/haberman's+survival. Accessed 01 April 2022
20. Pima Indians Diabetes Database. https://kaggle.com/uciml/pima-indians-diabetes-database. Accessed 16 May 2021
21. UCI Machine Learning Repository: Breast Cancer Wisconsin (Diagnostic) Data Set. https://archive.ics.uci.edu/ml/datasets/breast+cancer+wisconsin+(diagnostic). Accessed 01 April 2022
22. Bykov, M.M., et. al.: Research of neural network classifier in speaker recognition module for automated system of critical use. Presented at the Photonics Applications in Astronomy, Communications, Industry, and High-Energy Physics Experiments 2017 , Wilga, Poland August 7 (2017)

23. Mochurad, L., Hladun, Y.: Modeling of psychomotor reactions of a person based on modification of the tapping test. Int. J. Comput. **20**, 1–10 (in press) (2021)
24. Fernández, A., García, S., del Jesus, M.J., Herrera, F.: A study of the behaviour of linguistic fuzzy rule based classification systems in the framework of imbalanced data-sets. Fuzzy Sets Syst. **159**, 2378–2398 (2008). https://doi.org/10.1016/j.fss.2007.12.023
25. Izonin, I., et. al.: PNN-SVM approach of TI-based powder's properties evaluation for biomedical implants production. Comput. Mater. Continua **71**, 5933–5947 (2022). https://doi.org/10.32604/cmc.2022.022582

PBRE: A Rule Extraction Method from Trained Neural Networks Designed for Smart Home Services

Mingming Qiu[1,2](✉), Elie Najm[1], Rémi Sharrock[1], and Bruno Traverson[2]

[1] Télécom Paris, Palaiseau, France
{Mingming.Qiu,Elie.Najm,Remi.Sharrock}@telecom-paris.fr

[2] EDF R&D, Palaiseau, France
Bruno.Traverson@edf.fr

Abstract. Designing smart home services is a complex task when multiple services with a large number of sensors and actuators are deployed simultaneously. It may rely on knowledge-based or data-driven approaches. The former can use rule-based methods to design services statically, and the latter can use learning methods to discover inhabitants' preferences dynamically. However, neither of these approaches is entirely satisfactory because rules cannot cover all possible situations that may change, and learning methods may make decisions that are sometimes incomprehensible to the inhabitant. In this paper, PBRE (Pedagogic Based Rule Extractor) is proposed to extract rules from learning methods to realize dynamic rule generation for smart home systems. The expected advantage is that both the explainability of rule-based methods and the dynamicity of learning methods are adopted. We compare PBRE with an existing rule extraction method, and the results show better performance of PBRE. We also apply PBRE to extract rules from a smart home service represented by an NRL (Neural Network-based Reinforcement Learning). The results show that PBRE can help the NRL-simulated service to make understandable suggestions to the inhabitant.

Keywords: Rule extraction · Neural network · Reinforcement learning · Smart home

1 Introduction

Numerous smart home applications are rapidly emerging to provide various services to inhabitants. Most of these applications belong to the knowledge-based approaches. Expert systems [12] are one of the most well-known knowledge-based systems. They allow inhabitants to design their services based on a set of rules. However, despite the potential security and privacy risks [19], developing smart home services with knowledge-based approaches is usually a complicated manual process, especially when the services are complex or the actuators are diverse and tightly interconnected. Moreover, it is not easy to design rules when only ultimate

C. Strauss et al. (Eds.): DEXA 2022, LNCS 13427, pp. 158–173, 2022.
https://doi.org/10.1007/978-3-031-12426-6_13

objectives are known, e.g., it is cumbersome to design rules for an HVAC (heating, ventilation, and air conditioning) system when the desired indoor temperature is specified along with the energy consumption to be minimized.

Other applications to implement a smart home system mainly belong to data-driven approaches. These approaches make strategic decisions based on the analysis and interpretation of data. And learning methods that learn from data and make predictions based on it are at the forefront of data-driven decision making [5]. They try to automatically discover the patterns of the systems by analyzing the datasets provided. Thus, in a smart home, these approaches can figure out regulation solutions by studying inhabitants' activities. It is essential to consider the reactions of an inhabitant when attempting to design a user-friendly smart home system [6]. Reinforcement learning (RL) [8], whose basic idea is that an artificial agent learns the system's behavior patterns by interacting with the environment, can consider the inhabitant's reactions to the proposed actions to find out his habitual behaviors, and a group of habitual behaviors can be translated into a service. In this way, RL enables the inhabitant to participate in the control of smart home services. Moreover, neural network-based reinforcement learning (NRL), which integrates neural networks with RL, facilitates the modeling of high-dimensional systems for RL [8]. However, NRL works like a black box as it does not explain why it proposes new services or modifies existing ones.

To overcome the above shortcomings of the two approaches, we propose to extract rules from a trained NRL. The extracted rules allow showing the inhabitant in which situations the NRL suggested certain actions and enrich the knowledge base, which saves the inhabitant from manually creating rules. However, most of the existing rule extraction methods focus either only on neural networks with discrete inputs or on binary classification problems. Nevertheless, in a smart home, there are both discrete (window state: open or closed) and continuous states (light intensity or temperature). Moreover, the control of smart home services is not only a binary but also a multi-class classification problem. In this paper, a method called PBRE (Pedagogic Based Rule Extractor) is proposed to extract rules from a trained NRL that takes discrete and continuous states as input and proposes states for multiple actuators.

In the rest of the paper, Sect. 2 presents existing work on rule extraction. Section 3 explains the principle of PBRE. Section 4 evaluates PBRE and compares it with an existing method called RxNCM.[1] Section 5 shows how the NRL learning and rule extraction methods can be integrated into a smart home system. Section 6 simulates smart home light services with NRLs and evaluates the performance of PBRE in extracting rules from NRLs (see Footnote 1). Section 7 summarizes the main contributions and provides interesting perspectives.

2 Context and Related Work

The smart home is usually implemented by setting up various services. In our study, a set of possible operations performed by different devices can change

[1] The codes for the implementations of all experiments can be found in: https://github.com/mingming81/PBRE.git.

the value of a particular environment state. A service can be denoted by the name of that state, and the operations involved are means to implement that service. For example, if an inhabitant is at home, raising the heater to increase the temperature or opening a window to decrease the temperature are two ways to implement a temperature service.

To create services, we can use methods of knowledge-based [9,17,18] or data-driven approaches. However, knowledge-based systems usually require manual input from the inhabitant to design services, which hinders the creation of complex services. Although neural networks are more and more popular, and NRLs are used by many smart home systems [16,23,24] to create services that can interact with the environment and adapt to the inhabitant's activities, they are like black boxes, and we do not know why they suggest certain services.

To make NRLs understandable and enrich the knowledge base, we consider extracting rules from NRLs. There is a lot of work on extracting rules from trained neural networks. For example, in a tree-based machine learning approach [4,15] collects all formulas from the root to a leaf node with a decision value and conjugates all these formulas to obtain a rule. [3] proposes an algorithm called RxNCM. This algorithm first removes insignificant input attributes from the trained neural network. Then, it determines the ranges for each attribute by selecting its minimum and maximum values from the training samples. The rules are created by combining attributes with ranges of values and the corresponding outputs. These rules are then pruned by removing conditions from a rule if the accuracy of the rules can be increased in the test dataset. Finally, the pruned rules are updated by removing overlapping ranges of attribute values between rules if the accuracy of the new rules in the test dataset is increased. [22] proposes MOFN to extract rules. First, a neural network is created using KBANN [21] and then trained. Next, units with similar weighted connections are grouped. The average values replace the weight values in each group, and the groups with low link weights are deleted. The updated neural network is subsequently trained again by optimizing the bias values. Then rules with weights and biases are extracted by combining inputs and outputs. The final rules are obtained by removing the weights and biases. However, with the exception of RxNCM, the above work focuses on neural networks that either have specific structures, e.g., [4] is only suitable for tree-based machine learning methods, or they only accept categorical and limited integer inputs, e.g., [22] is only suitable for neural networks with limited integer inputs. In this work, we propose PBRE to extract rules from trained neural networks or NRLs by ignoring the input data types and structures of neural networks, and then evaluate it with RxNCM to prove its better performance.

3 The Proposed PBRE Method

The principle of PBRE is illustrated in Fig. 1: First, PBRE extracts an instance rule from a trained neural network, where an instance rule is a mapping between the inputs and outputs of a neural network at a given time step. Then, it generalizes the instance rule. Next, it combines the generalized rules by merging those whose conclusions are the same and the range values of the states in the conditions

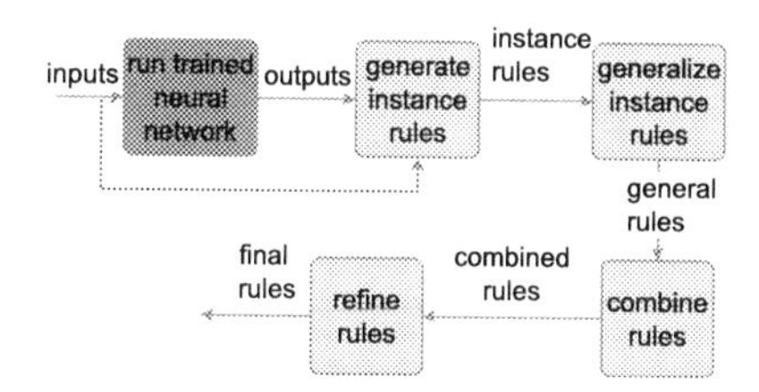

Fig. 1. PBRE rule extraction process

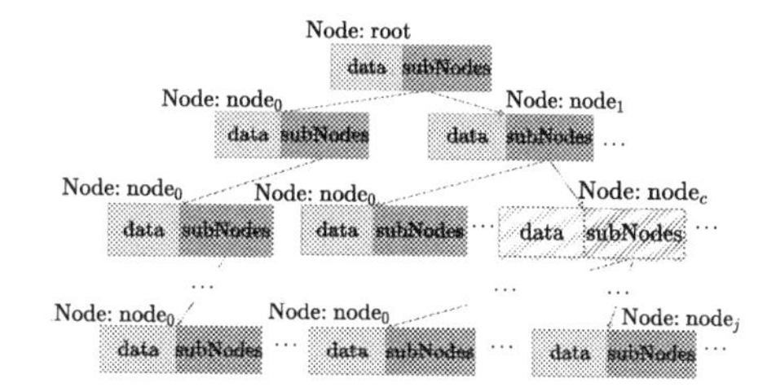

Fig. 2. Tree data structure

overlap. Finally, it refines the combined rules by removing insignificant states in the conditions based on the accuracy of the rules in the unseen dataset. The unseen dataset is a dataset that contains samples from which no rules are extracted, while the seen dataset is used to extract rules. The use of the unseen dataset allows the evaluation of the generalization capability [14, 20] of the extracted rules.

3.1 Generate Instance Rules

A neural network can directly output the desired action that an actuator will perform. Nevertheless, an NRL may provide the quality values for all possible states of an actuator, e.g., a deep neural network-based Q-learning (DQN) [10], which we use in this paper to implement smart home services. Therefore, we first introduce how to select the state of an actuator when the NRL generates quality values for all possible states of that actuator. Suppose that there is a trained NRL with an unknown structure, its inputs are $\mathbf{s}_t = [s_t^{i^0}, \cdots, s_t^{i^n}]$ where the superscript indicates the type of input state and the subscript denotes the time step. The outputs are action quality values $\mathbf{q}_t = [q_{0,t}^{act^0}, \cdots, q_{i,t}^{act^0}, \cdots, q_{0,t}^{act^m}, \cdots, q_{k,t}^{act^m}]$, where the superscript indicates the type of actuators and the subscript represents the actuator's certain state and time step. Thus, the proposed states are

$$s_{i',t}^{act^0} = argmax(q_{0,t}^{act^0}, \cdots, q_{i,t}^{act^0}), \cdots, s_{k',t}^{act^m} = argmax(q_{0,t}^{act^m}, \cdots, q_{k,t}^{act^m}). \quad (1)$$

Moreover, we can obtain and represent the instance rule ir_t at time step t as

$$\textit{if state } i^0 \textit{ has value } s_t^0, \cdots, \textit{ then actuator } act^0 \textit{ will have state } s_{i',t}^{act^0}, \cdots. \quad (2)$$

However, instance rules are not like rules for general situations. We should generalize them into rules by performing the following procedure.

3.2 Generalize Instance Rules

We generalize instance rules by first expressing instance rules in a tree structure with linked lists. This tree structure gives a concise and clear idea of an instance rule's compositions and generalization process. Then, we generalize an instance rule by merging it with another instance rule or rule whose conclusions are the same as those of the instance rule and whose certain conditions have close values or contain the values of those of the instance rule.

Figure 2 shows the structure of the linked list tree. Each branch is a linked list and stores an instance rule or a rule. Each branch node consists of a state value belonging to either conclusions or conditions. Except for the root node whose value is constant, the other nodes at the same level describe the values of the same state type of different instance rules or rules. The data structure of each node belonging to the *Node* class is shown in Fig. 2. It consists of two parts: *data* being the state value of the current node and *subNodes* storing the child nodes of the current node. The state value of the node is represented as

$$node.data = \{Float : mean, Float : min, Float : max\} \tag{3}$$

where *mean, min,* and *max* are the average, minimum, and maximum of the state value of all combined samples. The introduction of a range of values between the minimum and the maximum makes it possible to combine instance rules or an instance rule and a rule when their conclusions are the same, while the values of certain states in the conditions are the same or close or overlap. The data structure of *subNodes* is shown in Eq. 4. It is a vector consisting of all subnodes of the current node. Each element in this vector contains a subnode of class *Node* and an integer *count* indicating how many times the state value of the current subnode appears simultaneously with the state values of the nodes in the same branch and at higher levels.

$$node.subNodes = \{\{Node{:}node_0, Integer{:}count_0\}, \cdots\}. \tag{4}$$

To access a node and its subnodes, we use the dot notation like the property accessor in JavaScript, i.e. the node n_i in slashes in Fig. 2 can be described as $root.node_1.node_c$, which belongs to the class *Node*, its state value is $n_i.data$, and the subnodes are: $n_i.subNodes$. To access its average, minimum and maximum values, we use $n_i.data.mean, n_i.data.min$ and $n_i.data.max$; to query its jth subnode, we use $n_i.subNodes[j]$.

However, after generalizing an instance rule, there is a possibility that some particular state value ranges overlap between rules with the same conclusions. Therefore, we need to combine the obtained rules.

3.3 Combine Rules

To combine rules R_a whose conclusions are the same while the ranges of states in the conditions overlap, for each state c in the condition, we first sort rules of R_a in ascending order of the minimum values of c. Next, we name the first rule of R_a as r_m. Then we iterate R_a. For each rule r_a in R_a, its minimum value of state c is compared with the maximum value of the same state in r_m. If the former is not larger than the latter, we change $r_m.c.max$ to the larger value between $r_a.c.max$ and $r_m.c.max$. If there is no such rule r_a, we store r_m in a new rule set R_b, define the current r_a as r_m, and compare r_m with the remaining rules of R_a. After all rules in R_a have been tested, we store the last r_m in R_b to ensure that no r_m is forgotten. Finally, we need to delete duplicate rules from R_b because the above process is performed for each state in the conditions, therefore it may

Algorithm 1: Remove Insignificant States (RIS)

```
   Result: final rules
 1 D: dataset storing samples with each containing state values and target values
 2 D_unseen: unseen dataset with each containing state values and target values
 3 R = {r_1, ···, r_m}: extracted rules
 4 maxAcc: the maximum accuracy
 5 stateTypes: the available state types
 6 insignificantStates: vector storing insignificant state types
 7 Function RIS(D, D_unseen, R):
 8   corrState = D.corr().iloc[: -1, -1]; sort corrState in ascending order
 9   sort stateTypes with the same order of corrState
10   for stateType in stateTypes do
11     numCorrect = 0
12     for d_unseen in D_unseen do
13       create R_1 ∈ R without states with types in insignificantStates and
          stateType
14       select R_2 ∈ R_1, whose rules' states ranges contain those of d_unseen
15       create R_0 by adding deleted states to R_2
16       if size(R_2) equals 1 then inference = conclusions of R_2
17       else
18         arr=concatenate d_unseen.states and R_0.states.mean
19         arr, corrState2=remove states with types in
            insignificantStates and stateType from arr, corrState
20         inference = Inference(corrState2, arr)
21       end
22       if inference equals d_unseen.targets then numCorrect+ = 1
23     end
24     acc = numCorrect/length(D_unseen)
25     if acc ≥ maxAcc then add stateType to insignificantStates;
        maxAcc = acc
26   end
27   for stateType in insignificantStates do
28     for d_unseen in D_unseen do
29       arr=execute lines 13∼18
30       arr, corrState2=arr, corrState remove states with types in
          insignificantStates and add state with type stateType
31       execute lines 20∼24
32     end
33     if acc ≥ maxAcc then remove stateType from insignificantStates;
        maxAcc = acc
34   end
35   remove states with types in insignificantStates from R; return R
```

result in duplicate rules. The final R_b stores rules with the same conclusions but without states in the conditions whose ranges overlap with those of other rules.

3.4 Refine Rules

In this section, we focus on removing insignificant states in the conditions using Algorithm 1. First, we use Pearson product-moment correlation coefficients

Algorithm 2: Infer the unseen sample (Inference)

Result: Inference result
1 $corrState$: correlation vector between state types and targets
2 arr:matrix concatenating states of the unseen sample and averages of the states in the conditions of the rules
3 ϵ: predefined small number
4 **Function** `Inference`($corrState$, arr):
5 $corr = correlation(arr * corrStates)$
6 **if** $\forall i, j \in corr, (i - j) < \epsilon$ **then** select $conclusions$ of the rule which has the maximum sum of frequencies of occurrence of the conclusions
7 **else** select $conclusions$ of the rule with the maximum correlation
8 return $conclusions$

(PPMCC) [2] to calculates the states-targets correlation vector. This vector stores the correlation between available state types and the targets. Then, we sort this correlation vector in ascending order to ensure the least correlated state types come first (lines 8–9). For each *stateType* in available state types *stateTypes*, first, we define a rule set R_1 acquired from R by removing states with types in $insignificantStates$ and *stateType*. Then, we create a rule set R_2 containing rules from R_1, whose states' range values in the conditions contain the states' values of the unseen sample under study d_{unseen}. We also define a rule set R_0 equal to R_2 with the deleted states added (line 10–15). If the size of R_2 is 1, we select the conclusions of this rule as the targets of d_{unseen} (line 16). Otherwise, we remove states with types in $insignificantStates$ and *stateType* from the concatenated states matrix which combines the states of d_{unseen} and the averages of the states in the conditions of R_0. The states with types in $insignificantStates$ and *stateType* are also removed from the states-targets correlation vector (lines 17–19). Next, we use Algorithm 2 to derive the unseen sample d_{unseen} (line 20).

As shown in Algorithm 2, to obtain inference, we first multiply the updated concatenated matrix about states with the states-targets correlation vector, and then use PPMCC to calculate the correlation between d_{unseen} and rules in R_2 based on this weighted state matrix (line 5). If all rules have close correlations with d_{unseen}, we choose the conclusions of the rule which has the maximum sum of frequencies of occurrence (*count* in Eq. 4) of the conclusions; otherwise, we select the conclusions of the rule whose states are most strongly correlated with those of d_{unseen} (lines 6–7).

After all unseen samples are derived, we compute the accuracy of R without state with types in $insignificantStates$ and *stateType* (line 24). If the accuracy is not less than the maximum accuracy, the state with type *stateType* is not important for the rules to correctly make inference. It will be added to the $insignificantStates$ vector, and the current accuracy will be the maximum accuracy (line 25). Next, after having run through all *stateTypes* and obtained the final $insignificantStates$, we decide which *stateType* in $insignificantStates$ can be re-added to rules to maintain or improve the accuracy on the unseen dataset. If such a *stateType* exist, it will be removed from $insignificantStates$ vector, and the updated accuracy will be the new maximum accuracy as shown in lines 27–33. When the updated final *insignificantStates* is acquired, we remove states belonging to types in *insignificantStates* from R and return the updated R as the final extracted rules (line 35). An extracted rule after having been converted to "if-then" rule can be written as:

$$\textit{if state } i^0 \textit{ is between } s_0^0 \textit{ and } s_1^0 \textit{ and has average } s_m^0, \cdots, \textit{ then actuator } act^0 \textit{ will have state } s_0^{act^0} \textit{ with frequency of occurrence } count_0^{act^0}, \cdots. \quad (5)$$

4 Evaluation and Comparison with Existing Work

Before applying PBRE to extract rules from DQNs in a smart home, we compare and evaluate the performance of PBRE with that of RxNCM on six datasets from the machine learning repository of the University of California Irvine: the Iris dataset, the Wisconsin Breast Cancer (WBC) dataset, the Sonar dataset, the German Credit dataset, the Ionosphere dataset, and the Heart Disease dataset. Descriptions of the datasets are provided in Table 3 (Appendix 1).

4.1 Comparative Experiment

Metrics. We use the following metrics to evaluate PBRE and compare it with RxNCM: (1) The number of extracted rules. [7,14] (2) Accuracy describes the number of samples where the updated controllable environment states conform to the inhabitant's habitual behaviors as a percentage of the total samples. [1,22,25] (3) Similarity, or fidelity [1,22,25], is the number of samples where conclusions derived from the rules are the same with propositions proposed by the neural networks as a percentage of the total samples. (4) Inference is the number of samples that the extracted rules can derive as a percentage of the total samples. Metrics (2)–(4) are evaluated for both seen and unseen samples. The procedure for determining the metrics is shown in Fig. 9 (Appendix 2).

Results. Table 1, Fig. 3 and Fig. 4 show the metric results for PBRE and RxNCM. From Table 1, we see that the number of rules extracted from each dataset by PBRE or RxNCM is not large, which ensures that storing the extracted rules does not require large memory, which is an important metric for a high-dimensional smart home system. Figure 3 and Fig. 4 show that although RxNCM, like PBRE, can infer all seen and almost all unseen samples (see "PBRE Infe." and "RxNCM Infe."), the rules extracted with PBRE generally have higher accuracy and similarity than those extracted with RxNCM (see "PBRE Acc.", "RxNCM Acc.", "PBRE Sim." and "RxNCM Sim."). To illustrate the general performance, we calculate the average of metrics (2) to (4) and denote it as "PBRE Ave." and "RxNCM Ave.". The results show that PBRE has higher general performance than RxNCM for both datasets, which is consistent with the observations made above for metrics (2) to (4). Moreover, including general performance, RxNCM has higher metric results in the unseen datasets than in the seen datasets, and PBRE does the opposite and has higher metric results than RxNCM in both datasets. This is because

Table 1. Number of rules extracted with PBRE and RxNCM

	Iris	WBC	Sonar	German credit	Ionosphere	Heart disease
PBRE num. of rules	3	2	2	2	2	5
RxNCM num. of rules	3	2	2	2	2	5

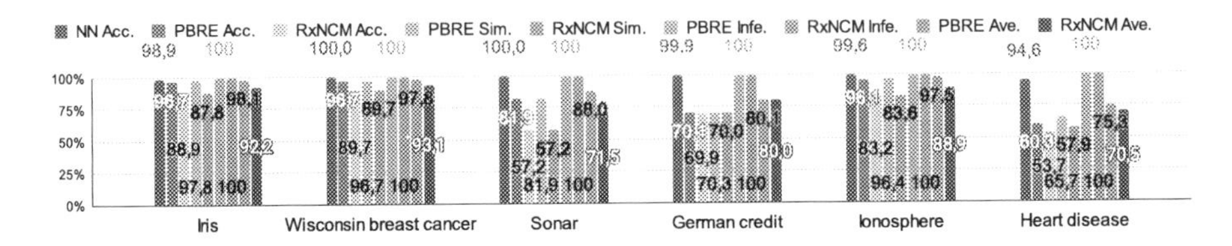

Fig. 3. PBRE and RxNCM experiment results by working on seen datasets

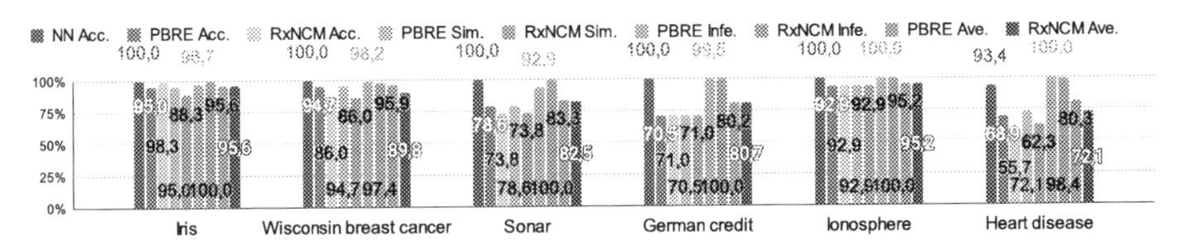

Fig. 4. PBRE and RxNCM experiment results by working on unseen datasets

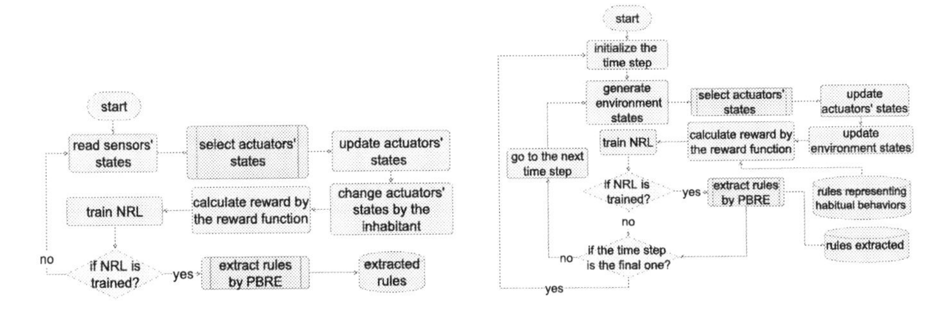

Fig. 5. A smart home system with one service in practice

Fig. 6. A smart home system with one service in simulation

to refine rules, PBRE not only deletes the states in the conditions but also adds them back, which ensures the number of states in the conditions and guarantees that PBRE achieves good performance in the unseen datasets and maintains the performance in the seen datasets.

5 NRL and Rule Extraction Methods in the Smart Home

5.1 Smart Home System in Practice

Figure 5 shows how the smart home system that uses NRL to implement a service and integrates rule extractions looks like in practice: When the system starts, it reads the values of the sensors associated with the service. Then the service selects the involved actuators' states. The actuators update their states to the selected ones. The executions of the actuators lead to changes in the controllable environment states, e.g., indoor light intensity. If the inhabitant is satisfied with the updated controllable states, he takes no action; otherwise, he can change some or all associated actuators' states to match his habitual behaviors. Considering the changes that the inhabitant makes to the actuators' states, the system can then obtain the reward calculated by the predefined reward function. It then trains the NRL using the transitions as input. Each transition contains the environment states detected by the sensors, the states of the actuators selected by the service, the rewards, and the updated environment states. If the NRL is well trained with high and stable accuracy, the system uses PBRE to extract rules and stores them in a database; otherwise, it repeats the above process.

5.2 Smart Home System in Simulation

Figure 6 shows how the smart home system looks like in simulation: First, the time step t is initialized. Next, the environment states, e.g., the inhabitant state, the indoor and outdoor light intensities at time step t, are generated by the predefined functions. Then, the service simulated by NRL selects the states that the actuators should take at time step t, and the actuators update their states to the selected ones. The controllable environment states are subsequently updated in terms of the actuators' executions. Depending on the predefined reward functions and simulated inhabitant's habitual behaviors, a reward is calculated to indicate whether the inhabitant is satisfied with the updated controllable environment states. After that, the system uses certain optimization algorithm to train the NRL based on the transitions. Once the NRL is well trained with high and stable accuracy, the rules are extracted using PBRE and stored in a database. The system then determines whether the current time step is the last time step. However, if the NRL is not well trained, the system directly checks if the current time step is the last one. If not, the system proceeds to the next time step; otherwise, it returns to the first time step and repeats the entire process.

Table 2. Number of rules extracted by PBRE from different DQNs

	DQN_v1	DQN_v2	DQN_v3
Num. of rules	4	4	19

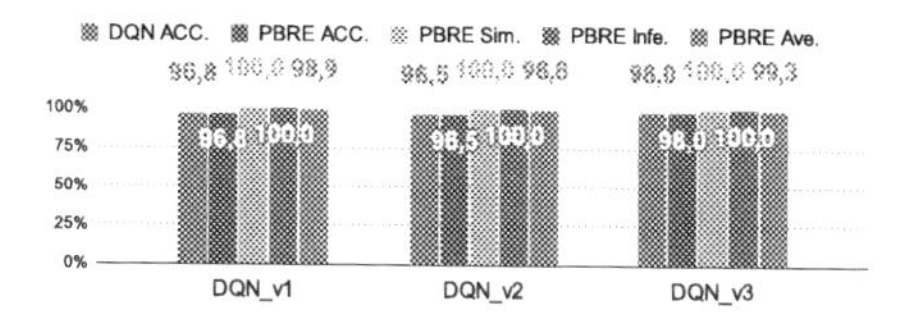

Fig. 7. PBRE with the seen datasets

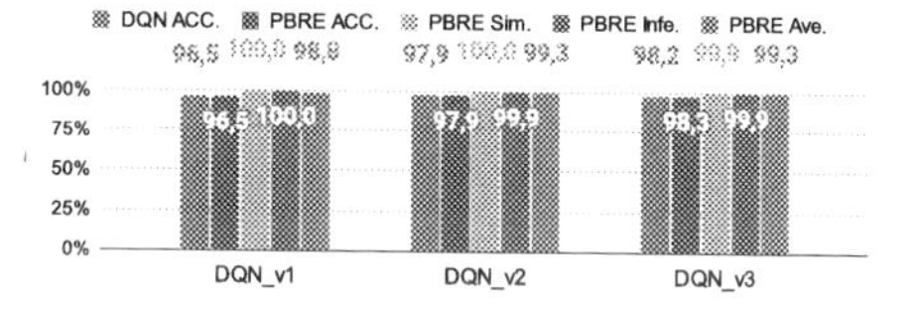

Fig. 8. PBRE with the unseen datasets

6 Experiment in the Smart Home Context

In this section, we run three tests (DQN_v1, DQN_v2, and DQN_v3), each using a DQN to simulate a smart home light service and evaluating PBRE. To perform these tests, we first introduce a simulated environment and then follow the process in Fig. 6, we use PBRE to extract rules from the light service.

6.1 Simulated Environment

The representations of the involved variables in the simulated smart home are: (1) s^{us}: state of the inhabitant; (2) s^{le}: outdoor light intensity; (3) s^{lr}: indoor light intensity;(4) s^{lp}: state of the lamp; (5) s^{cur}: state of the curtain.

Each light service first selects s^{lp} and s^{cur} as its outputs, and only considering s^{us} for DQN_v1 and also s^{le} for DQN_v2 and DQN_v3 as the inputs. The selected s^{lp} and s^{cur} are used to change s^{lr}. A reward r is then calculated by the predefined reward function with respect to the simulated inhabitant habitual behaviors and the obtained s^{lr}. According to Fig. 6, the system trains the DQN by using the transition at each time step. Each transition includes the current environment states s^{us} and s^{le}, the proposed actuators states s^{lp} and s^{cur}, the reward r and the new environment states s^{us} and s^{le} whose values have not been changed as a result of the actuators' executions. When the DQN is well trained, rules are extracted from it by using the seen dataset as its input, and the simulated smart home system starts a new iteration.

In this experiment, s_t^{us} at time step t is randomly generated by following the uniform distribution $U_{int}(0, n_{us})$ which generates an integer between 0 inclusive and n_{us} exclusive, where n_{us} is the total number of possible states of the inhabitant. s^{le} within a day is simulated with a Gaussian distribution [11,13]:

$$s_t^{le} = \mathcal{N}(amplitude = 600, mean = 12, stddev = 3) + 5 \cdot U(0,1) \qquad (6)$$

where some noise simulated in a uniform distribution $U(0,5)$ with a maximum value of 5 is added. s_t^{le} is generated every 5 min in one day.

The output of the light service is s_t^{lp} and s_t^{cur}. s_t^{lp} can be selected from multiple levels where level 0 indicates that the lamp is off and other levels represent that the lamp is on. The light intensity that s_t^{lp} can provide when it is on is $\beta \cdot s_t^{lp}$, where β is the light intensity that one level can provide. $s_t^{cur} \in \{0, 1/2, 1\}$ where 0 is closed, 1/2 is half-open, and 1 is fully open. s_t^{lr} thus is

$$s_t^{lr} = \beta \times s_t^{lp} + s_t^{le} \times s_t^{cur}. \qquad (7)$$

The inhabitant's habitual behaviors, which are the ultimate objectives that DQNs try to achieve, are simulated in "if-then" rules related to s^{us} and s^{lr}. For example: *if the inhabitant is absent, then the indoor light intensity is 0 lux.* We do not use specific actuators' states, which contributes to implementing a smart home service that can derive the regulation solutions when only the ultimate objectives of the inhabitant are given. The reward function (see Footnote 1) used defines rewards as constant numerical values when different habitual behaviors are satisfied.

6.2 Experiment Results

The metric results of PBRE in extracting rules from different DQNs for the seen and unseen datasets are shown in Table 2, Fig. 7 and Fig. 8. We can see that the number of extracted rules in Table 2 for each simulated service is not large, which ensures that storing these rules does not require large memory. Moreover, we can see that the number of rules in DQN_v3 has a larger value because the corresponding habitual behaviors of the inhabitant are more complex, as shown in Table 4 (Appendix 3). Furthermore, from Fig. 7 and Fig. 8, we see that PBRE can extract rules from DQNs with satisfactory general performance (see "PBRE Ave."), which can be further explained as follows: The extracted rules achieve the same and sometimes even higher accuracy than the DQNs; they have the same similarity to the DQNs in the seen datasets and almost the same similarity in the unseen datasets; moreover, they can infer all seen and almost all unseen samples. One of the extracted rules after having been deleted averages and frequencies of occurrence from Eq. 5 in DQN_v3 is: *if the inhabitant is working, and the outdoor light intensity is between 0.35 and 243.28, then the lamp is at level 3, and the curtain is closed.* More rules can be found in Table 4 (Appendix 3) where we approximate the lowest and highest values of each state's range to integers.

7 Conclusion

NRLs can implement smart home services by interacting with the inhabitant and adapting to his habitual behaviors. Yet, like other neural networks, NRLs are black boxes, and the inhabitant cannot know why they suggest certain services.

To address this problem, several contributions are made: (1) We propose PBRE to extract rules from trained neural networks or NRLs without considering their structures and input data types. And the comparison results with RxNCM prove the better performance of PBRE. (2) We show how the smart home working process, including using an NRL to implement a service and PBRE to extract rules, works in practice and in simulation. (3) We evaluate the performance of PBRE in extracting rules from a smart home light service simulated by a DQN. The results show that PBRE can satisfactorily extract rules from these DQNs.

In perspective, it is essential to evaluate the explainability of the extracted rules with qualitative results obtained through focus groups. Then it is promising to work on a proposal for a smart home system with multiple services.

Acknowledgements. This work is supported by Seido Laboratory, EDF R&D Saclay, Télécom Paris, and ANRT (Association Nationale Recherche Technologie) under grant number CIFRE n° 2018/1458.

Appendix 1 Datasets Descriptions

The descriptions of the six datasets are in Table 3.

Table 3. Datasets used for evaluating the performance of the extracted rules

Dataset	Num. of samples	Num. of attributes	Attribute characteristics	Num. of class
Iris dataset	150	4	Real	3
Wisconsin breast cancer	569	30	Real	2
Sonar dataset	208	60	Real	2
German credit dataset	1000	9	Categorical, Integer	2
Ionosphere dataset	350	34	Integer, Real	2
Heart disease dataset	303	13	Categorical, Integer, Real	5

Appendix 2 Metric Acquiring Procedure

We follow the process in Fig. 9 to obtain metrics (2)–(4). First, we simulate input sample 1 and input sample 2 which are the seen and unseen datasets. The neural network makes predictions and stores them in two databases for the two samples. Next, input sample 1 is used with the neural network to extract rules and obtain metric (1). These rules are used to derive the two samples. Finally, the derived conclusions are compared with the predictions to evaluate metrics (2)–(4).

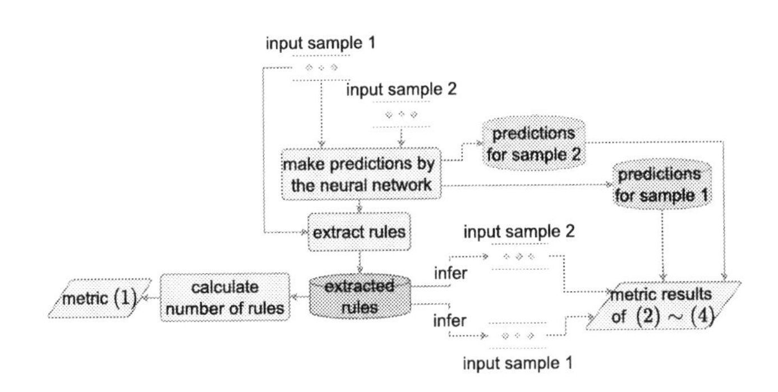

Fig. 9. Metrics acquiring procedure

Appendix 3 Extracted Rules for Light Services

Table 4 shows that DQN_v1 and DQN_v2 have the same extracted rules suggesting always closing the curtain. However, the rules in DQN_v3 have a curtain setting more variable as the energy saving is required in the inhabitant's habitual behaviors. Moreover, when expressing the rules, we only keep the states' ranges and forget their averages to make it easier to compare with habitual behaviors.

Table 4. Extracted rules for the light service simulated by different DQNs

DQN	Habitual behaviors	Extracted rules
v1	(1) If the inhabitant is absent, then the indoor light intensity is 0 lux; (2) If the inhabitant is working, then the indoor light intensity is between 250 lux and 350 lux; (3) If the inhabitant is seeing a movie, then the indoor light intensity is between 350 lux and 450 lux; (4) If the inhabitant is sleeping, then the indoor light intensity is 0 lux	(1) If the inhabitant is absent, then the lamp is off, and the curtain is closed; (2) If the inhabitant is working, then the lamp is at level 3, and the curtain is closed; (3) If the inhabitant is seeing a movie, then the lamp is at level 4, and the curtain is closed; (4) If the inhabitant is sleeping, then the lamp is off, and the curtain is closed
v2	(1) If the inhabitant is absent, then the indoor light intensity is 0 lux; (2) If the inhabitant is working, then the indoor light intensity is between 250 lux and 350 lux; (3) If the inhabitant is seeing a movie, then the indoor light intensity is between 350 lux and 450 lux; (4) If the inhabitant is sleeping, then the indoor light intensity is 0 lux	(1) If the inhabitant is absent, and the outdoor light intensity is between 0 and 605 lux, then the lamp is off, and the curtain is closed; (2) If the inhabitant is working, and the outdoor light intensity is between 0 and 605 lux, then the lamp is at level 3, and the curtain is closed; (3) If the inhabitant is seeing a movie, and the outdoor light intensity is between 0 and 605 lux, then the lamp is at level 4, and the curtain is closed; (4) If the inhabitant is sleeping, and the outdoor light intensity is between 0 and 605 lux, then the lamp is off, and the curtain is closed
v3	With the preference of decreasing the electricity consumption: (1) If the inhabitant is absent, then the indoor light intensity is 0 lux; (2) If the inhabitant is working, then the indoor light intensity is between 250 lux and 350 lux; (3) If the inhabitant is seeing a movie, then the indoor light intensity is between 350 lux and 450 lux; (4) If the inhabitant is sleeping, then the indoor light intensity is 0 lux	(1) If the inhabitant is absent, and the outdoor light intensity is between 0 and 605 lux, then the lamp is off, and the curtain is closed; (2) If the inhabitant is working, and the outdoor light intensity is between 246 and 357 lux, then the lamp is off, and the curtain is fully open; (3) If the inhabitant is working, and the outdoor light intensity is between 512 and 605 lux, then the lamp is off, and the curtain is half-open; (4) If the inhabitant is seeing a movie, and the outdoor light intensity is between 356 and 452 lux, then the lamp off, and the curtain is fully open; (5) If the inhabitant is sleeping, and the outdoor light intensity is between 0 and 605 lux, then the lamp is off, and the curtain is closed; $\cdots$

References

1. Arbatli, A.D., Akin, H.L.: Rule extraction from trained neural networks using genetic algorithms. Nonlinear Anal. Theory Methods Appl. **30**(3), 1639–1648 (1997)

2. Benesty, J., Chen, J., Huang, Y., Cohen, I.: Pearson correlation coefficient. In: Benesty, J., Chen, J., Huang, Y., Cohen, I. (eds.) Noise Reduction in Speech Processing, pp. 1–4. Springer, Heidelberg (2009). https://doi.org/10.1007/978-3-642-00296-0_5
3. Biswas, S.K., Chakraborty, M., Purkayastha, B., Roy, P., Thounaojam, D.M.: Rule extraction from training data using neural network. Int. J. Artif. Intell. Tools **26**(03), 1750006 (2017)
4. Bride, H., Dong, J., Dong, J.S., Hóu, Z.: Towards dependable and explainable machine learning using automated reasoning. In: Sun, J., Sun, M. (eds.) ICFEM 2018. LNCS, vol. 11232, pp. 412–416. Springer, Cham (2018). https://doi.org/10.1007/978-3-030-02450-5_25
5. Brunton, S.L., Kutz, J.N.: Data-Driven Science and Engineering: Machine Learning, Dynamical Systems, and Control. Cambridge University Press, Cambridge (2022)
6. Chan, M., Estève, D., Escriba, C., Campo, E.: A review of smart homes-present state and future challenges. Comput. Methods Programs Biomed. **91**(1), 55–81 (2008)
7. Craven, M.W., Shavlik, J.W.: Learning symbolic rules using artificial neural networks. In: Proceedings of the Tenth International Conference on Machine Learning, pp. 73–80 (2014)
8. François-Lavet, V., Henderson, P., Islam, R., Bellemare, M.G., Pineau, J.: An introduction to deep reinforcement learning. arXiv preprint arXiv:1811.12560 (2018)
9. García, C.G., G-Bustelo, B.C.P., Espada, J.P., Cueva-Fernandez, G.: Midgar: Generation of heterogeneous objects interconnecting applications. A domain specific language proposal for internet of things scenarios. Comput. Netw. **64**, 143–158 (2014)
10. Hester, T., et al.: Deep q-learning from demonstrations. In: Proceedings of the AAAI Conference on Artificial Intelligence, vol. 32 (2018)
11. Ilyas, S., et al.: The impact of revegetation on microclimate in coal mining areas in east kalimantan. J. Environ. Earth Sci. **2**(11), 90–97 (2012)
12. Jackson, P.: Introduction to expert systems (1986)
13. Juraimi, A.S., Saiful, M., Begum, M., Anuar, A., Azmi, M.: Influence of flooding intensity and duration on rice growth and yield. Pertanika J. Trop. Agric. Sci. **32**(2), 195–208 (2009)
14. Kamruzzaman, S., Islam, M., et al.: Extraction of symbolic rules from artificial neural networks. arXiv preprint arXiv:1009.4570 (2010)
15. Kern, C., Klausch, T., Kreuter, F.: Tree-based machine learning methods for survey research. In: Survey Research Methods, vol. 13, p. 73. NIH Public Access (2019)
16. Lee, S., Choi, D.H.: Reinforcement learning-based energy management of smart home with rooftop solar photovoltaic system, energy storage system, and home appliances. Sensors **19**(18), 3937 (2019)
17. Leong, C.Y., Ramli, A.R., Perumal, T.: A rule-based framework for heterogeneous subsystems management in smart home environment. IEEE Trans. Consum. Electron. **55**(3), 1208–1213 (2009)
18. Mainetti, L., Mighali, V., Patrono, L., Rametta, P.: A novel rule-based semantic architecture for IoT building automation systems. In: 2015 23rd International Conference on Software, Telecommunications and Computer Networks (SoftCOM), pp. 124–131. IEEE (2015)
19. Surbatovich, M., Aljuraidan, J., Bauer, L., Das, A., Jia, L.: Some recipes can do more than spoil your appetite: analyzing the security and privacy risks of IFTTT

recipes. In: Proceedings of the 26th International Conference on World Wide Web, pp. 1501–1510 (2017)
20. Taha, I.A., Ghosh, J.: Symbolic interpretation of artificial neural networks. IEEE Trans. Knowl. Data Eng. **11**(3), 448–463 (1999)
21. Towell, G.G.: Symbolic knowledge and neural networks: insertion, refinement and extraction (1993)
22. Towell, G.G., Shavlik, J.W.: Extracting refined rules from knowledge-based neural networks. Mach. Learn. **13**(1), 71–101 (1993)
23. Xu, X., Jia, Y., Xu, Y., Xu, Z., Chai, S., Lai, C.S.: A multi-agent reinforcement learning-based data-driven method for home energy management. IEEE Trans. Smart Grid **11**(4), 3201–3211 (2020)
24. Yu, L., Xie, W., Xie, D., Zou, Y., Zhang, D., Sun, Z., Zhang, L., Zhang, Y., Jiang, T.: Deep reinforcement learning for smart home energy management. IEEE Internet Things J. **7**(4), 2751–2762 (2019)
25. Zhou, Z.H.: Rule extraction: using neural networks or for neural networks? J. Comput. Sci. Technol. **19**(2), 249–253 (2004)

Effective and Robust Boundary-Based Outlier Detection Using Generative Adversarial Networks

Qiliang Liang[1], Ji Zhang[2], Mohamed Jaward Bah[3], Hongzhou Li[4](✉), Liang Chang[4], and Rage Uday Kiran[5]

[1] Nanjing University of Aeronautics and Astronautics, Nanjing, China
ql.liang@nuaa.edu.cn
[2] University of Southern Queensland, Toowoomba, Australia
Ji.Zhang@usq.edu.au
[3] Zhejiang Lab, Hangzhou, China
easybah@zhejianglab.com
[4] Guilin University of Electronic Technology, Guilin, China
homzh@163.com, changl@guet.edu.cn
[5] The University of Aizu, Aizuwakamatsu, Japan
udayrage@u-aizu.ac.jp

Abstract. Outlier detection aims to identify samples that do not match the expected patterns or major distribution of the dataset. It has played an important role in many domains such as credit card fraud identification, network intrusion detection, medical image processing and so on. The inherent class imbalance in datasets is one of the major reasons why this problem is difficult to solve. The small number of outliers are not adequate to characterize their own overall distribution, which makes it difficult for classifiers to effectively learn the demarcation (boundary) between normal samples and outliers. To address this problem, we introduce an effective and robust Boundary-based Outlier Detection method using Generative Adversarial Networks (BOD-GAN). Here, we extract the border data containing normal samples and outliers, expand them to form the initial reference boundary outliers. With the min-max game between a generator and two discriminators in GAN, the boundary outliers are further augmented by BOD-GAN, which, together with the boundary normal data, provides the valuable demarcation information for classifier. However, the increase of the data dimension may bring some gaps in the initial boundary, which are difficult to effectively fill by the augmentation method alone. To address this, we innovatively add density-loss to the loss function of the generator to explore these boundary gaps, making our model rather robust even with the high dimensional data. The extensive experimental evaluation demonstrates that our proposed method has achieved significant improvements compared with existing classic and emerging (i.e., GAN-based) outlier detection methods.

Keywords: Outlier detection · GAN · Boundary-based model

C. Strauss et al. (Eds.): DEXA 2022, LNCS 13427, pp. 174–187, 2022.
https://doi.org/10.1007/978-3-031-12426-6_14

1 Introduction

Outlier detection is an active research area in the field of machine learning. It aims to identify samples that do not match the expected patterns or major distribution of the dataset. Outliers are often difficult to obtain, but can be informative and critical for many applications. In order to detect and leverage outliers, many outlier detection technologies have been studied and applied to various applications, such as credit card fraud detection [9,11,20,27], network intrusion detection [8,16,22], and medical image processing [23].

Outliers are rare and distribute randomly in a more scattered data space than that of the normal data. The small amount of labeled outliers is not able to give an accurate and complete characterization of all outliers, making traditional classifiers not able to work well for outlier detection. Typically, the number of outliers can be increased simply using oversampling techniques. The synthetic minority oversampling technique (SMOTE) [5] and its variants such as Borderline-SMOTE [13] and the synthetic minority oversampling technique(ADASYN) [14] are representative of such methods. They use different oversampling strategies based on existing outlier, but at the same time may produce useless samples and increase the training time unnecessarily.

Recently, we have witnessed that Generative Adversarial Networks(GAN), thanks to their powerful modeling ability, have been widely used in various generative tasks, including outlier augmenting. On this issue, it is necessary to generate sufficient outliers, but such data augmentation based on the existing outliers is rather ineffective. The few existing outliers may be far from describing their overall distribution. In order for the outlier detector to learn the demarcation between normal samples and outliers, the most important outliers to be generated are often located on the outer edge of the area where the normal data are distributed and thus forming a boundary that well separates the normal samples from its vast outlier space. [19,21,25,26] stand at different angles adopting different model structures and training strategies, all of which are aimed at generating outliers at this critical position. However, existing GAN-based methods haven't yet been effective and accurate in generating the boundary outliers, requiring extensive parameter tuning efforts.

To effectively generate the boundary outliers, we propose in this paper a novel outlier detection method that utilizes boundary outliers and GAN (called BOD-GAN for short). In BOD-GAN, we first leverage the information contained in the dataset itself to form the initial boundary. More specifically, data clusters are first quickly generated from the dataset. For each cluster, we extract the border samples (which are treated as boundary normal samples) and expand them outward to form the initial boundary outliers. However, such extraction and expansion operations alone cannot generate the complete demarcation between the normal samples and outliers as the boundary outliers formed are usually sparse and cannot effectively gave a good coverage of the boundary.

To solve this problem, we propose a novel GAN-based model in BOD-GAN to further argument boundary outliers for a better coverage. Our model is characterized by a three-player adversarial game involving a generator and two discriminators, allowing the model to focus more on boundary outlier augmentation and correction.

In addition, due to the curse of dimensionality, we found that some uncovered blank areas in the boundary are likely to appear when data becomes sparser with increasing dimensionality. In order to mitigate this problem, we further modify the loss function of the generator and introduce a new penalty term, called density-loss, in BOD-GAN so that the model can heuristically explore the space and fill these gaps in a more effective matter.

Specifically, the main contributions of this paper are as follows:

1. We propose BOD-GAN, an effective and robust boundary-based outlier detection method using an improved structure of GAN. The initial boundary outliers are generated using clustering and distance extrapolation techniques, which provides a good foundation of the boundary for the subsequent GAN model, allowing it to focus more on augmentation and correction of the boundary so that the generation efficiency of the GAN framework can be significantly improved;
2. In BOD-GAN, we have modified the traditional structure of GAN to make our model characterized by a three-player adversarial game involving a generator and two discriminators. Our GAN model can effectively generate a high-quality boundary to assist the outlier detector to effectively detect unseen outliers;
3. In order to deal with the data sparseness problem caused by the curse of dimensionality, we further propose a novel penalty item, called density-loss, in training our GAN model to better generate the outlier detection boundary for high-dimensional datasets;
4. We have carried out extensive experimental evaluations on BOD-GAN using 12 real-world datasets, and compare the experimental results with an array of both traditional and emerging (i.e., deep-learning based) methods. The experiments show that our method has achieved a better performance in accuracy and AUC score.

The remaining of the paper is organized as follows. Section 2 reviews the related work involving the traditional and emerging outlier detection methods. Section 3 introduces our proposed method in detail. Section 4 presents our experiments design and results. Section 5 gives the conclusion of this paper.

2 Related Work

Outlier detection is an active research topic, over the years have emerged some excellent machine learning algorithms. One-class SVM [24], unlike traditional SVM, use a hypersphere instead of a hyperplane to do the division, so as to minimize the influence of outliers by minimizing the volume of the hypersphere. As a representative of lazy learning, kNN [1] uses a multi-nearest neighbor voting mechanism for classification. Ensemble learning algorithm [4,6,12,18] are constructed and combined with multiple learners to optimize the decision-making process and complete learning tasks. Clustering methods, such as classical DBSCAN [10], K-Means [28] and recently proposed Border-Peeling clustering [3], part of this algorithm is also used in our work, identify outliers by forming clusters of samples.

Class imbalance and especially rare outliers are a major problem in outlier detection. Typically, the number of outliers can be increased simply using oversampling techniques. The random oversampling technique (ROS) [15] is one of traditional methods which can achieve class balance by random sampling with replacement for outliers. However, repeated samples make the rules learned by the model too specific and reduce its generalization ability, thus introducing the problem of overfitting. Based on kNN, the synthetic minority oversampling technique (SMOTE) [5] uses linear interpolation skill to avoid overfitting, but, without filtering of outliers, smote may introduce more noise when the original ones are themselves noise. To solve this issue, Borderline-SMOTE [13] and the synthetic minority oversampling technique (ADASYN) [14] which can be seen as the variants of SMOTE use different strategies to select valuable outliers for oversampling. However, when the selected neighbor is rather close to the original outlier, both SMOTE and its variant methods may generate similar samples, and increase the unnecessary training time.

To address this problem, existing GAN-based outlier detection methods adopt different strategies to avoid the convergence of the generated outliers with the normal samples. For example, GAAL [19] and adGAN [26] choose to stop training earlier before they overlap with each other. In order to further reduce false positives, adGAN retrained the discriminator after earlystop to enable the classification to correctly distinguish between the two groups of data. However, their generative mechanisms for outliers rely on a large number of trial-and-error experiments, making it difficult to easily obtain the optimal demarcation between normal samples and outliers. By changing the loss function of GAN, FenceGAN [21] makes the generated samples directly on the boundary, but the model is complex with multiple parameters and requires extensive parameter tuning work for the respective datasets. [25] uses two discriminators with opposite goals which are trained against the generator. When the model converges, the opposite objective function allows the generated outliers to lie on the boundary, but the training of the generator is more difficult to converge in the face of two opposite discriminators, making the model less stable overall than the traditional GAN structure.

3 Our Method

3.1 Initial Reference Boundary

In real life, the distribution of data is diverse, but normal samples tend to appear in a specific pattern and are located in a relatively fixed position, thus presenting a multi-cluster distribution. While the between-cluster and the much wider out-of-cluster space are the locations outliers may appear, that is, the outlier space, which is extensive and unpredictable. Here, in order to remove the effect of between-cluster gaps on outliers generation, we use DBSCAN to pre-cluster the data, so as to separate the high-density areas where the normal samples are concentrated into small clusters.

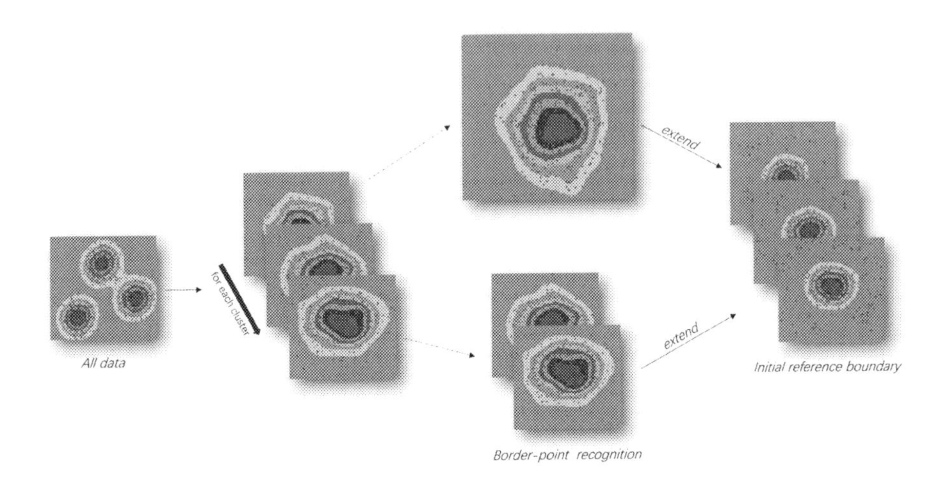

Fig. 1. Data processing: cluster the dataset and for each cluster, divide the data into border points and core points represented by the yellow and blue dots, respectively. Using the distance information between outliers and their closest border point, expand outwards to form the initial reference boundary, which is also shown in the figure as red dots. (Color figure online)

Border Recognition. As the [3] introduced, every cluster in the data is composed of a set of boundary samples around its core sample set, by identifying and peeling off the local border samples multiple times, it can gradually enter the core area of the cluster, thereby dividing it into a border sample set $\mathcal{X}_{bo}$ and core sample set $\mathcal{X}_{core}$. Based on the extracted border sample set, we perform an expansion operation towards outside direction, thus form the initial reference boundary.

Given a whole cluster $X = \{X_{nor}, X_{out}\}$, we separate its n normal data, denoted by $X_{nor} = \{x_1, x_2, ..., x_n\}$, and use $X_{nor}^{(t)}$ to indicate the sample set that has not yet been peeled at the beginning of the tth round. Denoted the reverse k neighbors of each $x_i \in X_{nor}^{(t)}$ as $RN_k^{(t)}$.

In order to estimate distance between samples, we adapt a Gaussian kernel with local scaling to the euclidean distance:

$$f(x_i, x_j) = exp\Big(-\frac{\|x_i - x_j\|_2^2}{\sigma_j^2}\Big) \tag{1}$$

and set the $\sigma_j = \left\|x_j - N_k^{(t)}(x_j)[k]\right\|_2$ which has proven to be effective in [3] and his prior work. For each samples $x_i \in X_{nor}^{(t)}$, apply f to its reverse k neighbors to get the border value $b_i^{(t)}$, where

$$b_i^{(t)} = \sum_{x_j \in RN_k^{(t)}(x_i)} f(x_i, x_j) \tag{2}$$

Samples located at the border of cluster, that is, the 10 percent samples with the lowest score, will be peeled at each iteration. At the same time, we will track the average border-value of peeled samples $\bar{b_p^{(t)}}$, and stop peeling when

$$\frac{\bar{b_p^{(t)}}}{\bar{b_p^{(t-1)}}} - \frac{\bar{b_p^{(t-1)}}}{\bar{b_p^{(t-2)}}} > \varepsilon \tag{3}$$

Expansion. Outliers are rare and distributed randomly, existing outliers are far from describing their overall distribution, thus cause the difficulty of obtaining information from them directly, this is also a reason why most of the related work ignores this part of the information which, however, can help us generate boundaries faster, more intuitively. Outliers located at the periphery of each cluster can provide expansion distance for the forming of initial reference boundary.

For each outliers in the cluster $x_j^{'} \in X_{out}$, we add the distance with its nearest normal sample, that is $x_i = N_1(x_j^{'})$ and $x_i \in X_{nor}$, to the calculation of the extension distance. N denote the number of outliers

$$dist(x_i, x_j^{'}) = \frac{1}{N} \sum_{\substack{x_j^{'} \in X_{abnor} \\ x_i = N(x_j^{'})}} ||x_i - x_j^{'}||_2 \tag{4}$$

In the mapping calculation, multiple outliers may be mapped to the same normal neighbor. In order to get accurate information, it is necessary to eliminate this duplicate situation and only take the distance between the normal sample and its nearest outlier into account. Using the calculated distance, expand the border samples towards the opposite direction to their nearest core sample to get $\mathcal{X}_{exp} = \{x_{exp_1}, x_{exp_2}, \cdot, x_{exp_k}\}$,

$$x_{exp_i} = \frac{dist}{||x_{bo_j} - x_{core_l}||_2}\left(x_{bo_j} - x_{core_l}\right) + x_{bo_j} \tag{5}$$

which $x_{bo_j} \in \mathcal{X}_{bo}, x_{core_l} \in \mathcal{X}_{core}$. $\mathcal{X}_{exp}$ is the set composed of informative outliers, shown as red dots in the right part of Fig. 1 which have formed the initial reference boundary.

3.2 Boundary-Based GAN

The Generative Adversarial Networks (GAN) have shown outstanding effects in modeling high-dimensional data and enabled it to achieve impressive results in many fields. But its objective does not completely meet the purpose of outlier detection and boundary outliers generating. For this reason, we propose the BOD-GAN using double discriminators, so that modified architecture can better complete the task of generating boundary outliers.

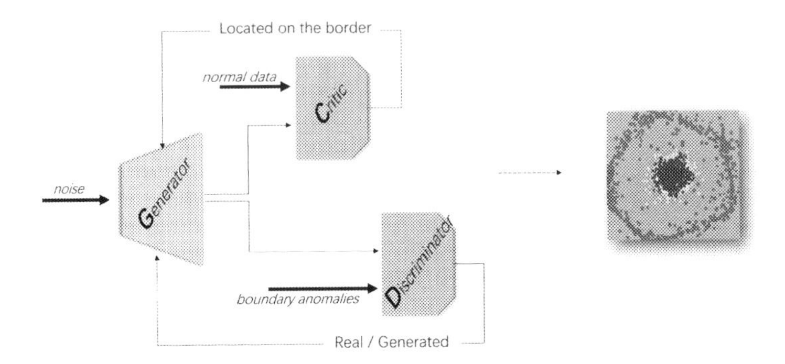

Fig. 2. The architecture of proposed BOD-GAN consists of one generator and double discriminators which named Discriminator (D) and Critic (C) respectively. The generator take noises as its input and output generated samples, C and D take the normal data and boundary outliers respectively as input, their job is to help the generator to augmentor the boundary outliers by the adversarial training process. Under the constraint of C, the boundary outliers are strictly limited to the boundary, like the right part of this figure, which provide a strong guarantee that the generated outliers are neither far away from data zone and thus loss the boundary information nor too close to the normal data which can possibly cause the increase of false positive rate.

Presented in Fig. 2 is the proposed architecture which consist of a generator G and double discriminators named as D and C respectively. D and C receive, in turn, boundary outliers and normal data respectively, as well as generated data for both of them. Their loss functions (i.e., goals) are different:

- D receives the generated samples as well as the initial boundary outliers as input. The training process between G and D is similar to traditional GAN, the training purpose of this part of BOD-GAN is to help G learn the distribution of boundary outliers, complete the task as a augmentor. We, thereby, adopt the similar loss function of GAN

$$L_D = \mathbb{E}_{x \sim p_{bo}(x)}\Big[\log D(x)\Big] + \mathbb{E}_{z \sim p_z(z)}\Big[\log(1 - D(G(z))\Big] \tag{6}$$

- The goal of C, different from D, is more about the correction of the boundary. C receives the generated samples from G and the normal samples of this cluster as its input. The training purpose of this part is to correct the generated outlier boundary, eliminate the uneven appearance of the boundary outliers, which is also means that the generated outliers are neither far away from the reference area cause the loss of boundary information, nor overlap with the normal samples cause the increase of false positive rate. We adopt the following formula as C's loss function, in addition, the corresponding part of G has also been modified for the goal of boundary correction, which will be described in detail later.

$$L_C = \mathbb{E}_{x \sim p_{nor}(x)}\Big[\log(D(x)] + \mathbb{E}_{z \sim p_z(z)}[\log(1 - D(G(z))\Big] \tag{7}$$

In order to generate the unified boundary outliers, as Formula (9) shows, G receives uniform noise z as its input, cooperate with both D and C by adding their output scores of generated outliers into loss function. The training goals with D follows the traditional way. The difference in training with C can be reflected in the introduction of α, which means only the α-level samples can reach the optimal generator loss, thus constrains the generated outliers to be located at the boundary position.

$$L_G = \mathbb{E}_{z \sim p_z(z)}\Big[\log(1 - D(G(z)) + \log(\alpha - C(G(z))\Big] + \beta \mathcal{L}_{den} + \gamma \mathcal{L}_{dispers} \tag{8}$$

Considering the possible impact of the curse of dimensionality, that is, the coverage of a certain size of data in the space decreases exponentially when their dimensionality increases, the boundaries extracted from sparse high-dimensional data do not necessarily contain complete information. In order to solve this drawback, we propose a novel heuristic extension method to the generator to encourage the generated samples explore the existed blank areas. The incompleteness of the boundary reflect in the unevenness of the sample density. Samples with higher density locate at the center position, surrounded by other low-density ones near the blank areas thus make them more suitable for exploring. Thereby, we calculate the density for each generated sample, denoted as

$$denRec = \{d_{x_{g_1}}, d_{x_{g_2}}, \cdots, d_{x_{g_n}}\}$$

where $\{x_{g_1}, x_{g_2}, \cdots, x_{g_n}\} \in X_g$, and X_g is the generate sample set. Join the sum of the density of the last 20% samples to the loss function of G, denoted as $\mathcal{L}_{den}$, force this part of samples fill the blank areas in the optimization process of the loss function. We set its weight (i.e., β) as the standard deviation of the selected sample density, which varies according to the blank situation of boundary outliers, and achieves parameter unification under different datasets.

It is worth noting that the structure of double discriminators in BOD-GAN also plays an important role in this part. The adversarial training among the three players in BOD-GAN constrains the extension behavior of these low-density samples and provides a strong guarantee for their explorations to be within the boundary range while maintaining the uniformity.

Besides, the situation of mode collapse which is common in GAN-based methods, could cause the generated samples only cover a small part of boundary and can hardly provide sufficient information. Here we avoid this from happening by maximize the distance between the generated samples and their center, which firstly proposed by [21]. We denote this dispersion loss in this article as $\mathcal{L}_{dispers}$, and set its weight (i.e., γ) according to previous work.

4 Experiments

4.1 Experiment Design

In our experiments, we test BOD-GAN on real-world datasets, comparing performance with classic, deep learning and GAN-based outlier detection methods. As presented in Table 1, we have selected 12 different datasets vary in dimension, number of samples and outlier ratios, but they all show the phenomenon of class imbalance which normal data accounted for a larger proportion.

Table 1. Summary of datasets used in our experiments

Dataset	Dim	Number	Outlier ratio
Pima	8	768	34.9%
Shuttle	9	58000	21.4%
PageBlocks	10	5393	9.4%
Annthyroid	21	7192	7.4%
Waveform	21	3443	2.9%
Ionosphere	33	351	35.9%
Satellite	36	6435	31.6%
Covtype	54	13000	8.3%
Spambase	57	4207	39.9%
Optdigits	64	5216	2.9%
KDD99	121	14200	15.5%
Musk	166	3062	3.2%

We compare our methods with classic algorithms such as OCSVM [24], LOF [17] and IForest [7], deep learning based method like AutoEncoder [29], and recently proposed FenceGAN [21], MO-GAAL [19] which using GAN as their basic structure. The addition of density-loss enables the model to explore the blank areas of the boundary while generating outliers, alleviating the data sparse problem caused by the curse of dimensionality. In order to verify the effectiveness of this novel methods, we also designed an ablation study to compare the effects of the model before and after adding density-loss.

In these experiments, the training set consists of both normal samples and outliers, we only use the distance information carried by outliers which will not participate in our subsequent training. We adopt fully connected network as the basic architecture for BOD-GAN. LeakyReLU and Sigmoid were chosen as activation function for the hidden and output layers respectively, we also set the optimizer as Adam with a learning rate of 0.001. After augmenting the data with BOD-GAN, we will train an outlier detector using the class-balanced dataset with informative boundary outliers, and verify its effect on the test set. Since the input and output forms of outlier detector are unified with D and C, we have adopted the same architecture for all three. The performance of outlier detector based on augmented data will be measured by the value of accuracy and the area under the ROC curve (AUC).

Besides, as mentioned in [2], a useful discriminator can provide the gradient information to generator, but can also cause gradient vanish if trained too frequently. This phenomenon is rather obvious in a model with additional discriminator. Therefore, we set $sleepTime$ for both D and C while updating the parameters of G per epoch, the choice of $sleepTime$ will also be evaluated in the experiment.

4.2 Experiment Results

Table 2 and 3 show the experiment results on the real-world datasets. For each dataset, we highlight the best result in bold and give the average ranking of all comparison methods on each dataset which can be seen that the proposed BOD-GAN has achieved better performance and higher rank on multiple datasets.

From the Table 2, it can be observed that the proposed BOD-GAN and BOD-GAN lite which is trained without density-loss have reached a higher accuracy in 8 out of 12 datasets, and this ratio raised to 9 out of 12 in the performance of AUC scores which can be observed in Table 3. Our methods also reach the top rank among all comparison methods after synthesizing the results of all datasets.

Table 2. Accuracy results on experimental datasets

Dataset	OCSVM	LOF	IForest	AutoEncoder	MO-GAAL	FenceGAN	BOD-GAN lite	BOD-GAN
Pima	0.6580	0.6407	0.6667	0.6320	0.6883	0.6926	0.6407	**0.7056**
Shuttle	0.7561	0.7382	0.8586	0.8514	0.7916	0.7482	**0.9289**	0.8529
PageBlocks	0.1334	0.6663	0.7010	0.7625	0.8538	0.7742	**0.8946**	0.8806
Annthyroid	0.8144	0.8237	0.8560	0.8303	0.8964	0.7438	0.8845	**0.9074**
Waveform	0.8722	0.8925	0.8887	0.7909	0.8887	0.7773	**0.9758**	0.9661
Ionosphere	0.7264	0.6887	**0.7398**	0.7264	0.5943	0.5755	0.6226	0.6226
Satellite	0.7721	0.6722	0.7587	0.7649	0.6810	0.4977	0.7923	**0.7991**
Covtype	0.8607	0.8300	0.8603	0.8003	0.8973	**0.9093**	0.8800	0.8827
SpamBase	0.5653	0.5796	0.5922	0.5590	0.6041	0.6342	0.6326	**0.6485**
Optdigits	0.8204	0.8195	0.8169	0.7280	0.8849	0.7073	0.8893	**0.8993**
KDD99	0.8459	0.8641	0.9068	0.9577	0.8923	**0.9418**	0.9118	0.9086
Musk	0.7449	0.7798	0.7510	0.6485	0.8540	**0.8601**	0.8535	0.8540
Average Ranks	5.67	5.92	4.50	5.58	3.92	4.92	2.92	**2.17**

Ablation Study. In the comparison between BOD-GAN and its lite version which is trained without density-loss, an interesting observation is that the augmented data under the training with density-loss does not always improves the performance in the accuracy, and even sometimes leads to a degradation. However, this part of the decline in accuracy does not affect its performance on AUC scores. Narrow the scope of comparison to the performance of BOD-GAN and its lite version on each dataset, the previous one outperformed in all datasets at AUC scores which lead to an improvement since the ratio in accuracy is 7 out of 12, and this advantage is more obvious in high-dimensional datasets.

Table 3. AUC scores on experimental datasets

Dataset	OCSVM	LOF	IForest	AutoEncoder	MO-GAAL	FenceGAN	BOD-GAN lite	BOD-GAN
Pima	0.6342	0.6246	0.7448	0.6061	0.6727	0.4540	0.6867	**0.7823**
Shuttle	0.6634	0.5525	0.8584	0.8122	0.5006	0.5379	0.9044	**0.9665**
PageBlocks	0.3457	0.6275	0.7770	0.8472	0.6259	0.4733	0.8388	**0.8478**
Annthyroid	0.4541	0.5784	0.6380	0.5255	0.5484	0.5379	0.7502	**0.7846**
Waveform	0.5231	0.7152	0.6877	0.6307	**0.7935**	0.4663	0.7496	0.7800
Ionosphere	0.7512	**0.9186**	0.8195	0.7985	0.4683	0.4315	0.8533	0.8878
Satellite	0.6458	0.5496	0.6686	0.5893	0.6690	0.6570	0.7697	**0.8135**
Covtype	**0.7698**	0.4880	0.7055	0.7715	0.5515	0.5000	0.7370	0.7585
SpamBase	0.5233	0.4327	0.5930	0.5407	0.4000	0.4259	0.7838	**0.8639**
Optdigits	0.3716	0.5682	0.3732	0.3560	0.4953	0.4903	0.8876	**0.9368**
KDD99	0.5783	0.6625	0.8775	0.9732	0.0781	0.4821	0.9354	**0.9852**
Musk	0.2069	0.4334	0.4036	0.2373	0.4332	0.5203	0.6612	**0.7325**
Average Ranks	6.08	5.00	4.08	5.00	5.42	6.50	2.58	**1.33**

In general, the addition of density-loss sacrifices a small part of the accuracy, but in exchange for better overall performance of the classifier, which is also in line with our goal of getting a more robust outlier detector by augmenting informative boundary outliers.

Parameter Choices. According to the experience of previous work, hyperparameters in the model are usually important, since their adjustment may directly affect the final result and thus increased workload when applying to new datasets. BOD-GAN involves four hyperparameters, and only $sleepTime$ and α need to be tuned actually. We select three datasets with different dimensions (dim), that is Pima (dim of 8), Waveform (dim of 21) and SpamBase (dim of 57), covering a large range of dimensions, for the purpose to get more representative result.

The value of α represents the target score for G, constraining the generated outliers to be located at the boundary position. Therefore, we tune it carefully and adjust its value to vary in the range of 0.46–0.50. $sleepTime$ is another strategy we chose to balance the three-player game of BOD-GAN in a more efficient way, we tested its value to vary from 5–25. Figure 3(a)–3(d) have shown the fluctuation trend of the final result (i.e., accuracy and AUC score) with the vary of α and $sleepTime$. As the parameters vary in a range, there are little fluctuate, and the result can still rank high according to the Table 2 and 3. Finding good hyperparameters of BOD-GAN is not complicated since our model is not sensitive to α or $sleepTime$.

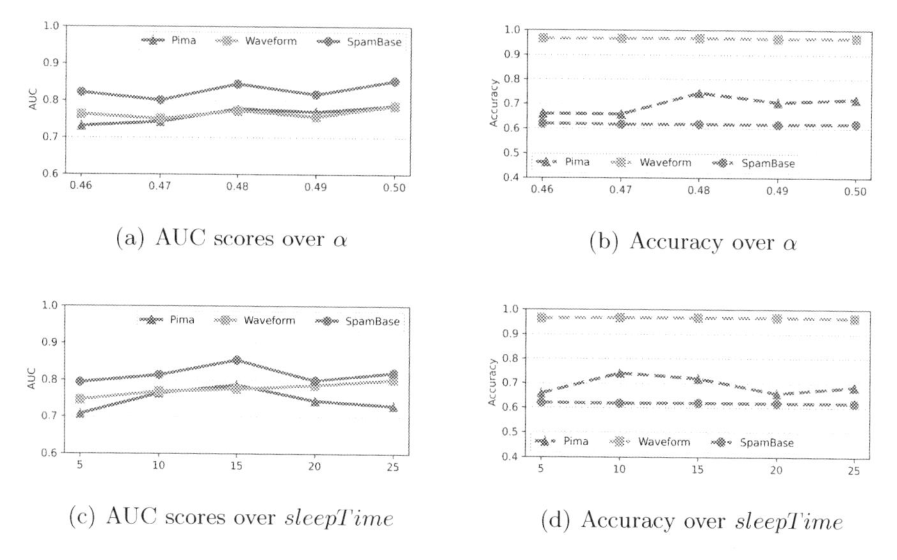

(a) AUC scores over α

(b) Accuracy over α

(c) AUC scores over $sleepTime$

(d) Accuracy over $sleepTime$

Fig. 3. Parameter experiments for both α and $sleepTime$. (a)–(b) show the degree of change in AUC and Accuracy of the three data sets with the change of α. As the value of α changes in the range of 0.46–0.50, the changes in AUC scores or Accuracy are small and still rank at a high level according to Table 2 and 3. Replace the tested hyperparameter to $sleepTime$, (c)–(d) also showed similar results which could lead to a conclusion that our method is not sensitive to both α and$sleepTime$.

5 Conclusion

In this paper, we have proposed a Boundary-based Outlier Detection using Generative Adversarial Networks (BOD-GAN), based on the initial reference boundary outliers by extracting the border samples and expanding. The structure of the double discriminators ensures that the generated outliers are strictly limited to the boundary position. Besides, considering the possible impact of the curse of dimensionality, we innovatively add the density-loss to the loss function of the generator so as to ensures the integrity of the boundary while generating the boundary. The result demonstrate that the proposed BOD-GAN outperforms than other competitive methods. However, the calculation of boundary recognition and exploration on blank areas, based on the nearest neighbors algorithms, are computationally expensive, especially in high-dimensional datasets. How to improve this situation and increase training efficiency requires further research in the future.

Acknowledgement. The authors would like to thank the support from Natural Science Foundation of China (No. 62172372), Zhejiang Provincial Natural Science Foundation (No. LZ21F030001), Postdoctoral Fund of Hangzhou City (No. 119001-UB2102QJ) and Henan Center for Outstanding Overseas Scientists (GZS2022011).

References

1. Abeywickrama, T., Cheema, M.A., Taniar, D.: k-nearest neighbors on road networks: a journey in experimentation and in-memory implementation (2016)
2. Arjovsky, M., Chintala, S., Bottou, L.: Wasserstein generative adversarial networks. In: International Conference on Machine Learning, pp. 214–223. PMLR (2017)
3. Averbuch-Elor, H., Bar, N., Cohen-Or, D.: Border-peeling clustering. IEEE Trans. Pattern Anal. Mach. Intell. **42**(7), 1791–1797 (2019)
4. Breiman, L.: Random forest. Mach. Learn. **45**, 5–32 (2001)
5. Chawla, N.V., Bowyer, K.W., Hall, L.O., Kegelmeyer, W.P.: Smote: synthetic minority over-sampling technique. J. Artif. Intell. Res. **16**, 321–357 (2002)
6. Chen, T., Guestrin, C.: XGBoost: a scalable tree boosting system. In: The 22nd ACM SIGKDD International Conference (2016)
7. Cheng, Z., Zou, C., Dong, J.: Outlier detection using isolation forest and local outlier factor. In: Proceedings of the Conference on Research in Adaptive and Convergent Systems, pp. 161–168 (2019)
8. Choi, H., Kim, M., Lee, G., Kim, W.: Unsupervised learning approach for network intrusion detection system using autoencoders. J. Supercomput. **75**(9), 5597–5621 (2019). https://doi.org/10.1007/s11227-019-02805-w
9. Dal Pozzolo, A., Boracchi, G., Caelen, O., Alippi, C., Bontempi, G.: Credit card fraud detection: a realistic modeling and a novel learning strategy. IEEE Trans. Neural Netw. Learn. Syst. **29**(8), 3784–3797 (2017)
10. Ester, M., Kriegel, H.P., Sander, J., Xu, X.: A density-based algorithm for discovering clusters in large spatial databases with noise. AAAI Press (1996)
11. Fiore, U., De Santis, A., Perla, F., Zanetti, P., Palmieri, F.: Using generative adversarial networks for improving classification effectiveness in credit card fraud detection. Inf. Sci. **479**, 448–455 (2019)
12. Friedman, J.H.: Greedy function approximation: a gradient boosting machine. Ann. Stat. 1189–1232 (2001)
13. Han, H., Wang, W.-Y., Mao, B.-H.: Borderline-SMOTE: a new over-sampling method in imbalanced data sets learning. In: Huang, D.-S., Zhang, X.-P., Huang, G.-B. (eds.) ICIC 2005. LNCS, vol. 3644, pp. 878–887. Springer, Heidelberg (2005). https://doi.org/10.1007/11538059_91
14. He, H., Bai, Y., Garcia, E.A., Li, S.: Adasyn: adaptive synthetic sampling approach for imbalanced learning. In: 2008 IEEE International Joint Conference on Neural Networks (IEEE World Congress on Computational Intelligence), pp. 1322–1328. IEEE (2008)
15. Kołcz, A., Chowdhury, A., Alspector, J.: Data duplication: an imbalance problem? (2003)
16. Kuypers, M.A., Maillart, T., Paté-Cornell, E.: An empirical analysis of cyber security incidents at a large organization. Department of Management Science and Engineering, Stanford University, School of Information, UC Berkeley 30 (2016)
17. Li, L., Huang, L., Yang, W., Yao, X., Liu, A.: Privacy-preserving LOF outlier detection. Knowl. Inf. Syst. **42**(3), 579–597 (2015)
18. Liu, F.T., Ting, K.M., Zhou, Z.H.: Isolation forest. In: 2008 Eighth IEEE International Conference on Data Mining, pp. 413–422. IEEE (2008)
19. Liu, Y., Li, Z., Zhou, C., Jiang, Y., Sun, J., Wang, M., He, X.: Generative adversarial active learning for unsupervised outlier detection. IEEE Trans. Knowl. Data Eng. **32**(8), 1517–1528 (2019)

20. Makki, S., Assaghir, Z., Taher, Y., Haque, R., Hacid, M.S., Zeineddine, H.: An experimental study with imbalanced classification approaches for credit card fraud detection. IEEE Access **7**, 93010–93022 (2019)
21. Ngo, P.C., Winarto, A.A., Kou, C.K.L., Park, S., Akram, F., Lee, H.K.: Fence GAN: towards better anomaly detection. In: 2019 IEEE 31St International Conference on tools with artificial intelligence (ICTAI), pp. 141–148. IEEE (2019)
22. Osada, G., Omote, K., Nishide, T.: Network intrusion detection based on semi-supervised variational auto-encoder. In: Foley, S.N., Gollmann, D., Snekkenes, E. (eds.) ESORICS 2017. LNCS, vol. 10493, pp. 344–361. Springer, Cham (2017). https://doi.org/10.1007/978-3-319-66399-9_19
23. Schlegl, T., Seeböck, P., Waldstein, S.M., Schmidt-Erfurth, U., Langs, G.: Unsupervised anomaly detection with generative adversarial networks to guide marker discovery. In: Niethammer, M., et al. (eds.) IPMI 2017. LNCS, vol. 10265, pp. 146–157. Springer, Cham (2017). https://doi.org/10.1007/978-3-319-59050-9_12
24. Schlkopf, B., Williamson, R.C., Smola, A.J., Shawe-Taylor, J., Platt, J.C.: Support vector method for novelty detection. In: Advances in Neural Information Processing Systems 12, NIPS Conference, Denver, Colorado, USA, 29 November–4 December 1999 (1999)
25. Schulze, J.P., Sperl, P., Böttinger, K.: Double-adversarial activation anomaly detection: adversarial autoencoders are anomaly generators. arXiv preprint arXiv:2101.04645 (2021)
26. Shen, H., Chen, J., Wang, R., Zhang, J.: Counterfeit anomaly using generative adversarial network for anomaly detection. IEEE Access **8**, 133051–133062 (2020)
27. Van Vlasselaer, V., Bravo, C., Caelen, O., Eliassi-Rad, T., Akoglu, L., Snoeck, M., Baesens, B.: APATE: a novel approach for automated credit card transaction fraud detection using network-based extensions. Decis. Support Syst. **75**, 38–48 (2015)
28. Zhang, X., He, Y., Jin, Y., Qin, H., Azhar, M., Huang, J.Z.: A robust k-means clustering algorithm based on observation point mechanism. Complexity **2020** (2020)
29. Zhou, C., Paffenroth, R.C.: Anomaly detection with robust deep autoencoders. In: Proceedings of the 23rd ACM SIGKDD International Conference on Knowledge Discovery and Data Mining, pp. 665–674 (2017)

Efficient Data Processing Techniques

Accelerated Parallel Hybrid GPU/CPU Hash Table Queries with String Keys

Tobias Groth[1(✉)], Sven Groppe[1], Thilo Pionteck[2], Franz Valdiek[2], and Martin Koppehel[2]

[1] Institute of Information Systems (IFIS), Universität zu Lübeck, Lübeck, Germany
groth@ifis.uni-luebeck.de

[2] Institute for Information Technology and Communications, Otto-von-Guericke Universität Magdeburg, Magdeburg, Germany

Abstract. GPUs can accelerate hash tables for fast storage and lookups utilizing their massive parallelism. State-of-the-art GPU hash tables use keys with fixed length like integers for optimal performance. Because strings are often used for structures like dictionaries, it is useful to support keys with variable length as well. Modern GPUs enable the combination of CPU and GPU compute power and we propose a hybrid approach, where keys on the GPU have a maximum length and longer keys are processed on the CPU. Therefore we develop a GPU accelerated approach of robin-map and libcuckoo based on string keys. We use OpenCL for GPU acceleration and threads for the CPU. Furthermore, we integrate the GPU approach into our benchmark framework H^2 and evaluate it against the CPU variants and the GPU approach adapted for the CPU. We evaluated our approach in the hybrid context by using longer keys on CPU and shorter keys on GPU. In comparison to the original libcuckoo algorithm our GPU approach achieves a speed-up of 2.1 and in comparison to the robin-map a speed-up of 1.5. For hybrid workloads our approach is efficient if long keys are processed on the CPU and short keys are processed on the GPU. By processing a mixture of 20% long keys on CPU and 80% short keys on GPU our hybrid approach has a 40% higher throughput than the CPU only approach.

Keywords: Hash table · Strings · Robin-map · Cuckoo hashing · GPU · CPU · Parallel · OpenCL

1 Introduction

Hash tables are widely used as data structure for storing key-value pairs. Many of these are designed for high throughput, but there are other characteristics which also matter. Hash tables have to be very flexible for different applications and scenarios. Modern systems have been evolved to highly parallel machines with dozens of cores. This performance can be used to increase the number of generated hashes and speed up the collision avoidance strategy of hash tables.

C. Strauss et al. (Eds.): DEXA 2022, LNCS 13427, pp. 191–203, 2022.
https://doi.org/10.1007/978-3-031-12426-6_15

Also modern specialized accelerators like GPUs and FPGAs can be used to further speed up the performance. These hardware accelerators are optimized for high parallel workloads, but their specialization on handling fixed size data for best performance rather than dynamic sized data strings can be a problem. Thus, handling variable long string keys directly in the GPU memory structure has the drawback of decreased performance, because the processing of each string takes as long as processing the longest of the currently processed strings. Hence it is not a surprise that current approaches for GPU hash tables focus on integer keys only and do not support string keys of variable size [11]. To overcome the discussed drawbacks, we propose a hybrid approach, such that string keys up to a maximum length are stored and handled on the GPU, and longer string keys are stored in main memory and processed by the CPU. Our approach takes advantage of both processing architectures, the CPU best suited to handle variable length data and the GPU tailored to massive parallelism of processing of fixed size data, for the overall goal of the best performance.

Our main contributions are

- a direct fast look-up of string queries on the GPU,
- hybrid scenarios handling shorter keys up to a fixed maximum length on GPU and longer keys on CPU, and
- an extensive evaluation of these approaches on a high performance computing system running recent many-core CPUs and GPUs for scientific calculations.

2 Related Work

	GPU Performance Criteria					GPU Hardware Features		
Hashing Technique	String Keys on GPU	Sufficient Parallelism	Memory Coalescing	Control Flow	CPU GPU Data Transfer	Shared Memory	Atomic Operations	Wrap-wide Voting
Open-addressing:								
CoherentHash	✗	✓	✓	✗	✗	✗	✓	✗
CuckooHash1	✗	✓	✗	✗	✗	✓	✓	✗
CuckooHash2	✗	✓	✗	✗	✗	✗	✓	✗
HortonHash	✗	✓	✗	✓	✓	✗	✗	✗
Our Hash Table Approach	✓	✓	✗	✗	✗	✗	✓	✗
MemcachedGPU	✗	✓	✗	✗	✓	✓	✓	✗
StadiumHash	✗	✓	✗	✗	✓	✓	✓	✓
Perfect Hashing:								
PerfectHash	✗	✓	✓	✓	✓	✗	✗	✗
Spatial Hashing:								
BiLevelLSH	✗	✓	✗	✗	✗	✓	✓	✗
EGSH	✗	✓	✗	✓	✓	✗	✗	✗
VoxelHash	✗	✓	✗	✗	✗	✗	✓	✗
Separate Chaining:								
SlabHash	✗	✓	✓	✓	✗	✗	✓	✓

Fig. 1. Overview of different GPU accelerated hash table algorithms and their features [11] compared with our approach.

Several papers related to hash tables focus on acceleration of hash table algorithms by using modern hardware and technologies. An overview of GPU accelerated hash tables is shown in Fig. 1, where we add the column about supporting string keys on GPUs and our approach for comparison matters. One important aspect of the hash table performance is efficient hashing. Therefore [2] describes efficient hashing with SIMD in OpenCL[1]. There are many papers with efficient GPU algorithms like Warpcore [8] which is a fast library of hash tables on GPU. It reaches 1.6 billion inserts and up to 4.3 billion retrievals per second on a Nvidia Quadro GV100. It is also faster than cuDPP [15], SlabHash [1] and NVIDIA RAPIDS cuDF[2]. HashGraph [6] uses sparse graph representations for scaleable hash tables. This highly parallel data structure uses information from their value-chain for a high performance collision management. DyCuckoo [13] is a dynamic cuckoo table which has a trade-off between GPU memory size and search performance. It is very efficient and enables fine-grained memory control. There are also hybrid hash tables on CPU and integrated GPU written modern program techniques like DPC++ [14]. [1] implements a fully concurrent dynamic hash table which runs on GPUs. Also they show a warp-cooperative work sharing strategy, which reduces the per-thread assignment and processing overhead.

To the best of our knowledge, all existing contributions to hash tables accelerated by GPUs do not support variable length string keys. Hence, our contribution is the first to investigate hash tables with string keys of variable length. Furthermore, we show that a parallel hybrid GPU/CPU hash table overcomes the drawbacks of GPUs designed to process data of fixed size and promises high performance.

3 Basics

In this chapter we look at basics of hash tables and CPU and GPU acceleration.

3.1 Hash Tables

Hashing is a widely used technique to quickly store data inside a structure like for example a hash table. Therefore we can use one or multiple different hash functions to translate a value into a hash. Hash tables often use a prime number for the size and the hash function includes a modulo operation with the prime number to shrink the range of values into the hash table size.

A hash table without collisions maps the key to a field in O(1) because we just calculate the hash value and access this position. The hash is ideally unique because collisions would lead to overwritten fields. However, hash functions are not collision free for arbitrary input, and we therefore need strategies to resolve these conflicts. There exist different strategies like for example linear probing [10], quadratic probing [10] or double hashing [7]. Another solution is e.g. to use cuckoo hashing with multiple tables.

Cuckoo hashing [16] uses two or more tables and a hash function for each table: First we try to insert the key into the first table by using the corresponding

[1] https://www.khronos.org/opencl/, accessed on 2022-03-08.
[2] https://github.com/rapidsai/cudf, accessed on 2022-03-13.

hash function. If the position is already occupied, we relocate the element which occupies the space from the first table into the second table and insert the element. If the position in the second table is already occupied, we insert the element there and move the previous element of this table in the next table and so on.

3.2 Hardware Acceleration

Modern CPUs and GPUs are designed for high parallelism and high bandwidth. There are many different vendors for graphics hardware and each vendor develops it's own API for their hardware, but some also support other hardware from other vendors. In addition cross-vendor APIs are developed from hardware independent developers. On modern multi-core and many-core systems there are a variety of different approaches for high performance computing. The first solution is to use multiple threads and distribute the workload onto the threads. A second idea is to use libraries like OpenMP[3] and Intel Threading Building Blocks[4] for loop and algorithm parallelism. Another technique uses asynchronous computation with tasks and coroutines. Each approach has different advantages and disadvantages like threads have an overhead and need synchronization, but allow manual optimization and custom parallelism. As an additional challenge there are multiple different implementations for the same concept in different languages. An interface for GPU acceleration is the widely used CUDA Toolkit[5], but it is limited to NVIDIA GPUs. CUDA offers high performance and extensive tooling and documentation. Another advantage is that many different frameworks and solutions are based on CUDA. For cross platform GPU acceleration there are OpenCL and Vulkan[6] which are both developed by the Khronos Group. OpenCL is designed for cross platform computing in contrast to Vulkan primarily designed for 3D-rendering, but can also be used for computing. Furthermore there exists other enterprise solutions like AMD's ROCm[7] for AMD's datacenter cards. Intel's OneAPI is a hybrid concept for CPU, APU, FPGA and GPU computing and also allows cross platform development for different hardware.

4 Parallel Hybrid GPU/CPU Hash Table for String Keys

In this chapter we show our acceleration design. We start with the concept idea followed by the data structure and how rollout and search-methods are designed. In the last subsection we look at hybrid approaches to combine CPU and GPU compute power.

---

[3] https://www.openmp.org/, accessed on 2022-03-08.
[4] https://www.intel.com/content/www/us/en/developer/tools/oneapi/onetbb.html, accessed on 2022-03-08.
[5] https://developer.nvidia.com/cuda-toolkit, accessed on 2022-03-08.
[6] https://www.vulkan.org/, accessed on 2022-03-08.
[7] https://rocmdocs.amd.com/en/latest/#, accessed on 2022-03-08.

4.1 Idea

For our design we choose an already implemented and well known hash table implementation. Therefore the robin-map[8] is used as a representative for a basic hash table. This implementation is based on robin hood hashing [3] and is also faster than the C++ build-in unordered_map. Robin hood hashing solves collisions by superseding existing elements which are closer to its original index position and this elements have to be relocated to indexes further away. As an example for a high performance well known variant of hashing, we choose the libcuckoo library which already has been considered by several different scientific contributions like [5,12]. This library implements the widely used cuckoo hashing.

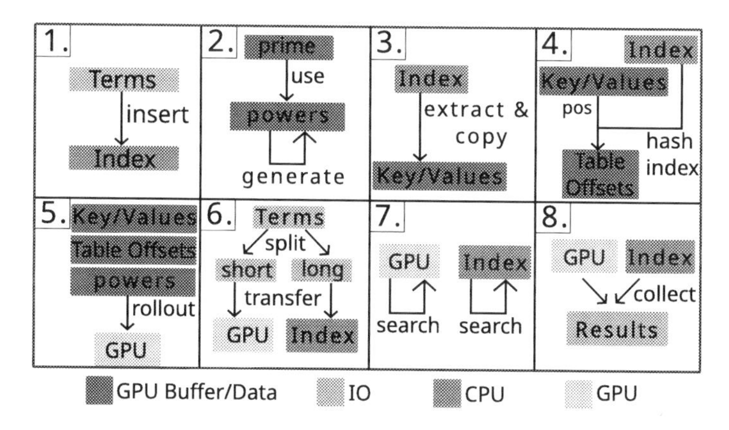

Fig. 2. This figure presents the steps that are necessary to process a search in a hybrid hash table. First terms are inserted in the index, then a prime number and powers are randomly generated. In the next step the index of the terms are copied to the OpenCL buffer before the hash is mapped to the buffer index and the content is the position in the key/values buffer. In step 5 all buffer content is transferred to the GPU. The terms are split into long and short keys, and are transferred to the corresponding computation device. Finally we start the search and transfer the results back to the host.

The main difference between the existing algorithms and our approach is the support of string keys. For these string keys we need an hash algorithm to convert the keys to a hash. The requirements for this hash algorithm are that it is simple, efficient and has a relative low collision rate. A simple approach is to use a polynomial rolling hash function like the Rabin-Karp-Algorithm [9]. Another advantage is that these kinds of algorithms can also be accelerated by GPUs [4]. Therefore we define and implement the polynomial rolling hash function. The following parameters are provided: $S = [s_0, ..., s_{n-1}]$ is the string key composed of the characters s_i, n is the number of characters in S, $P = [p_0, ..., p_{n-1}]$ is an

[8] https://github.com/Tessil/robin-map, accessed on 2022-03-08.

array of random powers and M is a prime number. We define the hash $H_M(S,P)$ of S as:

$$H_M(S,P) = \begin{cases} (s_0 * p_0) \ mod \ M & n = 1 \\ (H_M([s_0, ..., s_{n-2}], [p_0, ..., p_{n-2}]) + s_{n-1} * p_{n-1}) \ mod \ M & n > 1 \end{cases}$$

The operation in a hybrid search is shown in Fig. 2. We first fill our hash table and then generate an OpenCL data structure before transferring this structure. When we split the requests into batches and distribute the batches by different metrics on CPU and GPU, one metric can be the length of the keys. Afterwards we start the search on both accelerators and if these searches are finished, then we collect the results of both searches.

4.2 Data Structure and Search

Our GPU approach provides a generic interface for different hash tables. Therefore we can use the same OpenCL implementation for robin-map and libcuckoo. To support this, we also need a generic OpenCL data structure for the GPU memory. OpenCL uses 1-dimensional buffers for data transfer and each buffer element can store custom or built-in types. Furthermore OpenCL has an image type for 2- or 3-dimensional data. For the hash map itself we work with an 1d-buffer storing the key-value-pairs separated by a NULL byte. Because our values are 64 bit long values, every value takes up 8 bytes. We call this the terms-buffer as shown in Fig. 3. Now we have a buffer with data, but for efficiency we use an offset buffer for random access of each key-value-pair. This 1d-buffer for the offsets has a special order which is generated throughout a rollout. The indexes of this buffer correspond to the hash value of the key which is located behind the stored offset. The last buffer contains the powers for the hash algorithm and can be stored and transferred in different ways. We need enough space for the maximum string length multiplied by the number of tables. For this memory, we can use a simple flat 1d-buffer like is shown in Fig. 3. Alternatively, we can use the OpenCL built-in vector types for the number of tables, which has the advantage that we can calculate the hash for a single character in all tables with one vector operation. A disadvantage is that the OpenCL kernel has to be constructed during initialization because we do not know the number of tables beforehand. Furthermore, we also need a prime number for the hash algorithm which is randomly generated and transferred to the OpenCL kernel as an argument alongside the number of tables and the tables size.

After we transferred the buffers to the GPU, we use the hash function to calculate the hash for each key and look in the offsets buffer to find the position in the terms buffer. Then we compare the keys with each other. If we have a match, then we return the result and otherwise we continue with the next table. If the last table does not have the key, it is not in the structure and we return not found. A performance benefit can be achieved by the host OpenCL control structure. We are processing the requests in batches and use a simple form of

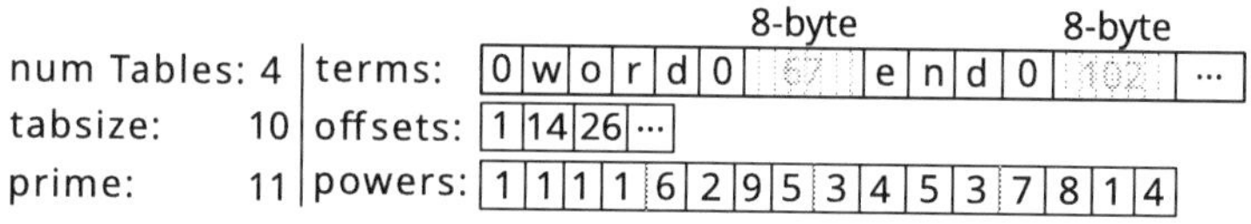

num Tables:	4	terms:	0 w o r d 0 [8-byte: 67] e n d 0 [8-byte: 102] ...
tabsize:	10	offsets:	1 14 26 ...
prime:	11	powers:	1 1 1 1 6 2 9 5 3 4 5 3 7 8 1 4

Fig. 3. This is the memory layout of the hash table on the GPU. We use three OpenCL buffers and three values. The values are the number of tables, the maximum size of the table and the prime number for hashing. The first buffer terms contains the key-value pairs separated by null bytes. The second buffer contains the start offsets for the key value pairs and the index corresponds to the hash of a key. The third and final buffer contains the generated powers for the hash algorithm.

pipelining by splitting the batch into two halves. We start by converting the keys of the first half to a GPU structure which consists of a character buffer and an offset buffer for the start position of each key. Then we transfer these data and start the search by inserting the write, execute and read operations into a queue. Meanwhile the second half is prepared and transferred and the OpenCL operations are inserted into a second queue. Afterwards we wait until both queues are finished.

5 Evaluation

In this chapter we evaluate our approach. We start with the environment and the different benchmark mods, then we look at the results.

5.1 Benchmark Framework

For benchmarking and comparing different hash implementations and other data structures, we develop an hybrid multi-threaded framework called H^2. The reason for this framework is to load or generate random workloads for testing and measure the execution times over multiple runs in different data structures. It is possible to switch between different data structures during runtime to compare different implementations in one run. Also the H^2 framework can process complete series where each point can be executed multiple times to reduce statistic variance.

For performance purposes, we use modern C++ with Boost for the H^2 framework. For acceleration H^2 can support different devices and APIs. Currently CPU, GPU and FPGA are supported devices and OpenCL, CUDA and OneAPI[9] are supported APIs. The general architecture is shown in Fig. 4. The key-value pairs are either generated inside the framework or are loaded from a flat text file with one pair per line. H^2 is optimized for batch processing and supports different hybrid and non-hybrid scenarios. The current supported scenarios are

[9] https://www.intel.com/content/www/us/en/developer/tools/oneapi/overview.html, accessed on 2022-03-08.

all batches on CPU, all batches on GPU and split batches according to a certain percentage or a certain maximum length. The H^2 framework has the ability to validate the results and to log resources like RAM, GPU memory and utilization usage. H^2 supports Linux and Windows operating systems.

Each benchmark run is split into two different phases, the loading or fill phase and the run phase. First the key-value store is initially filled with a certain number of pairs and an initial rollout to the accelerator is done. In the next phase batch requests are sent multi-threaded to the different accelerators and H^2 waits for completion of each thread. Afterwards, the execution time and throughput is written to an output file and the key-value store and accelerator are reset to the initial state. Now another pass-through or a different configuration can be executed. Also we use a Node.js[10] application written in TypeScript[11] for result aggregation and diagram generation. The basic configuration and benchmark setup can be handled via one single JSON file. Each file can store multiple series and these are combined into a single diagram.

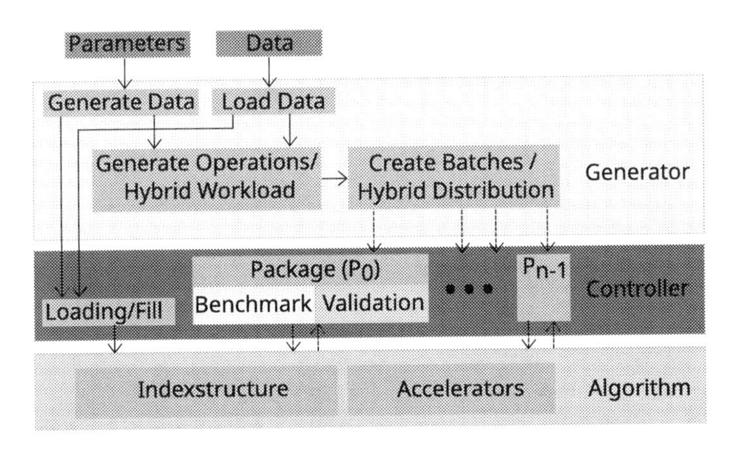

Fig. 4. Benchmark Framework Structure: the framework consists of three different parts, first the data and batch generators followed by the controller for each individual hybrid configuration (package) and then specific algorithms with index and accelerators.

5.2 Benchmark Environment

We use a modern many core high performance computing NUMA system for scientific computations. This system contains two AMD EPYC 7542 processors with 32-Cores each and a core can handle two threads (Hyper-Threading). Furthermore the system has 2 TiB of high bandwidth RAM for Big Data computing. For acceleration we use a NVIDIA A100 graphics card with 40 GiB memory. For real world data we use the complete triples data (btc2019-triples.nt.gz) from

[10] https://nodejs.org/en/, accessed on 2022-03-13.
[11] https://www.typescriptlang.org/.

the 2019 Billion Triple Challenge[12] (BTC) and filter the data after key length requirements.

5.3 Benchmark Results

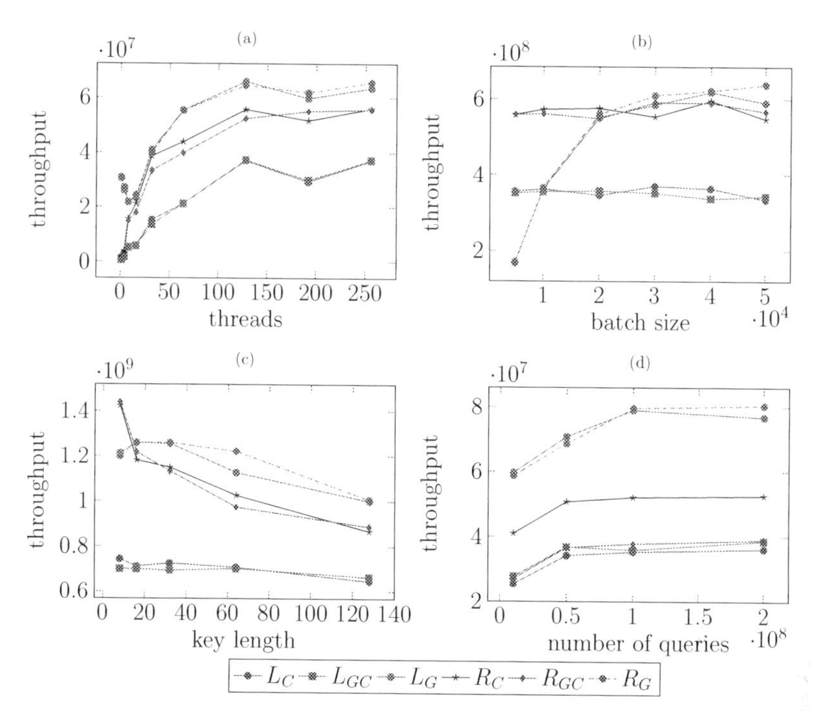

Fig. 5. Benchmark with throughput in case of (a) increasing number of threads, (b) batch size, (c) key length and (d) number of queries with BTC data.

We start by comparing the variants against each other in Fig. 5. Therefore we use abbreviations to avoid ambiguity as follows:

- libcuckoo CPU/libcuckoo GPU (L_C/L_G)
- robin-map CPU/robin-map GPU (R_C/R_G)
- libcuckoo GPU approach on CPU (L_{GC})
- robin-map GPU approach on CPU (R_{GC})

In Fig. 5a, we compare different numbers of threads. We use a thread pool with the same size as the threads, but capped it at 128 threads, which is the maximum number of parallel threads. All approaches benefit of more threads, but L_G and R_G have the highest throughput. L_C is on the same level as L_{GC}

[12] https://zenodo.org/record/2634588, accessed on 2022-03-13.

and the same is true for R_C and R_{GC}. In Fig. 5b, we consider the batch size. We see that a high batch size is important for GPU acceleration because L_G and R_G increase their throughput significantly with larger batches and reach a plateau by 30,000 operations. We can see that we need a certain amount of operations to be efficient, because we have some overhead at the start and the end of the pipeline. The third diagram (Fig. 5c) shows the effect of the key length. If the keys get longer, all approaches need more time for comparing the keys and the throughput is decreasing. We can also see that for short strings R_C and R_{GC} are faster. In the fourth diagram (Fig. 5d) we look at the real world BTC data with key length 32. We increase the number of queries for better utilization and see that both L_G and R_G are the fastest and that there is also a gap to R_C. L_{GC}, R_{GC} and L_C are even slower as R_C and on the same level. The speed-up is 2.1 between L_C and L_G and 1.5 between R_C and R_G.

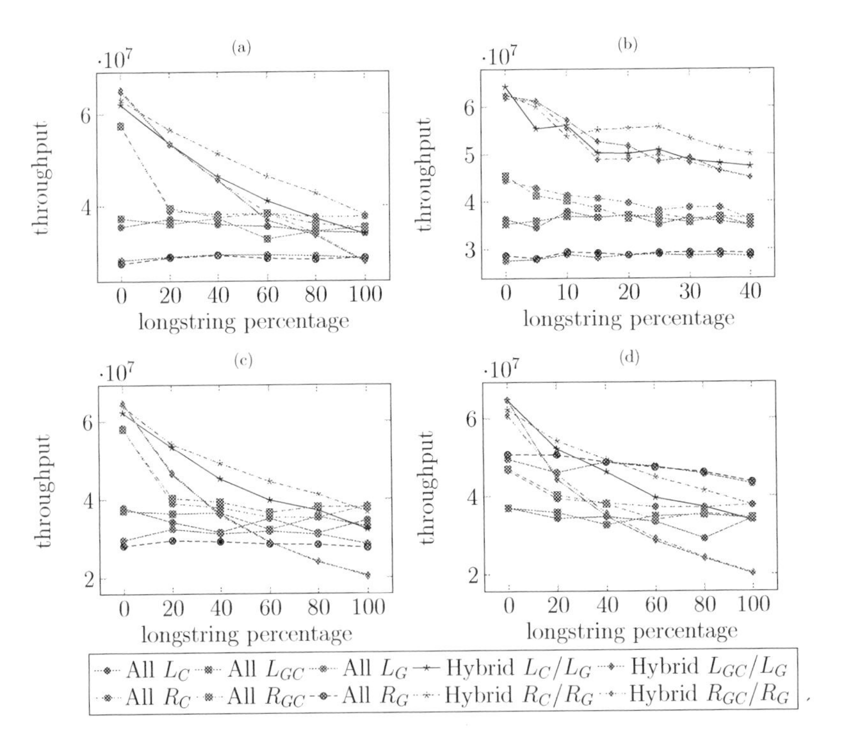

Fig. 6. Hybrid throughput trend with increasing long string amount, 128 Threads GPU, 128 Threads CPU, (a,c) short string length 8 and (b,d) 32, (a,b) long string length 128 and (c,d) 256, (b) shows the long key range between 0–40% in detail.

Now we compare different query splittings and thread counts in a hybrid scenario. The parameter for splitting is the key length and we have a limit after which we distribute the terms to the CPU approach. In the following chapter, we define long strings as strings which are longer than a certain threshold and are moved to the CPU if not specified differently. We start with zero percent long strings in our query set and increase the amount until all keys are long strings. In Fig. 6, we compare the performance of a single approach with all queries against the hybrid combinations of two approaches split between them. We can see in the graphs that with increasing amount of long strings all hybrid implementations lose throughput and tend towards the all CPU and GPU performances. But for medium percentage, the combination of both approaches is faster than CPU or GPU approach alone.

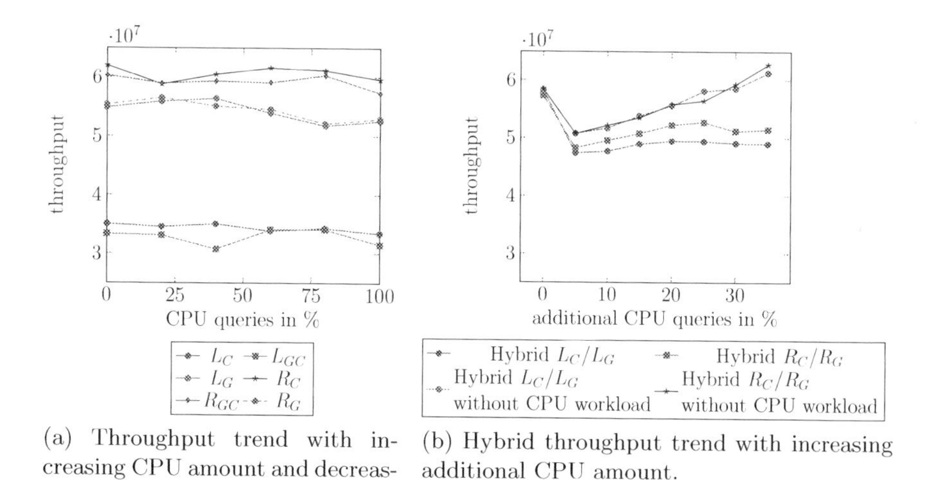

(a) Throughput trend with increasing CPU amount and decreasing GPU amount.

(b) Hybrid throughput trend with increasing additional CPU amount.

Fig. 7. Further hybrid results with percentage of (a) queries process on the CPU and with constant load on the GPU and (b) additional queries added to the CPU.

In Fig. 7a, we simply increase the share of CPU queries in percent, but leave the total number of queries constant. We see that this doesn't change the performance very much, but the throughput of all approaches drops with all queries on the CPU. In Fig. 7b, we have a constant amount of short length keys and add additional longer keys. The throughput of the hybrid variants L_C/L_G and R_C/R_G increase at the beginning, but they reach a maximum at 25% additional queries and make a slight drop after. This is the case because L_C and R_C slow the algorithm down. We check this by simply removing the workload on L_C and R_C and keeping the workload on L_G and R_G. A reason is that the parallelism of GPUs is much more efficient than a high number of CPU threads.

6 Summary and Conclusions

We design an GPU acceleration string based GPU implementation for robin-map and libcuckoo and integrate it into our hybrid benchmark framework H^2.

Furthermore, we propose a parallel hybrid GPU/CPU hash table for string keys of variable length. OpenCL is used for GPU acceleration and we use traditional thread in a thread pool for CPU parallelism. Then we evaluate our implementation against the original CPU variants. Our approach has a higher throughput than the CPU variant and achieves a speed-up of 2.1 compared to libcuckoo and 1.5 compared to robin-map. For long strings on the CPU and short strings on the GPU, our proposed hybrid approach is faster and has a 40% higher throughput with 20% long keys on the CPU.

In the future, update and delete operations could be supported in combination with a GPU based look-up. Also the modified hash tables could be merged with the already existing GPU structure to minimize the amount of data which has to be transferred. Another direction is the use of different accelerators like FPGA and APU, and to investigate different types of basic indices for hybrid index structures.

Acknowledgments. This work is funded by the Deutsche Forschungsgemeinschaft (DFG, German Research Foundation) – Project-ID 422742661.

References

1. Ashkiani, S., Farach-Colton, M., Owens, J.D.: A dynamic hash table for the GPU. In: 2018 IEEE International Parallel and Distributed Processing Symposium (IPDPS), pp. 419–429 (2018). https://doi.org/10.1109/IPDPS.2018.00052
2. Behrens, T., Rosenfeld, V., Traub, J., Breß, S., Markl, V.: Efficient SIMD vectorization for hashing in OpenCL. In: EDBT, pp. 489–492 (2018)
3. Celis, P., Larson, P.A., Munro, J.I.: Robin hood hashing. In: 26th Annual Symposium on Foundations of Computer Science (SFCS 1985), pp. 281–288 (1985). https://doi.org/10.1109/SFCS.1985.48
4. Dayarathne, N., Ragel, R.: Accelerating rabin karp on a graphics processing unit (GPU) using compute unified device architecture (CUDA). In: 7th International Conference on Information and Automation for Sustainability, pp. 1–6 (2014). https://doi.org/10.1109/ICIAFS.2014.7069589
5. Fan, B., Andersen, D.G., Kaminsky, M.: Memc3: Compact and concurrent memcache with dumber caching and smarter hashing. In: 10th USENIX Symposium on Networked Systems Design and Implementation (NSDI 2013), pp. 371–384. USENIX Association, Lombard, IL, April 2013. https://www.usenix.org/conference/nsdi13/technical-sessions/presentation/fan
6. Green, O.: Hashgraph-scalable hash tables using a sparse graph data structure. ACM Trans. Parallel Comput. **8**(2) (2021). https://doi.org/10.1145/3460872

7. Guibas, L.J., Szemeredi, E.: The analysis of double hashing. J. Comput. Syst. Sci. **16**(2), 226–274 (1978). https://doi.org/10.1016/0022-0000(78)90046-6, https://www.sciencedirect.com/science/article/pii/0022000078900466
8. Jünger, D., et al.: Warpcore: a library for fast hash tables on gpus. In: 2020 IEEE 27th International Conference on High Performance Computing, Data, and Analytics (HiPC), pp. 11–20 (2020). https://doi.org/10.1109/HiPC50609.2020.00015
9. Karp, R.M., Rabin, M.O.: Efficient randomized pattern-matching algorithms. IBM J. Res. Dev. **31**(2), 249–260 (1987). https://doi.org/10.1147/rd.312.0249
10. Knuth, D.E.: The Art of Computer Programming, Volume. 3: Sorting and Searching. Addison Wesley Longman Publishing, Redwood City (1974)
11. Lessley, B., Childs, H.: Data-parallel hashing techniques for GPU architectures. IEEE Trans. Parallel Distrib. Syst. **31**(1), 237–250 (2020). https://doi.org/10.1109/TPDS.2019.2929768
12. Li, X., Andersen, D.G., Kaminsky, M., Freedman, M.J.: Algorithmic improvements for fast concurrent cuckoo hashing. In: Proceedings of the Ninth European Conference on Computer Systems. EuroSys 2014, Association for Computing Machinery, New York, NY, USA (2014). https://doi.org/10.1145/2592798.2592820
13. Li, Y., Zhu, Q., Lyu, Z., Huang, Z., Sun, J.: Dycuckoo: dynamic hash tables on GPUs. In: 2021 IEEE 37th International Conference on Data Engineering (ICDE), pp. 744–755 (2021). https://doi.org/10.1109/ICDE51399.2021.00070
14. Lupescu, G., Ţăpuş, N.: Design of hashtable for heterogeneous architectures. In: 2021 23rd International Conference on Control Systems and Computer Science (CSCS), pp. 172–177 (2021). https://doi.org/10.1109/CSCS52396.2021.00035
15. Merrill, D.G., Grimshaw, A.S.: Revisiting sorting for GPGPU stream architectures. In: Proceedings of the 19th International Conference on Parallel Architectures and Compilation Techniques, PACT 2010, pp. 545–546. Association for Computing Machinery, New York, NY, USA (2010). https://doi.org/10.1145/1854273.1854344
16. Pagh, R., Rodler, F.F.: Cuckoo Hashing. In: auf der Heide, F.M. (ed.) ESA 2001. LNCS, vol. 2161, pp. 121–133. Springer, Heidelberg (2001). https://doi.org/10.1007/3-540-44676-1_10

Towards Efficient Discovery of Periodic-Frequent Patterns in Dense Temporal Databases Using Complements

P. Veena[1], Sreepada Tarun[2(✉)], R. Uday Kiran[2,3], Minh-Son Dao[3], Koji Zettsu[3], Yutaka Watanobe[2], and Ji Zhang[4]

[1] SBRIT, Ananthapur, Andhra Pradesh, India
[2] The University of Aizu, Aizu-Wakamatsu, Fukushima, Japan
{s1272001,udayrage,yutaka}@u-aizu.ac.jp
[3] National Institute of Information and Communications Technology, Tokyo, Japan
{dao,zettsu}@nict.go.jp
[4] University of Southern Queensland, Toowoomba, Australia
ji.zhang@usq.edu.au

Abstract. Periodic-frequent pattern mining involves finding all periodically occurring patterns in a temporal database. Most previous studies found these patterns by storing the temporal occurrence information of an item in a list structure. Unfortunately, this approach makes pattern mining computationally expensive on dense databases due to increased list sizes. With this motivation, this paper explores the concept of *complements,* and proposes an efficient algorithm that records non-occurrence information of an item to find all desired patterns in a dense database. Experimental results demonstrate that our algorithm is efficient.

Keywords: Data mining · Periodic patterns · Temporal databases

1 Introduction

Periodic-frequent pattern mining is a popular analytical technique that aims to discover all regularly occurring patterns in a temporal database. A classic application of these patterns is air pollution analytics. It involves finding the locations in which people were regularly exposed to harmful levels of air pollution. An example of a periodic-frequent pattern is as follows:

$$\{S1, S4, S5\}\ [support = 10\%, periodicity = 3\,\text{h}]$$

The above pattern informs us that 10% of the time, the people living in the vicinity of the sensors, S1, S4, and S5, were regularly (i.e., at least once in every 3 h) exposed to harmful levels of PM2.5. Such a piece of information may be helpful to the users in introducing policies to control emissions.

Several algorithms [2–4,7] have been described in the literature to find periodic-frequent patterns. The basic approach followed by these algorithms, however, remains the same. It involves the following two steps:

C. Strauss et al. (Eds.): DEXA 2022, LNCS 13427, pp. 204–215, 2022.
https://doi.org/10.1007/978-3-031-12426-6_16

1. Construct a list for each pattern by recording its temporal occurrences in the database, and
2. Determine a pattern is periodic-frequent or not by traversing through the generated list.

The above approach of maintaining the temporal lists for a pattern makes the periodic-frequent pattern mining possible on sparse databases. However, it also raises performance issues while dealing with dense databases. It is because items have to maintain very long temporal lists, which leads to performance issues. This paper tackles this challenging problem by exploring the concept of "*(set) complements*" and proposing an efficient algorithm to find the patterns in dense databases.

Please note that finding periodic-frequent patterns in dense databases is a non-trivial and challenging task due to the following reasons:

1. Several algorithms [1,5] have been described in the literature to find frequent patterns in a dense transactional database. However, since these algorithms capture only *frequency* and completely disregard the temporal occurrence information of a pattern in a database, they cannot be utilized to find periodic-frequent patterns in dense temporal databases.
2. Zaki et al. [8] explored the concept of *set difference*[1] to calculate the *frequency* of a pattern in a database. They have not described any methodology to determine the *periodicity* of a pattern. In this paper, we propose a novel method to calculate the *periodicity* of a pattern using *complements.*

The contributions of this paper are as follows. First, we introduce a novel methodology to determine the *periodicity* of a pattern given its non-occurrence temporal information. Second, we propose a novel depth-first search algorithm to find all desired patterns in a dense temporal database. We call our algorithm as Periodic-Frequent Pattern Miner with Complements (PFPM-C). Third, we empirically show that our algorithm is efficient.

The remainder of this study is organized as follows. Section 2 discusses the literature on frequent pattern mining and periodic-frequent pattern mining. In Sect. 3, we describe the model of a periodic-frequent pattern. Section 4 introduces the basic idea and presents our PFPM-C algorithm. In Sect. 5, we present the experimental results. Finally, in Sect. 6, we provide a conclusion of this study and discuss future research directions.

2 Related Work

Agrawal et al. [1] introduced the concept of frequent pattern mining to extract useful information in transactional databases. Luna et al. [5] conducted a detailed survey on frequent pattern mining and presented the improvements that happened in the past 25 years. Zaki et al. [8] first explored the concept of set difference and proposed a depth-first search algorithm to find all frequent patterns

[1] The term *set difference* refers to relative complement, whereas the term *complement* typically refers to absolute complement.

in a dense transactional database. This algorithm uses the **size of set difference list** to determine whether a pattern is frequent or infrequent in the database. This algorithm does not describe any methodology to determine the *periodicity* of a pattern from the set difference list. Henceforth, this algorithm cannot be directly extended to find periodic-frequent patterns. This paper proposed a novel methodology to calculate the *periodicity* of a pattern given its set difference information.

Tanbeer et al. [7] introduced the idea of periodic-frequent pattern mining. A highly compacted periodic frequent-tree (PF-tree) was constructed. A pattern growth technique was applied to generate all periodic-frequent patterns in a database based on the user-specified $minSup$ and $maxPer$ constraints. Amphawan et al. [2] designed an efficient depth-first search-based algorithm named for mining top-K periodic-frequent patterns without using the user-specified $minSup$ constraint. Uday et al. [4] introduced a novel greedy approach to discover periodic-frequent patterns. Authors have designed a two-phase architecture named expanding and shrinking phases to efficiently store all the patterns with support and periodicity. Where these phases have effectively utilized the newly introduced local periodicity concept. Finally, the authors have created a PF-tree++ and applied a pattern growth technique to generate periodic-frequent patterns in a database based on the user-specified $minSup$ and $maxPer$. Anirudh et al. [3] introduced a novel concept of periodic summaries to find the periodic-frequent patterns in temporal databases. Authors have introduced a novel concept called periodic summaries-tree to maintain the time stamp information of the patterns in a database and designed a pattern growth algorithm to generate a complete set of periodic-frequent patterns. Unfortunately, all of the above algorithms store the temporal occurrence information of a pattern in a list structure and thus, suffer from computational issues while dealing with dense databases.

Overall, our algorithm to find periodic-frequent patterns in dense databases using set complements is novel and has not been explored in the literature.

3 Periodic-Frequent Pattern Model

Let I be the set of items. Let $X \subseteq I$ be a **pattern** (or an itemset). A pattern containing β, $\beta \geq 1$, number of items is called a **β-pattern.** A **transaction**, $t_k = (ts,\ Y)$ is a tuple, where $ts \in R^+$ represents the timestamp at which the pattern Y has occurred. A **temporal database** TDB over I is a set of transactions, i.e., $TDB = \{t_1, \cdots,\ t_m\}$, $m = |TDB|$, where $|TDB|$ can be defined as the number of transactions in TDB. For a transaction $t_k = (ts,\ Y)$, $k \geq 1$, such that $X \subseteq Y$, it is said that X occurs in t_k (or t_k contains X) and such a timestamp is denoted as ts^X. Let $TS^X = \{ts_j^X, \cdots, ts_k^X\}$, $j,\ k \in [1, m]$ and $j \leq k$, be an **ordered set of timestamps** where X has occurred in TDB. The number of transactions containing X in TDB is defined as the **support** of X and denoted as $sup(X)$. That is, $sup(X) = |TS^X|$. The pattern X is said to be a **frequent pattern** if $sup(X) \geq minSup$, where $minSup$ refers to the user-specified *minimum support* value. Let ts_c^X and ts_d^X, $j \leq c < d \leq k$, be the two consecutive timestamps in TS^X. The time difference (or an inter-arrival time) between ts_d^X and ts_c^X is defined as a ***period*** of X, say p_e^X. That is, $p_e^X = ts_d^X - ts_c^X$. Let $P^X = (p_1^X, p_2^X, \cdots, p_d^X)$ be the set of all *periods* for pattern X. The ***periodicity*** of X, denoted as $per(X) = maximum(p_1^X, p_2^X, \cdots, p_d^X)$. The frequent pattern X is said to be a **periodic-frequent pattern** if $per(X) \leq maxPer$, where $maxPer$ refers to the user-specified *maximum periodicity* value. Given a temporal database (TDB) and the user-specified *minimum support* ($minSup$) and *maximum periodicity* ($maxPer$) constraints, the problem definition of periodic-frequent pattern mining is to discover the complete set of periodic-frequent patterns that have *support* no less than $minSup$ and *periodicity* no more than the $maxPer$ constraints.

Example 1. Let $I = \{p, q, r, s, t, u\}$ be the set of items. A hypothetical row temporal database generated from I is shown in Table 1. Without loss of generality, this row temporal database can be represented as a columnar temporal database as shown in Table 2. The temporal occurrences of each item in the entire database is shown in Table 3. The set of items 'q' and 'r', i.e., $\{q, r\}$ is a pattern. For brevity, we represent this pattern as 'qr'. This pattern contains two items. Therefore, it is 2-pattern. The pattern 'qr' appears at the timestamps of 1, 2, 3, 4, 6, 9, and 10. Therefore, the list of timestamps containing 'qr', i.e., $TS^{qr} = \{1, 2, 3, 4, 6, 9, 10\}$. The *support* of '$qr$,' i.e., $sup(qr) = |TS^{qr}| = 7$. If the user-specified $minSup = 5$, then qr is said to be a frequent pattern because of $sup(qr) \geq minSup$. The periods for this pattern are: $p_1^{qr} = 1\ (= 1 - ts_{initial})$, $p_2^{qr} = 1\ (= 2 - 1)$, $p_3^{qr} = 1\ (= 3 - 2)$, $p_4^{qr} = 1\ (= 4 - 3)$, $p_5^{qr} = 2\ (= 6 - 4)$, $p_6^{qr} = 3\ (= 9 - 6)$, $p_7^{qr} = 1\ (= 10 - 9)$, and $p_8^{qr} = 0\ (= ts_{final} - 10)$, where $ts_{initial} = 0$ represents the timestamp of initial transaction and $ts_{final} = |TDB| = 10$ represents the timestamp of final transaction in the database. The *periodicity* of qr, i.e., $per(qr) = maximum(1, 1, 1, 1, 2, 3, 1, 0) = 3$. If the user-defined $maxPer = 3$, then the frequent pattern 'qr' is said to be a periodic-frequent pattern because $per(qr) \leq maxPer$.

4 Proposed Algorithm

4.1 Basic Idea: Calculating the *Periodicity* of a Pattern Using Complements

Definition 1 ***(The complement list of a pattern*** X***).*** *The* ***compliment list*** *of* X*, denoted as* $\widehat{TS^X} = \{TS - TS^X\}$*, where* TS *denotes the universe of timestamps contained in a database and* TS^X *represents the timestamps at which* X *has appeared in a database.*

Example 2. In Table 1, the pattern qr has occurred at the timestamps of 1, 2, 3, 4, 6, 9, and 10. Thus, $TS = \{1, 2, 3, 4, 5, 6, 7, 8, 9, 10\}$ and $TS^{qr} = \{1, 2, 3, 4, 6, 9, 10\}$. The complement list of qr, denoted as $\widehat{TS^{qr}} = \{1, 2, 3, 4, 5, 6, 7, 8, 9, 10\} - \{1, 2, 3, 4, 6, 9, 10\} = \{5, 7, 8\}$.

Table 1. Row database

ts	Items	ts	Items
1	pqru	6	pqrs
2	qrst	7	pq
3	pqrs	8	rsu
4	pqrt	9	pqrs
5	rtu	10	qrsu

Table 2. Columnar database

	Items							Items					
ts	p	q	r	s	t	u	ts	p	q	r	s	t	u
1	1	1	1	0	0	1	6	1	1	1	1	0	0
2	0	1	1	1	1	0	7	1	1	0	0	0	0
3	1	1	1	1	0	0	8	0	0	1	1	0	1
4	1	1	1	0	1	0	9	1	1	1	1	0	0
5	0	0	1	0	1	1	10	0	1	1	1	0	1

Table 3. Items list

Item	TS-list
p	1, 3, 4, 6, 7, 9
q	1, 2, 3, 4, 6, 7, 9, 10
r	1, 2, 3, 4, 5, 6, 8, 9, 10
s	2, 3, 6, 8, 9, 10
t	2, 4, 5
u	1, 5, 8, 10

Typically, $|TS^X| \leq |\widehat{TS^X}|$ in sparse databases. Henceforth, finding periodic-frequent patterns using the TS^X of a pattern is recommended for sparse databases. However, $|TS^X| \geq |\widehat{TS^X}|$ in dense database. Thus, finding periodic-frequent patterns using the TS^X information of a pattern is not recommended for dense databases. We need to explore the *complement* of TS^X, i.e., $\widehat{TS^X}$, to find periodic-frequent patterns in dense databases effectively. To do so, we first need to address the following two key challenges:

1. How to derive the *complement* information of a pattern given the *complement* information of its subsets? That is, how to calculate $\widehat{TS^Z}$ from $\widehat{TS^X}$ and $\widehat{TS^Y}$, where $Z = X \cup Y$ and $X \cap Y = \emptyset$?
2. How to determine the *periodicity* of a pattern from its *complement* list? That is, how to calculate $per(Z)$ given $\widehat{TS^Z}$?

We solve these challenges using the Properties 1 and 2.

Property 1. If X, Y and Z be three patterns such that $Z = X \cup Y$ and $X \cap Y = \emptyset$, then $\widehat{TS^Z} = \widehat{TS^X} \cup \widehat{TS^Y}$.

Example 3. The $\widehat{TS^{qr}} = \widehat{TS^q} \cup \widehat{TS^r} = \{5, 8\} \cup \{7\} = \{5, 7, 8\}$. It means the pattern qr does not appear at the timestamps of 5, 7, and 8.

Property 2. Let $\widehat{MTS^Z_k} \subseteq \widehat{TS^Z}$, where $k \geq 1$, be a maximal set of consequence timestamps in $\widehat{TS^Z}$ such that $\widehat{TS^Z} = \{\widehat{MTS^Z_1} \cup \widehat{MTS^Z_2} \cup \cdots \cup \widehat{MTS^Z_k}\} = \cup_{j=1}^{k} \widehat{MTS^Z_j}$ and $\widehat{MTS^Z_i} \cap \widehat{MTS^Z_j} = \emptyset, 1 \leq i \leq j \leq k$. The *periodicity* of Z given $\widehat{TS^Z}$, i.e., $per(Z) = max(|\widehat{MTS^Z_i}| \forall \widehat{MTS^Z_i} \in \widehat{TS^Z}) + 1$.

Example 4. Continuing with the previous example, the $\widehat{TS^{qr}}$ can be split into two maximal consecutive subsets, $\widehat{MTS^{qr}_1} = \{5\}$ and $\widehat{MTS^{qr}_2} = \{7, 8\}$. Thus, periodicity of qr, denoted as $per(qr) = max(|\{5\}|, |\{7, 8\}|) + 1 = 3$.

4.2 PFPM-C

The space items in a database represent an itemset lattice. This lattice represents the search space of periodic-frequent pattern mining. Using the *downward closure property* (see Property 3), our PFPM-C algorithm finds all desired patterns by performing a depth-first search on the itemset lattice.

Property 3 **(Downward closure property** [1]**):** All non-empty sets of a periodic-frequent pattern must also be a periodic-frequent pattern.

The PFPM-C algorithm involves the following steps: (*i*) Scan the database and find all periodic-frequent items (or 1-patterns) in the database, (*ii*) Construct complement set, i.e., $\widehat{TS^{i_j}}$ for each periodic-frequent item i_j, and (*iii*) using the Properties 1 and 2, perform the depth-first search on the lattice and find all periodic-frequent patterns from the database. Algorithms 1 and 2 describe the procedures to find periodic-frequent patterns from the database. We will now illustrate these algorithms using the temporal database shown in Table 1. Let $minSup = 5$ and $maxPer = 3$.

Scan the database and construct the list of timestamps for each item (Lines 1 to 8 in Algorithm 1). Figure 2(a)–(c) shows the TS-list generated after scanning the first, second, and every transaction in the database. Using the *downward closure property*, we prune all items whose *support* is less than $minSup$ or *periodicity* is more than $maxPer$ (Lines 9 to 15 in Algorithm 1). The remaining items are considered periodic-frequent items and sorted in *support* descending order (Line 16 in Algorithm 1). Figure 2(d) shows the sorted list of periodic-frequent items. Let L denote this sorted list of periodic-frequent items. We replace the list of timestamps of periodic-frequent items with their *compliments* (Lines 17 and 18 in Algorithm 1). Figure 2(e) shows the TS-list generated after applying set complement. We now call this list a CTS-list for brevity. Next, we perform a depth-first search on the lattice generated by L and find all periodic-frequent patterns in a database (Line 18 in Algorithm 1). The optimization is achieved during the depth-first search by preventing the search on child nodes if a parent node fails to be a periodic-frequent pattern (Algorithm 2). Figure 3 shows the depth-first search performed on the lattice. We start with the first item in L, i.e., r. Since r is a periodic-frequent pattern (or item), we concatenate r with the second item in L, i.e., q. The result is a new pattern rq. Using Property 1, we construct its complement list, and determine whether is periodic-frequent pattern or not using Property 2. We repeat this process until the child node fails to be a periodic-frequent pattern or lattice is traversed.

5 Experimental Results

In the literature, researchers have described PFP-growth [7], PFP-growth++ [4] and PF-ECLAT [6] algorithms to find periodic-frequent patterns. We compare our PFPM-C algorithm against these algorithms and show that ours is efficient.

5.1 Experimental Setup

The algorithms, PFP-growth, PFP-growth++, PF-ECLAT, and PFPM-C, were written in Python and executed on a GIGABYTE R282-Z93-00 rack server containing two AMD EPYC 7452 CPUs running at 2.35 GHz with an NVIDIA

Algorithm 1. PeriodicFrequentItems(TDB: temporal database, $minSup$: minimum Support, $maxPer$: maximum periodicity

1: Let $TS\text{-}list = (i, ts\text{-}list(i))$ be a dictionary that records the temporal occurrence information of an item i in a TDB. Let TS_l be a temporary list to record the *timestamp* of the last occurrence of an item in the database. Let Per be a temporary list to record the *periodicity* of an item in the database.
2: **for** each transaction $t_{cur} \in TDB$ **do**
3: Set $ts_{cur} = t_{cur}.ts$;
4: **for** each item $i \in t_{cur}.X$ **do**
5: **if** i does not exit in TS-list **then**
6: Insert i and its timestamp into the TS-list. Set $TS_l[i] = ts_{cur}$ and $Per[i] = (ts_{cur} - ts_{initial})$;
7: **else**
8: Add i's timestamp in the TS-list. Update $TS_l[i] = ts_{cur}$ and $Per[i] = max(Per[i], (ts_{cur} - TS_l[i]))$;
9: **for** each item i in TS-list **do**
10: **if** $length(ts\text{-}list(i)) < minSup$ **then**
11: Prune i from the TS-list;
12: **else**
13: Calculate $Per[i] = max(Per[i], (ts_{final} - TS_l[i]))$;
14: **if** $Per[i] > maxPer$ **then**
15: Prune i from the TS-list.
16: Consider the remaining items in the TS-list as periodic-frequent items (or 1-patterns) and sort them in *support* descending order. Let L denote this sorted list of items.
17: **for** each item i in TS-list **do**
18: Replace i's timestamp information with its complement information. That is, set $ts\text{-}list(i) = TS - ts\text{-}list$. Let us call this new TS-list as CTS-list for brevity.
19: Call PFPM-C(CTS-List).

Algorithm 2. PFPM-C(CTS-List)

1: **for** each item i in PFP-List **do**
2: Set $pi = \emptyset$ and $X = i$;
3: **for** each item j that comes after i in the CTS-list **do**
4: Set $Y = X \cup j$. Determine $ts\text{-}list(Y) = ts\text{-}list(X) \cup ts\text{-}list(j)$; Calculate $sup(Y) = |TS| - |ts\text{-}list(Y)|$. Set $per(Y) = 0$, $periods = \emptyset$ and $tempPer = 0$;
5: **if** ts_{final} is in ts-list **then**
6: Remove ts_{final} from ts-list
7: **for** $(i = 0; i < ts\text{-}list(Y).size() - 1; + + i)$ **do**
8: **if** $(ts\text{-}list(Y)[i+1] - ts\text{-}list(Y)[i]) == 1$ **then**
9: $tempPer+ = 1$;
10: **else**
11: Append $tempPer$ in to periods. Set $tempPer = 0$;
12: $per(Y) = max(periods) + 1$
13: **if** $sup(Y) \geq minSup$ and $per(Y) \leq maxPer$ **then**
14: Add Y to pi and Y is considered as periodic-frequent pattern;
15: Call $PFPM\text{-}C(pi)$;

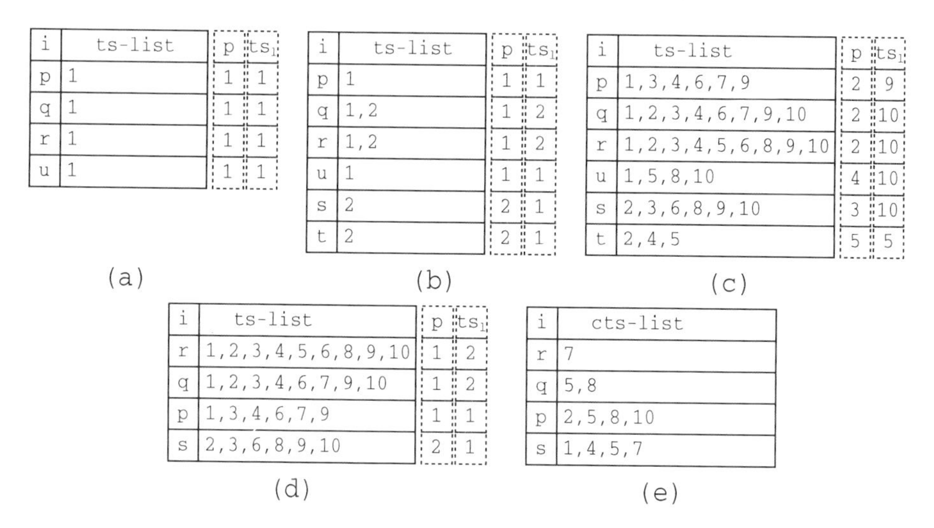

Fig. 1. Finding periodic-frequent items. (a) TS-list generated after scanning first transaction, (b) TS-list generated after scanning second transaction, (c) TS-list generated after scanning entire database, (d) Sorted list of periodic-frequent items, and (e) The complement list generated for each periodic-frequent item.

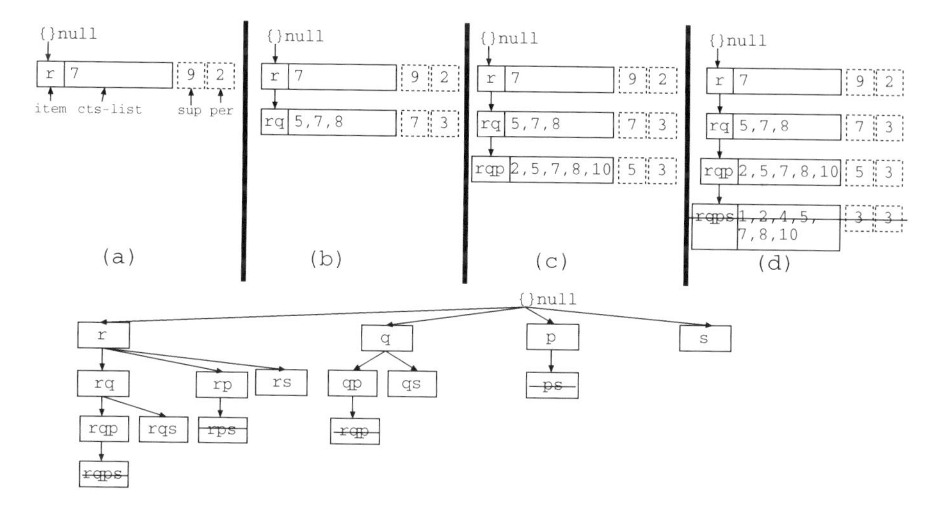

Fig. 2. Mining periodic-frequent patterns using depth-first search on the itemset lattice. (a) Start from the periodic-frequent item r. (b) As r is a periodic-frequent item, we visit its child node rq, constructs its cts-list, and determine its *support* and *periodicity*. (c) As rq is a periodic-frequent pattern, the algorithm visits its child node rqp, and determines rqp as a periodic-frequent patter. (d) $rqps$ child node is visited and pruned as it fails to be a periodic-frequent pattern. (e) The depth-first search on the itemset lattice of r, q, p and s items is shown.

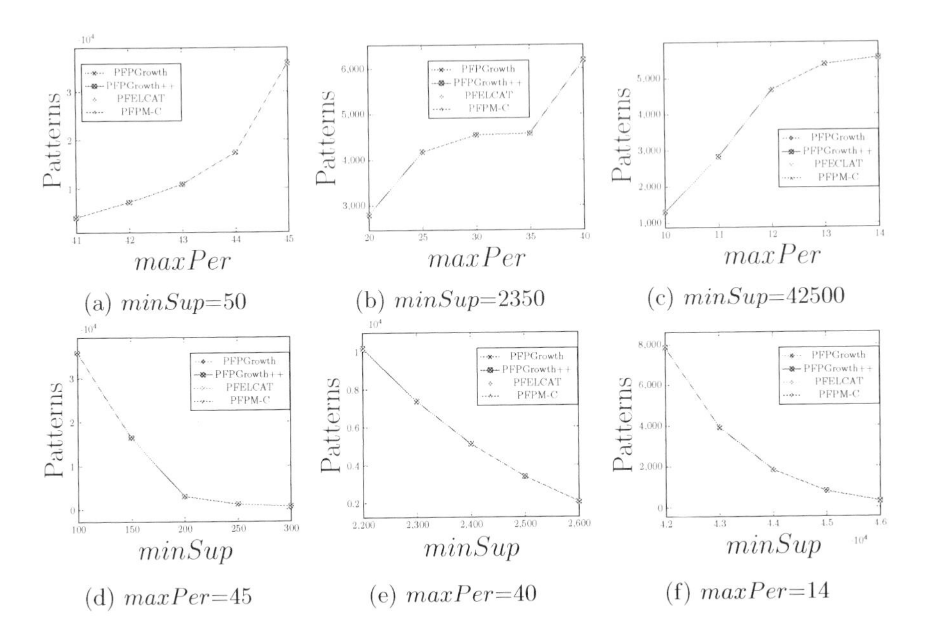

(a) $minSup$=50 (b) $minSup$=2350 (c) $minSup$=42500

(d) $maxPer$=45 (e) $maxPer$=40 (f) $maxPer$=14

Fig. 3. Number of periodic-frequent patterns generated in various databases

Titan RTX with 4608 CUDA cores and 24 GB GDDR6 dedicated GPU memory. This server machine has 64 GB of memory and runs on Ubuntu 20.04.4. The experiments have been conducted on real-world (Pollution, Chess, and Pumsb) databases (Fig. 1).

The **Pollution** database contained 1029 items (or stations) with 719 transactions. The *minimum*, *average*, and *maximum* transaction lengths are 11, 460 and 971, respectively. The **Chess** database is a high dimensional real-world database containing 75 items and 3196 transactions. The *minimum*, *average*, and *maximum* transaction lengths are 37, 37, and 37, respectively. The **Pumsb** database is a real-world database containing 2,113 items and 49,046 transactions. The *minimum*, *average*, and *maximum* transaction lengths are 74, 74, and 74 respectively.

5.2 Generation of Periodic-Frequent Patterns

Figure 3a, 3b, and 3c respectively show the number of periodic-frequent patterns generated in Pollution, Chess and Pumsb databases at different $maxPer$ values. The $minSup$ in Pollution, Chess and Pumsb databases has been set at 50, 2350, and 42500, respectively. It can be observed that $maxPer$ has a positive effect on the generation of periodic-frequent patterns as many patterns which were earlier considered as aperiodic were being considered as periodic-frequent patterns. **Please note that all the algorithms generate the same periodic-frequent patterns at any given** $minSup$ **and** $maxPer$.

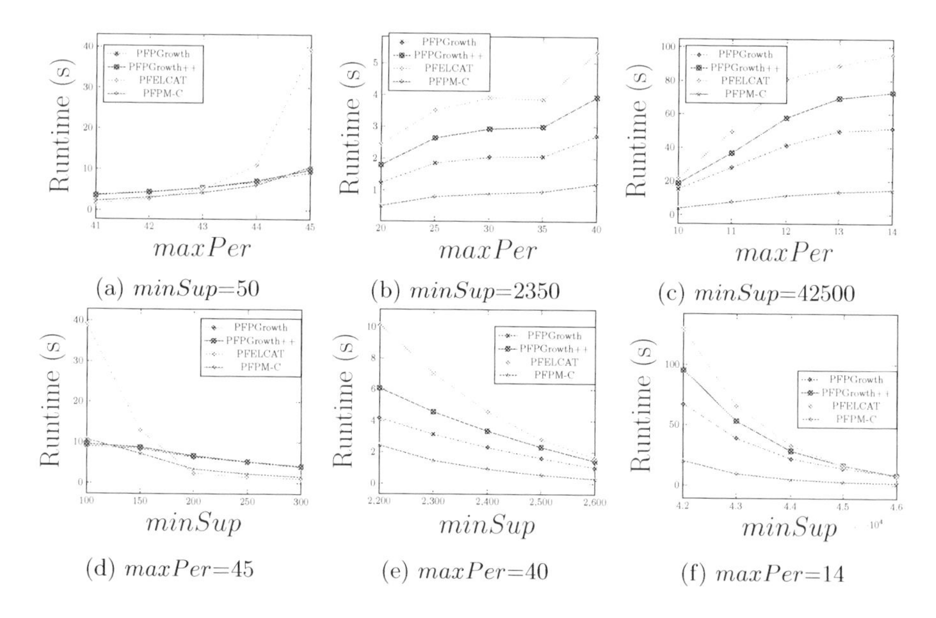

(a) $minSup$=50 (b) $minSup$=2350 (c) $minSup$=42500

(d) $maxPer$=45 (e) $maxPer$=40 (f) $maxPer$=14

Fig. 4. Runtime comparison of various algorithms on different databases

Figure 3d, 3e, and 3f respectively show the number of periodic-frequent patterns generated in Pollution, Chess and Pumsb databases at different $minSup$ values. The $maxPer$ in Pollution, Chess and Pumsb databases has been set at 45, 40, and 14, respectively. It can be observed that an increase in $minSup$ has a negative effect on the generation of periodic-frequent patterns. It is because many patterns fail to satisfy the increased $minSup$.

5.3 Runtime Evaluation of the Algorithms

Figure 4a, 4b, and 4c respectively show the runtime requirements of PFP-growth, PFP-growth++, PF-ECLAT, and PFPM-C algorithms in Pollution, Chess and Pumsb databases at different $maxPer$ values. It can be observed that even though the runtime requirements of all the algorithms increase with the increase in $maxPer$, the PFPM-C algorithm completed the mining process much faster than the remaining algorithms in dense databases at any given $minSup$.

Figure 4d, 4e, and 4f respectively show the runtime requirements of PFP-growth, PFP-growth++, PF-ECLAT, and PFPM-C algorithms in Pollution, Chess and Pumsb databases at different $minSup$ values. It can be observed that even though the runtime requirements of all the algorithms decrease with the increase in $minSup$, the PFPM-C algorithm completed the mining process much faster than the remaining algorithms in dense databases at any given $minSup$. The PFPM-C algorithm was an order of magnitude time faster than the remaining algorithms in the Pollution database. More importantly, the PFPM-C algorithm was several times faster than the remaining algorithms, especially at low $minSup$ values.

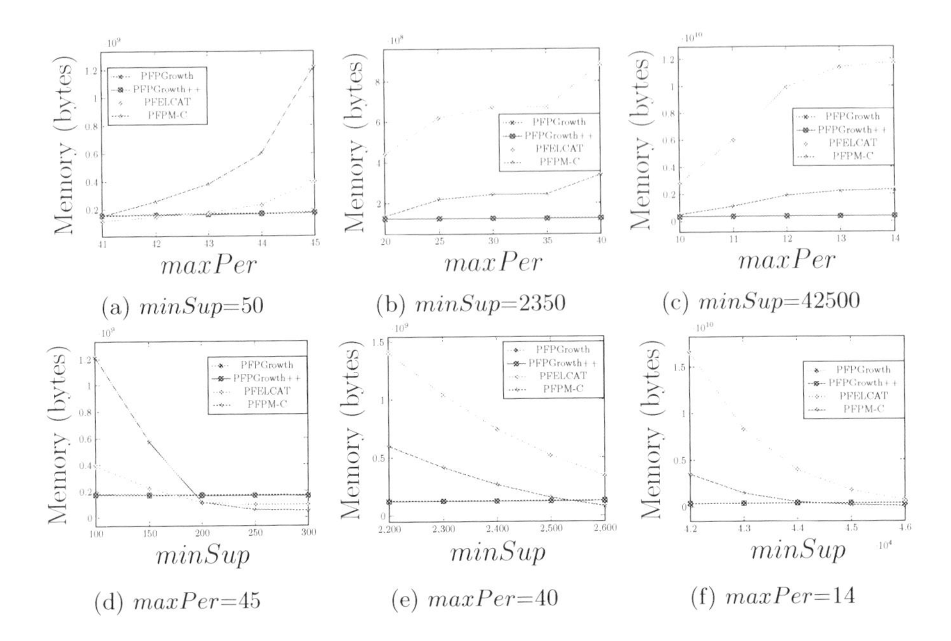

(a) $minSup$=50 (b) $minSup$=2350 (c) $minSup$=42500

(d) $maxPer$=45 (e) $maxPer$=40 (f) $maxPer$=14

Fig. 5. Memory comparison of various algorithms on different databases

5.4 Memory Evaluation of the Algorithms

Figure 5a, 5b, and 5c respectively show the memory requirements of PFP-growth, PFP-growth++, PF-ECLAT, and PFPM-C algorithms in Pollution, Chess and Pumsb databases at different $maxPer$ values. The following observations can be drawn from these figures: (i) increase in $maxPer$ increases the memory requirements of all the algorithms. It is because the algorithms will be generate and storing more number of periodic-frequent patterns. (ii) Our algorithm has consumed more memory in the Pollution database it is because it is a sparse database and not a dense database (unlike Chess and Pumsb databases).

Figure 5d, 5e, and 5f respectively show the memory requirements of PFP-growth, PFP-growth++, PF-ECLAT, and PFPM-C algorithms in Pollution, Chess and Pumsb databases at different $minSup$ values. It can be observed that though an increase in $minSup$ resulted in the decrease of memory requirements for all the algorithm.

Overall, our algorithm with moderate memory requirements was observed to be 2 to 4 times faster than the state-of-the-art algorithms.

6 Conclusions and Future Work

This paper explored the notion of set complements and proposed an efficient algorithm, called PFPM-C, to find periodic-frequent patterns in a dense database. A novel property has been introduced in this paper to determine the *periodicity* of a pattern from its complement temporal information. The performance of the PFPM-C was verified by comparing it with other algorithms on different real-world databases. Experimental analysis shows that PFPM-C exhibits high performance in periodic-frequent pattern mining and can obtain all periodic-frequent patterns faster and with moderate memory usage against the state-of-the-art algorithms. In particular, our algorithm was found to be 2 to 4 times faster than the existing algorithms.

Future work may be expanded as follows, but the scope is not limited: We would like to extend our algorithm to the distributed environment to find periodic-, partial- and fuzzy periodic-frequent patterns in very large temporal databases. In addition, we would like to investigate novel measures or techniques to reduce further the computational cost of mining the periodic-frequent patterns.

References

1. Agrawal, R., Imieliński, T., Swami, A.: Mining association rules between sets of items in large databases. In: SIGMOD, pp. 207–216 (1993)
2. Amphawan, K., Lenca, P., Surarerks, A.: Mining top-K periodic-frequent pattern from transactional databases without support threshold. In: Papasratorn, B., Chutimaskul, W., Porkaew, K., Vanijja, V. (eds.) IAIT 2009. CCIS, vol. 55, pp. 18–29. Springer, Heidelberg (2009). https://doi.org/10.1007/978-3-642-10392-6_3
3. Anirudh, A., Kiran, R.U., Reddy, P.K., Kitsuregawa, M.: Memory efficient mining of periodic-frequent patterns in transactional databases. In: 2016 IEEE Symposium Series on Computational Intelligence, pp. 1–8 (2016)
4. Kiran, R.U., Kitsuregawa, M.: Novel techniques to reduce search space in periodic-frequent pattern mining. In: Bhowmick, S.S., Dyreson, C.E., Jensen, C.S., Lee, M.L., Muliantara, A., Thalheim, B. (eds.) DASFAA 2014. LNCS, vol. 8422, pp. 377–391. Springer, Cham (2014). https://doi.org/10.1007/978-3-319-05813-9_25
5. Luna, J.M., Fournier-Viger, P., Ventura, S.: Frequent itemset mining: a 25 years review. Wiley Interdiscip. Rev. Data Min. Knowl. Discov. **9**(6), e1329 (2019)
6. Ravikumar, P., Likhitha, P., Venus Vikranth Raj, B., Uday Kiran, R., Watanobe, Y., Zettsu, K.: Efficient discovery of periodic-frequent patterns in columnar temporal databases. Electronics **10**(12), 1478 (2021)
7. Tanbeer, S.K., Ahmed, C.F., Jeong, B.-S., Lee, Y.-K.: Discovering periodic-frequent patterns in transactional databases. In: Theeramunkong, T., Kijsirikul, B., Cercone, N., Ho, T.-B. (eds.) PAKDD 2009. LNCS (LNAI), vol. 5476, pp. 242–253. Springer, Heidelberg (2009). https://doi.org/10.1007/978-3-642-01307-2_24
8. Zaki, M.J., Gouda, K.: Fast vertical mining using diffsets. In: SIG KDD, pp. 326–335 (2003)

An Error-Bounded Space-Efficient Hybrid Learned Index with High Lookup Performance

Yuquan Ding[1], Xujian Zhao[1(✉)], and Peiquan Jin[2]

[1] Southwest University of Science and Technology, Mianyang, China
yuquanding@mails.swust.edu.cn, jasonzhaoxj@gmail.com
[2] University of Science and Technology of China, Hefei, China
jpq@ustc.edu.cn

Abstract. Learned index is a novel index structure that enhances traditional index structures like B+-tree with machine learning models. It uses a trained model to locate a key position directly without the costly root-to-leaf traversing in the conventional B+-tree. However, the performance of existing learned indices relies on the prediction accuracy of the model. Recently, to address this problem, some existing learned indices, such as PGM-index, set a threshold to ensure that the prediction error is below the threshold. However, the PGM-index introduced additional space costs because it used a bottom-up strategy to build a tree structure to maintain an error threshold at each level. In this paper, we aim to find out a better trade-off between prediction accuracy and space consumption of learned indices. In particular, we propose a new learned index called HLI (Hybrid Learned Index) that offers high prediction accuracy and searching performance with limited space costs. HLI constructs error-bounded leaf nodes to provide high prediction accuracy and searching performance. Meanwhile, it organizes inner nodes without error bounds to reduce extra space costs. Consequently, we demonstrate that HLI can deliver high searching performance with limited space overhead. We conduct experiments on various datasets to compare HLI with several existing indices, including PGM-index, RMI, and B+-tree. The results show that HLI outperforms PGM-index in lookup performance on all datasets by 1.5× on average. In addition, compared to PGM-index, HLI reduces the index size by up to 13× without decaying the lookup performance.

Keywords: Learned index · Hybrid index · Data distribution · Piecewise linear approximation

1 Introduction

Recently, learned indices have been proposed to map keys to their storage positions based on a pre-trained machine learning model [6]. The early learned index, RMI [6], used a normalization method to divide the data into several subsets

C. Strauss et al. (Eds.): DEXA 2022, LNCS 13427, pp. 216–228, 2022.
https://doi.org/10.1007/978-3-031-12426-6_17

and train the corresponding models on these subsets, respectively. However, RMI does not have an error bound for predicting key positions, which will lead to unstable searching performance on the index. To address this issue, FITing-tree [4] and PGM-index [3] introduce an error bound to specify the acceptable amount of the prediction error, namely the maximum distance between the predicted and the actual position of the lookup key. However, these indices require a fine-grained fitting of the data distribution to provide a sufficiently small error bound, which means that more linear models are required. Furthermore, these two learned indices generate a bottom-up tree structure. FIT treats the PLA-model as a leaf node of the B+-tree, while the PGM-index utilizes a multi-level structure consisting of linear models as the inner node to index the PLA-model. Therefore, a small error bound leads to a rise in the height of these tree-structured learned indices, which results in a large space cost.

On the other hand, the trade-off between the prediction accuracy and the space cost of the tree structure is difficult to control. In a tree-structured learned index, each query needs to traverse from the root node to the leaf node, which inevitably introduces additional overhead at query time because the selection of lower-level paths requires multiple key comparison operations. Although the space cost can be reduced by increasing the error threshold, it also requires determining the correct position of the lookup key over a larger range. Thus, the time spends to return the correct position using the traditional algorithm dominates the overall query time.

In this paper, we propose a new error-bounded Hybrid Learned Index (HLI) that can deliver high searching performance with limited space overhead. HLI uses a two-level Recursive Model Index (RMI) as the routing structure for locating the PLA-model. Since our routing structure has no error bound, there is no need to recursively build the upper-level structure to share the overhead caused by the error bound of the previous level, just like in other data-aware learned indices. For any lookup operation, this structure requires only one key comparison operation to locate the underlying model with a small error bound corresponding to the lookup key. Overall, HLI provides a better trade-off between prediction accuracy and space consumption. The main contribution of this paper can be summarized as follows.

- We propose a hybrid learned index structure, HLI, that adopts error-bounded leaf nodes to provide high prediction accuracy and searching performance but organizes inner nodes without error bound to reduce extra space costs. With such a hybrid node design, HLI can reduce space overhead without sacrificing prediction accuracy and lookup performance.
- We demonstrate that HLI can achieve better trade-offs between space costs and time performance because the error bound of HLI is easy to be tuned to meet the space or time requirement in real applications.
- We conduct experiments on various datasets and compare HLI with several existing indices, including the conventional B+-tree and two learned indices, namely PGM-index and RMI. The results show that HLI outperforms all the competitors in both space and time efficiency. Particularly, HLI achieves 1.5×

faster lookup time than PGM-index. Also, HLI reduces the index size by up to 13× compared to PGM-index without worsening lookup performance.

2 Related Work

The introduction of the learned index [6] has opened up new research directions in the field of indices, which inspires a series of works. ASLM [8] divides data based on the similarity between data points to train more accurate models. AIDEL [7] eliminates inter-model dependencies of RMI by assigning liner models adaptively according to the data distribution. FITing-tree [4] uses a piecewise linear model to approximate the data distribution and indices these models by B+-tree, compressing the index size with a performance similar to B+-tree. Inspired by the FITing-tree, the PGM-index [3] uses the optimal algorithm to generate the minimum number of linear models and replaces the non-leaf node with linear models.

Since the original Learned Index does not support insert operations, the subsequent proposed learned indices are dedicated to solving the issue. ASLM, AIDEL, and FITing-tree all use extra buffer to store new data, while the PGM-index uses the idea of LSM-tree [11], where each insertion may lead to a series of subsets merging. ALEX [2] supports the insert operations by using a model-based insertion method that inserts directly into the gapped array based on the predicted location. In addition, to support dynamic workloads, ALEX uses an adaptive node splitting mechanism. To address the problems caused by inaccurate predictions in ALEX and the PGM-index, LIPP [14] uses a model-based insert but eliminates the last mile search by creating a new node to hold the conflicting keys that are mapped into the same position. Meanwhile, COLIN [16] supports query and insert operations in a cache-friendly way.

In addition, there are some learned indices that have attracted considerable attention. CDFShop [10] is a tool for exploring and optimizing RMI. XIndex [12] proposes a two-phase compaction scheme to handle concurrent operations. SIndex [13] is a concurrent learned index specifically optimized for variable-length strings. RadixSpline [5] is a single-pass learned index dedicated to building index structures quickly. Additionally, Marcus et al. [9] proposed a unified benchmark for comparing learned indices with traditional indices.

3 Hybrid Learned Index

3.1 Overview

The HLI is a pure learned index that consists of two modules (see Fig. 1), namely a PLA-model with an error bound that ultimately gives the predicted positions of the lookup key and an RMI without an error bound used to index the PLA-model. The characteristic of this hybrid learned index is that it combines the respective advantages of models with and without error bound.

The first module of the HLI is a Piecewise Linear Approximation model (PLA-model). Since real-world datasets often have complex data distributions, a single linear model is clearly insufficient to fit them. Therefore, we compactly capture the trends in the data distribution by using a series of linear regression models. Meanwhile, we use a greedy algorithm discussed in the PGM-index [3] to compute the optimal PLA-model with the minimum number of models. This algorithm ensures that the maximum prediction error of each model does not exceed a given threshold. Thus the prediction error can be efficiently corrected to return the actual position of the lookup key by using a binary search.

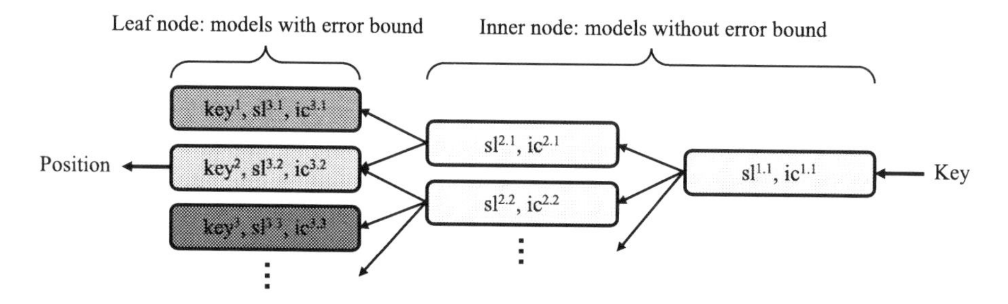

Fig. 1. The hybrid learned index.

The second module of HLI is a two-level Recursive Model Index (RMI) that indices the optimal PLA-model. To minimize the overhead caused by traversing the tree structure, we train RMI on the sorted keys derived from each segment in the optimal PLA-model. By adopting this structure without error bound, we avoid the requirement of recursively constructing a tree-like structure to maintain the error bound at each level, thus greatly reducing the space occupancy.

Generally, we choose the two-level RMI as the root node and the PLA-model as the leaf node to predict the position of the lookup key. There are three major benefits to using this new-type learned index. First, HLI only needs a single compare operation to reach the leaf node, thus greatly reducing the traversal time in internal levels. Second, since the number of additional operations of HLI is fixed to one, there is no significant degradation in its performance as the data amount increases. Third, HLI can provide a clear trade-off between lookup time and index size due to its simple structure. In the next two subsections, we will discuss these two basic modules of HLI in detail.

3.2 Leaf Nodes with Error Bound

In this section, we describe how to efficiently build an optimal PLA-model for predicting an approximate location of the lookup key, and how this approximation model provides fast query operation and concise storage.

A segment s is a triple $(key, slope, intercept)$ consisting of two floats and one key, which computes a predicted position of a lookup key through the function $f_s(key) = key \times slope + intercept$. It can be seen that a segment provides a $O(1)$ query time by direct computation and $O(1)$ space occupancy. Besides, the distribution of real-world datasets may be extremely complex, so it is difficult to learn the exact mapping from the key to its position. To this end, our goal is to approximate a function rather than learn a precise one. Therefore, the prediction of each model is considered as an approximation, i.e., there may be a certain distance between the predicted position and the actual position.

Note that a linear segment is not a valid approximation of the whole complex data distribution. Although any complex function can be used to approximate the index, the cost of modeling the accurate function is too expensive, and the computation of query operation is more complex. Besides, the piecewise linear function is efficient enough and dramatically reduce the cost of building such an approximation. Thus we train a set of linear models to approximate any complex function.

There exists a greedy algorithm [15] that computes the minimum number of segments in linear time and space, i.e., the optimal PLA-model. The core idea of this algorithm is to represent as many keys as possible with a linear function whose prediction error does not exceed a given error threshold ε. For example, if the preset error is 32, then the distance between the predicted position and the actual position of the lookup key does not exceed 32. In the construction process, new keys are continuously read into a subset until the newly added keys make such a function non-existent in that subset. Eventually, the algorithm generates a set of piecewise linear models to approximate the data distribution. By ensuring that the distance between the predicted position and the actual position of any key does not exceed the given error threshold, the range of the final binary search is effectively limited to 2ε.

3.3 Inner Nodes Without Error Bound

In order to minimize the time spent on traversing the tree structure before reaching the leaf node, there are two factors that need to be taken into account. First, the learned indices are the optimal choice for indexing the optimal PLA-model, because the query performance of a learned index is much better than that of a traditional tree index as mentioned above. Second, since any supervised learning method is only an approximation of the desired result, the predicted result at each level of the learned index needs to be corrected immediately. Therefore, we must minimize these additional operations.

Aimed at the two factors, RMI has significant advantages. Intuitively, each model in RMI provides $O(1)$ query time. Furthermore, without additional comparison operations, the model in the upper level directly selects the model in the lower level. More importantly, this structure only needs to correct the prediction result given by the last level, which means it needs only one key comparison process.

Therefore, we use the two-level RMI to replace the original recursive index structure in the PGM-index. In the process of building such an index (see Fig. 2), the minimum number of segments is first generated over the whole dataset. As mentioned in Sect. 3.2, each segment stores the first key of the subset it covers. We first extract these keys to create a training set, and then train a single linear model as the first level of HLI on the entire training set, this model $f(x)$ gives a predicted

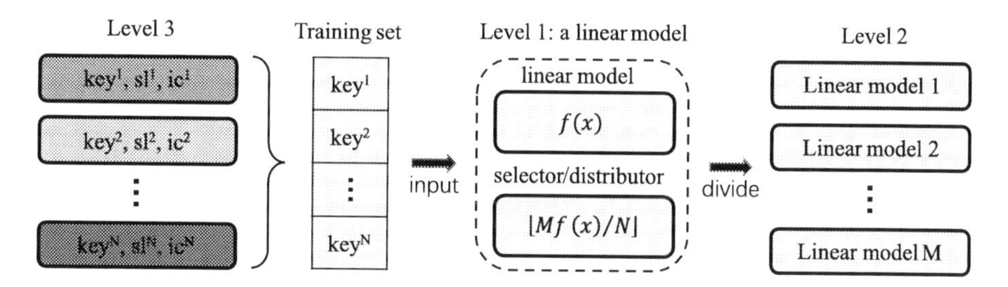

Fig. 2. Constructing the three-level structure of HLI.

position, in which x is the lookup key. The first level uses a normalization method to divide the dataset. This method can be described as follows:

$$\lfloor Mf(x)/N \rfloor \tag{1}$$

where N represents the total amount of keys, and there are M models in level 2. Given this equation, the model in the first level provides a prediction in the range from 0 to $M - 1$ and divides the dataset into some subsets based on this prediction. After that, the linear models are trained on each subset, respectively, to form the second level of RMI. Each model in this layer predicts the position of the segment corresponding to a given key. Eventually, we find the most reasonable number of models on the second level by a grid search to ensure that the trained RMI structure has the minimum error. Each model in our design only stores two parameters, namely slope and intercept.

Algorithm 1. Lookup Algorithm

Input: k: the lookup key, segments: the array of segment, rmi: the root of HLI index
Output: the subscript of the key in the dataset array

```
 1: function LOOKUP(key)
 2:     segment_i ← get_segment(key)
 3:     pred_pos ← segments[segment_i].predict(key)
 4:     act_pos ← binray_search(pred_pos, pred_pos − ε, pred_pos + ε)
 5:     return act_pos
 6: end function
 7:
 8: function GET_SEGMENT(key)
 9:     pred_s ← rmi.predict(key)
10:     act_s ← exponential_search(pred_s)
11:     return act_s
12: end function
```

Now, a query operation of a lookup key k over HLI works as Algorithm 1 shows. Firstly, the first level of RMI, which is a single linear model, accepts a lookup key and directly selects a model in the second level. Secondly, the model in the second level predicts an approximate position of a segment which is responsible for

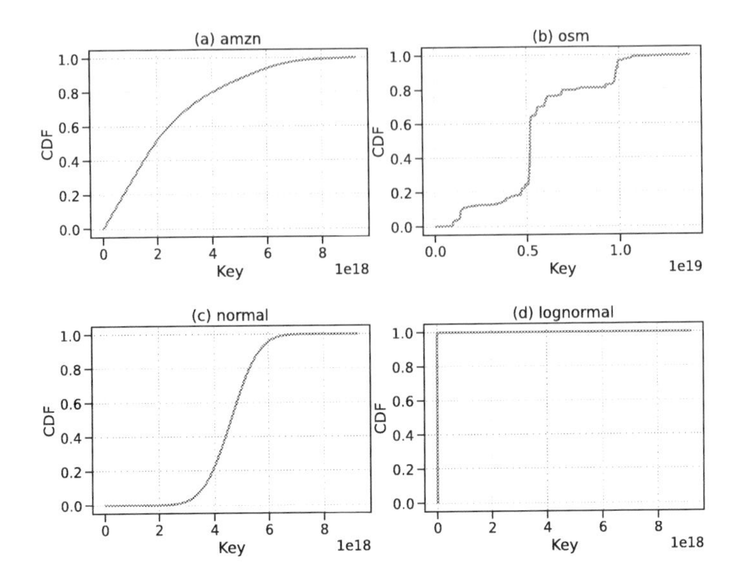

Fig. 3. CDF plots of each testing datasets.

the given key on the third level. Since the location given by this routing structure is an approximation, we need to perform an exponential search around the predicted position to find the correct one, namely, the rightmost segment s_j such that $s_j.key \leq k$. Eventually, a certain segment in the third level predicts the position of the lookup key k in the dataset. The real position is then found by a binary search in a range of size 2ε centered around the predicted position.

4 Performance Evaluation

4.1 Experimental Settings

We implement HLI in C++. Experiments are conducted on an Ubuntu Linux machine with a 2.1 GHz Intel Xeon and 64 GiB memory, only using a single thread. We compare HLI against three baselines.

- **PGM-index** [3]. The state-of-the-art data-aware learned index is implemented in C++.
- **RMI** [6]. The two-level Recursive Model Index (RMI) structure consists entirely of linear models. Note that the design of RMI is also used as an internal level to index the PLA-model in our proposal.
- **B+-tree** [1]. A standard B+-tree structure in traditional work.

As the PGM-index and HLI use a threshold ε that determines the error bound and impacts the lookup performance, the lookup time for an error threshold of ε can be modeled as $Latency(\varepsilon)$. Then, the ε' that makes the PGM-index and

Table 1. The error threshold ε' of the F-PGM index on all datasets

Dataset	Size			
	200M	400M	600M	800M
amzn	2^4	2^5	2^6	2^6
osm	2^4	2^5	2^4	2^7
normal	2^4	2^4	2^5	2^5
lognormal	2^4	2^4	2^4	2^4

Table 2. The error threshold ε' of HLI on all datasets

Dataset	Size			
	200M	400M	600M	800M
amzn	2^4	2^5	2^5	2^6
osm	2^4	2^4	2^5	2^5
normal	2^3	2^3	2^3	2^3
lognormal	2^3	2^3	2^3	2^3

HLI have the fastest lookup time is given by Eq. 2, where E is the set of ε and its range is $\{2^3, 2^4, ..., 2^{13}\}$.

$$\varepsilon' = \arg\min_{\varepsilon \in E} Latency(\varepsilon) \tag{2}$$

For a complete demonstration of the performance of the PGM-index, we included two versions of the PGM-index in all experiments. The PGM-index with the fastest lookup time is denoted as F-PGM ($\varepsilon = \varepsilon'$) while the one with a smallest threshold at the bottom level is denoted as S-PGM ($\varepsilon = 2^3$). In addition, we set $\varepsilon = 4$ for internal level as in [3]. Meanwhile, we tune the number of models in the second level of RMI to achieve the best lookup time on each dataset. And for the B+-tree, we use the default configuration as Bingmann's work [1].

In the paper, we use the following real-world datasets and synthetic datasets used in [9] for evaluation. Each dataset consists of unsigned 64-bit integer keys, and we generate 8-byte payloads for each key. To further evaluate the performance of each index on datasets of different sizes, we configure four different sizes for each dataset, ranging from 200 million keys to 800 million keys.

- amzn: book popularity data from Amazon. Each key represents the popularity of a particular book.
- osm: cell IDs from Open Street Map. Each key represents an embedded location.
- normal: a dataset is generated artificially according to the normal distribution.
- lognormal: a dataset is generated artificially according to the log-normal distribution.

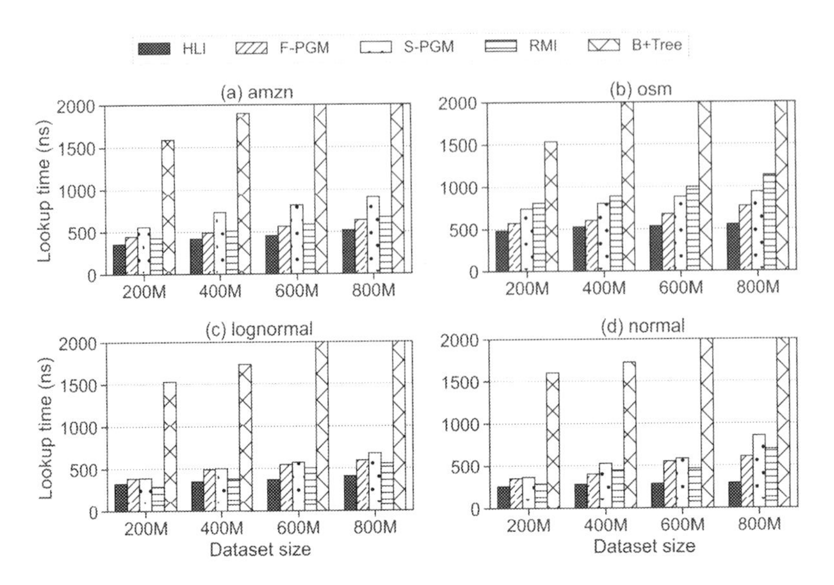

Fig. 4. Comparison of lookup performance.

The CDFs of each of these datasets are plotted in Fig. 3. For each datasets, we generate 10M random lookup keys. Indices are required to return search bounds that contain the lower bound of each lookup key. We perform this procedure five times and report the average lookup time.

4.2 Query Performance

For query performance evaluation, first, we compare our proposal with the traditional index. Then, we will discuss the performance evaluation of HLI with F-PGM, S-PGM, and RMI. Here, the parameter ε' of F-PGM and HLI on all datasets is shown in Table 1 and 2.

HLI vs. B+-Tree. We compared the HLI against the B+-tree, and the evaluation result is given in Fig. 4. Obviously, the HLI dominates this traditional index on all datasets. For example, on all four datasets with 200M keys, HLI achieves up to 4.4× higher lookup performance than the B+-tree on average. This is because the HLI uses the linear models as routing tables and only needs constant time to predict the position of the lookup key. Meanwhile, HLI uses the efficient computational method instead of traditional indirect key comparison operations, thus greatly speeding up query performance. Besides, HLI effectively limits the search range to a small area of size 2ε, while the nodes of the B+-tree need to pay an additional search cost as they grow larger. In terms of the index height, HLI has a much lower height than B+-tree. Specifically, the former has three levels, whereas the latter has seven levels on amzn-200M dataset, thus HLI experiences a shorter traversal time which is induced by a higher branching factor.

Table 3. Space costs of HLI and F-PGM when offering similar lookup performance.

Dataset	HLI			F-PGM		
	ε	Lookup time (ns)	Size (MiB)	ε	Lookup time (ns)	Size (MiB)
amzn	2^{10}	663.2	**0.1**	2^{6}	640	12.2
osm	2^{11}	749.6	**0.8**	2^{7}	771	10.5
normal	2^{10}	635.2	**0.009**	2^{5}	608	0.042
lognormal	2^{8}	596.8	**0.036**	2^{4}	595.4	0.083

Table 4. Space costs of HLI and S-PGM when offering similar lookup performance.

Dataset	HLI			S-PGM		
	ε	Lookup time (ns)	Size (MiB)	ε	Lookup time (ns)	Size (MiB)
amzn	2^{13}	924	**0.003**	2^{3}	909.8	259.7
osm	2^{13}	946.2	**0.27**	2^{3}	937	209.6
normal	2^{12}	808.4	**0.004**	2^{3}	853.6	0.084
lognormal	2^{9}	652.4	**0.025**	2^{3}	678.2	0.118

HLI vs. PGM-Index and RMI. Figure 4(a) to 4(d) show that HLI, on average, achieves up to 1.24×, 1.26×, 1.41×, 1.47× lower lookup latency compared with F-PGM on all four datasets at 200M, 400M, 600M and 800M, respectively. Because HLI requires only one error correction process before reaching the leaf node while PGM-index requires multiple error corrections of the predicted result using binary search at each level, HLI saves considerable time in traversing the tree structure.

In addition, we find that S-PGM, which intuitively has the fastest query time, actually works not well. The performance of S-PGM with the highest number of levels tends to be poor on real-world datasets. As shown in Fig. 4, the F-PGM outperforms the S-PGM in query performance on all datasets. For example, on amzn-800M dataset, F-PGM with $\varepsilon = 2^6$ achieves 1.42× better query performance than S-PGM with $\varepsilon = 2^3$. This is because S-PGM requires one more level than F-PGM to maintain a smaller error threshold. Therefore, S-PGM takes more time for traversing the tree structure.

Further, we conduct comparative experiments with RMI, namely, the two-level RMI with a linear model at each node. Figure 4 shows that HLI dominates RMI almost on all datasets. On average, HLI achieves up to 1.27×, 1.41×, 1.59×, 1.72× lower lookup latency than RMI on all four datasets at 200M, 400M, 600M, and 800M, respectively. Because RMI divides the data that do not satisfy the same distribution into a subset, the model error trained on the subset is large, so RMI needs a significant overhead in the final binary search. Unlike RMI, HLI effectively limits the range of binary search by capturing regularity trends in the data distribution.

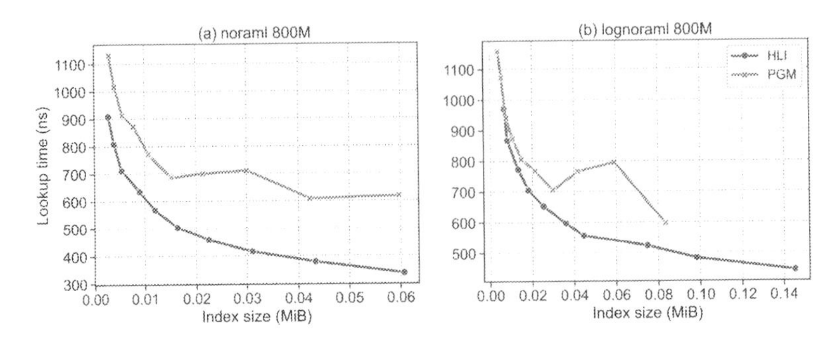

Fig. 5. Space-time trade-offs of HLI and PGM on two synthetic datasets.

4.3 Space Cost

Table 3 and 4 shows that on all datasets with 800M keys, HLI has a much lower space occupancy than F-PGM and S-PGM with similar lookup time. For example, on the osm-800M dataset, which is the largest and most complex dataset, the average lookup time of F-PGM that occupied 10.5 MiB is 771 ns, while HLI with a average lookup time of 750 ns occupied only 0.8 MiB, a 12.4× reduction in space footprint with a faster query time. Although F-PGM has a smaller error threshold, before reaching the bottom level, it requires four times key comparison operations to locate the path to the next level. In contrast, HLI only needs to perform these operations once. Basically, the performance bottleneck of the tree-structured learned index is not dependent on the ε, but on the number of key comparison operations between levels. Therefore, due to fewer additional operations, the HLI can select a larger error threshold, which greatly reduces the number of segments.

For RMI, on the same dataset, the average query time of the fastest RMI is 1135 ns that occupied 4.6 MiB while HLI with a average query time of 1062.4 ns occupied only 147 KiB, a 31.8× reduction in space footprint. The reason for the smaller space occupancy of HLI is that it adaptively assigns a minimum number of models according to the data distribution instead of presetting the number of models and inspecting the errors afterward.

For a traditional index, even the largest HLI achieves up to 69× smaller index size than the B+-tree. Because the B+-tree ignores the underlying data distribution, it misses the compression opportunity. In contrast, HLI uses a piecewise linear function to approximate the data distribution, each model only needs to store the parameters of the trained model.

4.4 Space-Time Trade-Off

As shown in Fig. 5, although both HLI and the PGM-index introduce a trade-off between lookup time and space occupancy, HLI has a clearer relationship between lookup time and space occupancy, whereas PGM-index has the opposite. For example, on the lognormal dataset with 800M keys, the lookup performance of the PGM-index becomes better as the index size increases, but this trend disappears when the size of the index increases to 0.03 MB. And when the index size increases from 0.06 MiB to 0.08 MiB, the lookup performance of the PGM-index shows a significant improvement again. This is because the PGM-index requires many additional operations during the query process, and these additional operations introduce a lot of uncertainty, so multiple factors that jointly affect the lookup performance of the PGM-index. On the contrary, HLI simplifies the steps of query operations as much as possible, such as reducing the height of the tree structure or removing redundant search operations between two levels, which makes the relationship between lookup performance and space occupancy much simpler, so that we can more easily control such trade-offs.

5 Conclusions and Future Work

In this paper, we propose a hybrid learned index that offers high prediction accuracy and searching performance with limited space costs. By combining the respective advantages of models with and without error bound, we decrease the number of additional operations required by other data-aware learned indices before reaching a leaf node from numerous to one. The experimental results on real-world and synthetic datasets show a great improvement of our proposal in lookup performance and space costs. In addition, HLI can offer a better trade-off between lookup time and space occupancy than existing learned indices.

For the future work, we plan to find a more efficient way to build the index to reduce the training overhead. Further, we will also consider extending HLI to support update operations.

Acknowledgements. This paper is supported by the Humanities and Social Sciences Foundation of the Ministry of Education (17YJCZH260), the National Science Foundation of China (62072419), the Sichuan Science and Technology Program (2020YFS0057).

References

1. Bingmann, T.: STX B+ Tree (2013). https://panthema.net/2007/stx-btree
2. Ding, J., et al.: ALEX: an updatable adaptive learned index. In: SIGMOD, pp. 969–984 (2020)
3. Ferragina, P., Vinciguerra, G.: The PGM-index: a fully-dynamic compressed learned index with provable worst-case bounds. Proc. VLDB Endow. **13**, 1162–1175 (2020)
4. Galakatos, A., Markovitch, M., Binnig, C., Fonseca, R., Kraska, T.: Fiting-tree: a data-aware index structure. In: SIGMOD, pp. 1189–1206 (2019)
5. Kipf, A., et al.: RadixSpline: a single-pass learned index. In: aiDM@SIGMOD, pp. 1–5 (2020)
6. Kraska, T., Beutel, A., Chi, E.H., Dean, J., Polyzotis, N.: The case for learned index structures. In: SIGMOD, pp. 489–504 (2018)

7. Li, P., Hua, Y., Zuo, P., Jia, J.: A scalable learned index scheme in storage systems. ArXiv abs/1905.06256 (2019)
8. Li, X., Li, J., Wang, X.: ASLM: adaptive single layer model for learned index. In: Li, G., Yang, J., Gama, J., Natwichai, J., Tong, Y. (eds.) DASFAA 2019. LNCS, vol. 11448, pp. 80–95. Springer, Cham (2019). https://doi.org/10.1007/978-3-030-18590-9_6
9. Marcus, R., et al.: Benchmarking learned indexes. Proc. VLDB Endow. **14**, 1–13 (2020)
10. Marcus, R., Zhang, E., Kraska, T.: CDFShop: exploring and optimizing learned index structures. In: SIGMOD, pp. 2789–2792 (2020)
11. O'Neil, P.E., Cheng, E.Y.C., Gawlick, D., O'Neil, E.J.: The log-structured merge-tree (LSM-tree). Acta Informatica **33**, 351–385 (2009). https://doi.org/10.1007/s002360050048
12. Tang, C., et al.: XIndex: a scalable learned index for multicore data storage. In: PPoPP, pp. 308–320 (2020)
13. Wang, Y., Tang, C., Wang, Z., Chen, H.: SIndex: a scalable learned index for string keys. In: APSys, pp. 17–24 (2020)
14. Wu, J., Zhang, Y., Chen, S., Wang, J., Chen, Y., Xing, C.: Updatable learned index with precise positions. Proc. VLDB Endow. **14**, 1276–1288 (2021)
15. Xie, Q., Pang, C., Zhou, X., Zhang, X., Deng, K.: Maximum error-bounded piecewise linear representation for online stream approximation. VLDB J. **23**, 915–937 (2014). https://doi.org/10.1007/s00778-014-0355-0
16. Zhang, Z., Jin, P., Wang, X., Lv, Y., Wan, S., Xie, X.: COLIN: a cache-conscious dynamic learned index with high read/write performance. J. Comput. Sci. Technol. **36**, 721–740 (2021). https://doi.org/10.1007/s11390-021-1348-2

Continuous Similarity Search for Text Sets

Yuma Tsuchida, Kohei Kubo, and Hisashi Koga(✉)

University of Electro-Communications, Tokyo 182-8585, Japan
{tsuchida,kubo,koga}@sd.is.uec.ac.jp

Abstract. We focus on searching similar data streams. Recent works regard the latest W items in a data stream as a set and reduce the problem to set similarity search. This paper uniquely studies similarity search for text streams and treats evolving sets composed of texts. We formulate a new continuous range search problem named the CTS problem to find all the text streams from the database whose similarity to the query becomes larger than a threshold ϵ. The CTS is challenging because it allows both the query and the database to change dynamically.

Keywords: Data stream · Similarity search · Text sets · Inverted index · Pruning

1 Introduction

Similarity search for data streams has become significant [5] in the area of information recommendation. Recent works search similar data streams by regarding a data stream as a set of data and find similar data streams by means of set similarity search. For instance, [7] characterizes a POI (Point of Interest) as a set of visitors.

Xu *et al.* [6] studied the "Continuous similarity search for Evolving Queries" (CEQ). Given a query set which consists of the latest W elements of a data stream, the CEQ tries to find the top-k most similar sets out of the n static sets in the database. Because the query set evolves dynamically, the top-k most similar sets must be updated continuously. By contrast, Koga and Noguchi [4] studied the problem "Continuous similarity search for Evolving Database" (CED). There, the query set Q becomes static and the database consists of n data streams. In the CEQ and the CED both, sets consist of items.

This paper formulates a new problem named "Continuous similarity search for Text Sets" (CTS). The CTS is quite different from the CED and the CEQ, since set elements are texts rather than items. In the CTS, the query set consists of the W texts in the sliding window (SW) of the query text stream X_Q. The CTS demands to search all the text streams in the database whose SWs are more similar to the query set than a threshold ϵ. The CTS models a scenario in which a user-based recommendation system searches users who contribute similar posts to the query user Q on some SNS. Here, one text stream X_A abstracts the posts from a user A. As time elapses, A adds a new post (i.e., a new text) to X_A.

C. Strauss et al. (Eds.): DEXA 2022, LNCS 13427, pp. 229–234, 2022.
https://doi.org/10.1007/978-3-031-12426-6_18

Thus, the query set is associated with the recent posts from Q. Thus, the CTS can discover users whose recent posts are similar to that of Q. The CTS problem is noteworthy, as it allows both the query and the database to evolve dynamically for the first time. We develop a fast pruning-based algorithm for the CTS.

2 Problem Statement

We formulate our new CTS problem. It characterizes a user A with a set of texts posted to A's text stream, where a text consists of more than one word. Let A_T be the SW of A's text stream at time T. A_T stores the latest W texts posted by A at T. To all the streams, exactly one new text is passed every time. Let us denote the text added to A's text stream at time t by a_t. Then, $A_T = \{a_{T-W+1}, a_{T-W+2}, \cdots, a_T\}$. When the time advances from T to $T+1$, the oldest a_{T-W+1} vanishes and a_{T+1} joins A_{T+1}.

In the CTS, the SW of the query user Q at T serves as the query set Q_T. The database D holds n text streams, i.e., n users. The CTS must find all the users U at T for whom the similarity $\mathrm{sim}(Q_T, U_T) \geq$ a threshold ϵ. Because Q_T and U_T evolve with T, we must update the search results continuously.

Let us define $\mathrm{sim}(Q_T, U_T)$ rigorously: Two texts o and o' are called *matchable* if their word sets satisfy $\frac{|o \cap o'|}{|o \cup o'|} \geq \beta$. To compute $\mathrm{sim}(Q_T, U_T)$, we first make a bipartite graph $G(Q_T, U_T)$ in which one text grows one node. In $G(Q_T, U_T)$, an edge connects a text $o_Q \in Q_T$ to a text $o_U \in U_T$, if they are matchable. Though it is convincing to define $\mathrm{sim}(Q_T, U_T)$ as the size of the maximum matching, the maximum matching is too heavy to compute every time. Thus, the CTS approximates it with the size of a maximal matching. By doing so, a text pair (o_Q, o_U) in the matching remains valid, until o_Q or o_U leaves the SW.

In the subsequence, to shrink the exposition, we only discuss the procedure to judge if a SINGLE USER U in D satisfies $\mathrm{sim}(Q_T, U_T) > \epsilon$. One straightforward method to solve the CTS is to construct a maximal matching $M(Q_T, U_T)$ completely and to check whether $|M(Q_T, U_T)| \geq \epsilon$ at every time T, while exploiting $M(Q_{T-1}, U_{T-1})$ as much as possible so as to create $M(Q_T, U_T)$ promptly. We refer to this simple algorithm as SIMPLE.

3 Our Algorithm for CTS

We present our fast pruning-based algorithm. At every time T, whereas SIMPLE constructs a maximal matching $M(Q_T, U_T)$ fully, our algorithm only creates a part of it which is just enough to judge if $\mathrm{sim}(Q_T, U_T) \geq \epsilon$. Our algorithm always categorizes all the texts in Q's SW into one of the three classes:

1. Q_M : A text in Q_M is confirmed to belong to the maximal matching.
2. Q_{NM} : A text in Q_{NM} is assured not to belong to the maximal matching.
3. Q_{UM} : A text in Q_{UM} is unsure to belong to the maximal matching.

Let o_Q be a text in Q_T. After our algorithm searches a matchable text for o_Q, o_Q is assigned to either Q_M or Q_{NM} depending on whether a matchable text for o_Q

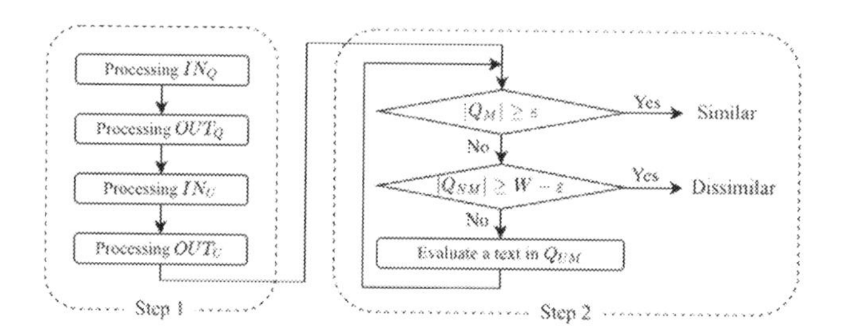

Fig. 1. Lazy evaluation algorithm

is found or not. We remark that it is quite time-consuming to verify $o_Q \in Q_{NM}$, since we must prove that any text in $U_T \setminus U_M$ is not matchable to o_Q. Here, U_M symbolizes a set of texts in U's SW that belong to the current matching. o_Q stays in Q_{UM}, if our algorithm has not searched a matchable text for o_Q yet. Hereafter, we often rewrite the phrase "search a matchable text for o_Q" with a shorter phrase "evaluate o_Q".

When the time advances from $T-1$ to T, $Q_T = \{q_{T-W+1}, q_{T-W+2}, \cdots, q_{T-1}, q_T\}$ is derived by adding q_T to Q_{T-1} and ejecting q_{T-W} from Q_{T-1}. Denote q_T by IN_Q and q_{T-W} by OUT_Q. IN_U and OUT_U are defined analogously for U. W.l.o.g., we assume that Q_{T-1} transforms to Q_T before U_{T-1} changes to U_T. Since SIMPLE builds the maximal matching fully, it always holds that $Q_{UM} = \emptyset$. Especially, SIMPLE evaluates IN_Q, as soon as IN_Q emerges. By contrast, our algorithm defers evaluating IN_Q as much as possible. Due to this nature, we name our algorithm LE (Lazy Evaluation).

The flowchart in Fig. 1 outlines how our LE operates when the time advances from $T-1$ to T. LE runs in (Step 1) and (Step 2).

Step 1: Update of the Text Classes

(Step 1) handles the four events, i.e., (1) arrival of IN_Q, (2) removal of OUT_Q, (3) arrival of IN_U and (4) removal of OUT_U and updates the classes of texts in Q_T. Due to space limitations, we only explain the first three events. One important task in Step 1 is to check if the texts which belonged to Q_{NM} at $T-1$ remain there at T with a little overhead by utilizing the result of past computations.

Processing of IN_Q: When IN_Q emerges, LE merely puts IN_Q into Q_{UM} without evaluating it immediately unlike SIMPLE.

Processing of OUT_Q: LE first discards OUT_Q. If $OUT_Q \in M(Q_{T-1}, U_{T-1})$, let o_U be a matching partner of OUT_Q. After discarding OUT_Q, o_U becomes unmatched. At this point, the texts in Q_{NM} at $T-1$ are uncertain to remain in Q_{NM} at T, because they may match to o_U. Therefore, LE compares o_U with each of them one by one. If a text o_Q kept in Q_{NM} at $T-1$ is not matchable to o_U, we have $o_Q \in Q_{NM}$ at T. Else if o_Q is matchable to o_U, the text pair (o_Q, o_U) joins $M(Q_T, U_T)$ and $o_Q \in Q_M$.

Processing of IN_U: The texts in Q_{NM} at $T-1$ are uncertain to remain in Q_{NM} at T, because they may match to IN_U. Therefore, LE compares IN_U with them. The rest of this procedure is exactly the same as the processing of OUT_Q.

Step 2: Judgment of Similarity between Q_T and U_T
After (Step 1), (Step 2) evaluates the texts in Q_{UM} incrementally until U_T is judged as similar or dissimilar, that is to say, until either $|Q_M| \geq \epsilon$ or $|Q_{NM}| > W - \epsilon$ holds true. Concretely, LE iteratively chooses a text, say o_Q, in Q_{UM} and searches a text matchable to o_Q from $U_T \backslash U_M$. Then,

- if o_Q did not match any text in $U_T \backslash U_M$, $o_Q \in Q_{NM}$ and $|Q_{NM}|$ increments.
- if o_Q matched to some text o_U in $U_T \backslash U_M$, (o_Q, o_U) joins in $M(Q_T, U_T)$ and $|Q_M|$ augments by 1.

To optimize Step 2, we should consider which text in Q_{UM} to evaluate first. Regarding this issue, we recommend the LIFO (Last-In-First-Out) order which chooses the latest text in Q_{UM} first. The LIFO order enjoys an advantage: Let o_Q (respectively o_U) be a text posted by Q (respectively U). Once the text pair (o_Q, o_U) belong to the matching, the matching keeps holding (o_Q, o_U) until either o_Q or o_U leaves the SW. Thus, the LIFO order leads to extending the lifetime of a text pair in the matching, so that it reduces the frequency of extra processing incurred whenever a text in the current maximal matching leaves the SW.

4 Usage of Inverted Indices

This section accelerates LE with the inverted index.

For the CTS, we must insert the newest text to and delete the oldest text from the inverted index, every time the SW moves forward. Thus, the cost to update the inverted index cannot be ignored. On the other hand, the inverted index has the advantage to shorten the time to evaluate a text by skipping text pairs with no common words.

First, let us recall that LE executes two types of text evaluation.
(Type 1): In processing IN_U and OUT_Q, Step 1 searches a text from Q_{NM} that is matchable to some text $o_U \in U_T$
(Type 2): Step 2 searches a text from $U_T \backslash U_M$ that is matchable to some text in Q_{UM}

Because (Type 1) searches from Q_{NM} and (Type 2) searches from $U_T \backslash U_M$, it appears natural to support (Type 1) with the inverted index for the query user Q and (Type 2) with the inverted index for U. However, the latter is almost unfeasible, as we must maintain n inverted indices for the database D composed of n users. Therefore, we decide to create only one inverted index for Q and speed up the text evaluation of (Type 1) only. We name this algorithm LE-Q (LE with the inverted index for Query).

For a word w, LE-Q arranges one inverted list l_w which records the IDs of Q's texts with w. To evaluate texts in the LIFO order, we realize l_w as a queue and insert the newest text with w to the head of l_w, which means that the oldest text with w is to be deleted from the tail of l_w.

LE-Q evaluates a text $o_U \in U_T$ in Step 1 as follows. Assume that o_U consists of m words $\{w_1, w_2, \cdots, w_m\}$. First, LE-Q collects the m inverted lists corresponding to the m words. Since all of them manage the text IDs in the decreasing

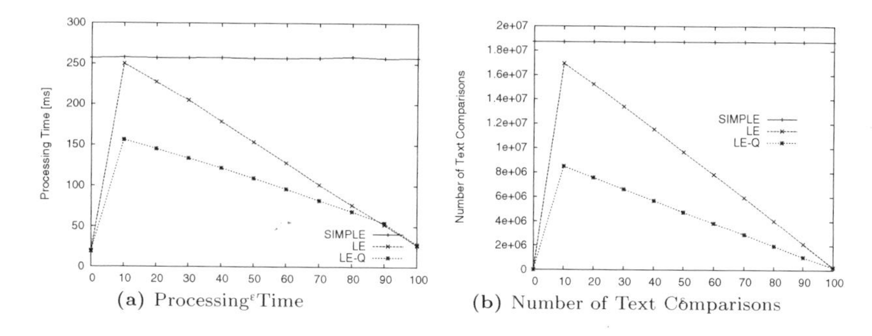

(a) Processing Time
(b) Number of Text Comparisons

Fig. 2. Processing time and number of text comparisons for cophir dataset

order of arrival times, LE-Q can retrieve texts from the m lists in chronological order one by one with DAAT [3].

5 Experiments

We evaluate our algorithms experimentally with a real dataset Cophir (Content-based Photo Image Retrieval) [1,2] which collects the metadata for the archive of Flickr image database. By grouping the tags attached to an image, we generate one text per image. Then, we characterize a user A with the texts associated with the images posted by A. We specify one user as the query user Q, while another 100 users constitute the database D.

First, we compare our LE and our LE-Q with the baseline algorithm SIMPLE. As for the experimental parameters, we change the user similarity threshold ϵ in the range $[0, 100]$, while fixing W to 100 and β to 0.1. Fig. 2(a) displays the total processing time and Fig. 2(b) shows the total number of text comparisons to compute the similarity between two texts, when the continuous range search repeats 1000 times. For any ϵ, LE-Q runs the fastest and LE runs faster than SIMPLE, which tells that the pruning in LE works effectively and that LE-Q shrinks the processing time surely owing to the inverted index for Q.

Next, we examine the effect of the two novel ideas in designing LE-Q: (1) We create the inverted index only for the query user Q to support Step 1, but not for the users in D to help Step 2. and (2) we pay attention to the order to evaluate texts. We give the highest priority to the latest text in Q_{UM}.

To confirm the effect of the first idea, we develop an algorithm which supplies the inverted indices not only for Q but also for all the users in D. This algorithm is referred to as LE-QD. LE-QD exploits the inverted indices for the users in D to evaluate the texts in Q_{UM} at Step 2. Figure 3 compares the processing time and the number of text comparisons between LE-Q and LE-QD. Interestingly, despite LE-QD does not compare text pairs frequently at all, LE-QD runs slower than LE-Q. This is because LE-QD keeps too many inverted indices so that the overhead to update them becomes too enormous. Thus, our strategy to build the inverted index only for Q is smart.

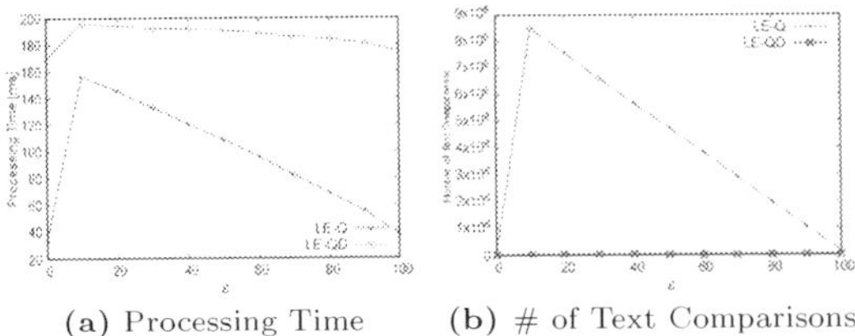

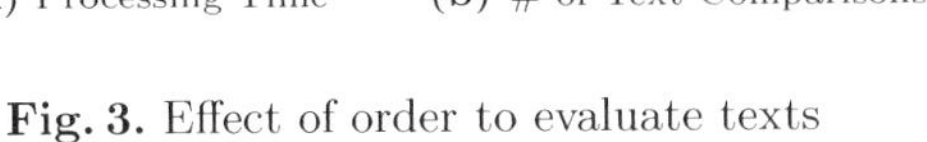

(a) Processing Time (b) # of Text Comparisons

Fig. 3. Effect of order to evaluate texts

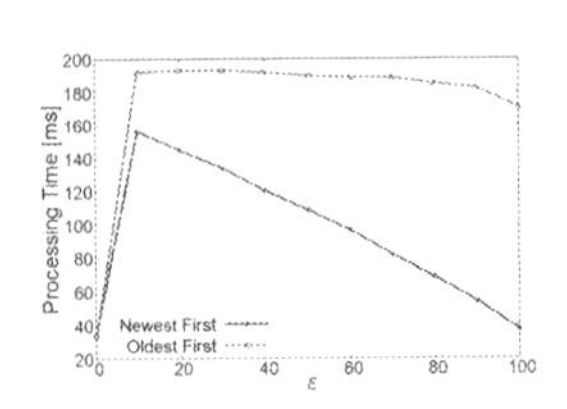

Fig. 4. Effect of order to evaluate texts

Next, we examine how the order to evaluate texts influences the processing time. While LE-Q gives the highest priority to the newest text, we prepare another variant of LE-Q which evaluates the oldest text first. Figure 4 compares their processing time, where we assign the legend "Newest-First" to the original LE-Q and the legend "Oldest-First" to the variant of LE-Q. The Newest-First runs much faster than the Oldest-First. Their gap enlarges as ϵ gets larger. The reason for this result can be inferred as follows: As ϵ gets larger, more users in D are judged as dissimilar, so finding matching text pairs becomes more useless. Nevertheless, the Oldest-First is forced to retrieve more matching text pairs in vain, as the matched text pairs expire in a short time.

Acknowledgments. This work was supported by JSPS KAKENHI Grant Number JP21K11901, 2021.

References

1. Bolettieri, P., et al.: CoPhIR: a test collection for content-based image retrieval. CoRR abs/0905.4627v2 (2009). http://cophir.isti.cnr.it
2. Bolettieri, P., Eusli, A., Falchi, F., Lucchese, C., Perego, R., Rabitti, F.: Enabling content-based image retrieval in very large digital libraries. In: Proceedings of the Second Workshop on Very Large Digital Libraries, pp. 43–50 (2009)
3. Fontoura, M., Josifovski, V., Liu, J., Venkatesan, S., Zhu, X., Zien, J.: Evaluation strategies for top-k queries over memory-resident inverted indexes. Proc. VLDB Endow. **4**(12), 1213–1224 (2011)
4. Koga, H., Noguchi, D.: Continuous similarity search for evolving database. In: Proceedings of the 13th SISAP, pp. 155–167 (2020)
5. Leong Hou, U., Zhang, J., Moruatidis, K., Li, Y.: Continuous top-k monitoring on document streams. IEEE Trans. Knowl. Data Eng. **29**(5), 991–1003 (2017)
6. Xu, X., Gao, C., Pei, J., Wang, K., Al-Barakati, A.: Continuous similarity search for evolving queries. Knowl. Inf. Syst. **48**(3), 649–678 (2015). https://doi.org/10.1007/s10115-015-0892-x
7. Yang, D., Li, B., Rettig, L., Cudré-Mauroux, P.: Histosketch: fast similarity-preserving sketching of streaming histograms with concept drift. In: 2017 IEEE International Conference on Data Mining (ICDM), pp. 545–554 (2017)

Exploiting Embedded Synopsis for Exact and Approximate Query Processing

Hiroki Yuasa, Kazuo Goda(✉), and Masaru Kitsuregawa

The University of Tokyo, Tokyo, Japan
{yuasa,kgoda,kitsure}@tkl.iis.u-tokyo.ac.jp

Abstract. Organizing data in structured forms is a mainstream approach to improve the efficiency of query processing on database. For example, a B+Tree index is widely employed to limit the search space for selective queries, speeding up the query processing. In this direction, this paper proposes *synopsis embedment*, an idea of incorporating synopsis (i.e. precomputed summary or abstraction information) into an existing data structure. Using the embedded synopsis, the query processing may obtain an exact or approximate answer without accessing all the concerned records in the data structure. Thus, the use of the synopsis may reduce the access cost, improving the efficiency of the query processing. This paper presents a design of efficient search algorithms on the synopsis-embedded B+Tree structure and experimentally clarifies its performance superiority.

Keywords: Synopsis embedment · Data structure · Approximate query processing

1 Introduction

Organizing data in structured forms has been a traditional but still mainstream approach to improve the efficiency of query processing on database. Suppose B+Tree [1,2], a data structure which is widely employed to organize a relational table or a secondary index in relational database. B+Tree is composed of two different types of nodes; leaf nodes store table records or index entries in the designated key order, whereas internal nodes store the navigation information to the leaf nodes. The navigation information allows the query processing to limit the search space, thus speeding up the processing if the query constraint is substantially selective along the key attribute.

In this technical direction, this paper proposes *synopsis embedment*, an idea of incorporating synopsis (i.e. precomputed summary or abstraction information) into an existing data structure. Using the embedded synopsis, the query processing may obtain an exact or approximate [3,5] answer without accessing all the concerned records stored in the leaf nodes. Thus, the use of the synopsis may reduce the access cost, improving the efficiency of the query processing. This paper presents a design of efficient search algorithms on the synopsis-embedded B+Tree structure and clarifies its performance superiority by showing our experiments in scenarios of exact and approximate query processing.

C. Strauss et al. (Eds.): DEXA 2022, LNCS 13427, pp. 235–240, 2022.
https://doi.org/10.1007/978-3-031-12426-6_19

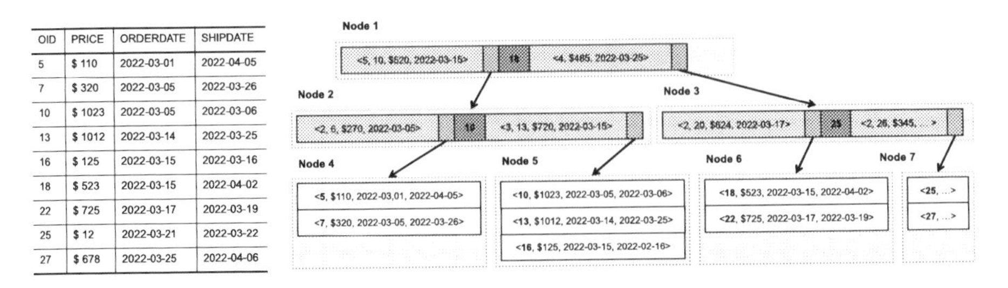

OID	PRICE	ORDERDATE	SHIPDATE
5	$ 110	2022-03-01	2022-04-05
7	$ 320	2022-03-05	2022-03-26
10	$ 1023	2022-03-05	2022-03-06
13	$ 1012	2022-03-14	2022-03-25
16	$ 125	2022-03-15	2022-03-16
18	$ 523	2022-03-15	2022-04-02
22	$ 725	2022-03-17	2022-03-19
25	$ 12	2022-03-21	2022-03-22
27	$ 678	2022-03-25	2022-04-06

Fig. 1. Example of synopsis embedment. A B+Tree structure organizes a given relational table with synopsis. The B+tree structure is composed of Nodes 1 to 7. The internal nodes, Node 1 to 3, contain synopses (colored in orange) as well as separation values (colored in blue) and references to its child nodes.

2 Synopsis Embedment and Synopsis-Aware Search

Synopsis Embedment. This section presents *synopsis embedment*, an idea of incorporating synopsis into an existing data structure, and search algorithms that exploit the embedded synopsis to improve the processing efficiency of queries. We take B+Tree as an example, but the same idea can be extended to other structures. B+tree is a major data structure employed in relational database. According to the standard design, a B+Tree structure is composed of multiple nodes (i.e., fixed-length pages) that are structured in a balanced tree form. Records are stored only in the leaf nodes in an sorted order of the designated key attribute. Each of other (internal) nodes contains one or more separation values of the designated key attribute and references to its child nodes (internal nodes or leaf nodes), each being in charge of one of the separated key space. Query processing starts at the root (top internal) node, and then recursively visits each of its child node only if its descendant nodes may hold records satisfying the query constraint until fetching all such records. This approach limits the search space, reducing the query processing cost.

Synopsis embedment extends the information that the internal node holds regarding its descendant nodes. Figure 1 illustrates an example case of synopsis embedment in B+Tree. Suppose a relational table presented in the figure, which mimics a sales order history; each record contains an order identifier (`ORDERID` as the primary key), price (`PRICE`), order date (`ORDERDATE`) and ship date (`SHIPDATE`) of a respective order. The figure also illustrates a B+Tree structure that embeds synopsis information. Each entry of the internal node contains synopsis (colored in orange) regarding the records stored in its descendant nodes. For example, the first synopsis in Node 2 covers all the records stored in Node 4, whereas the first synopsis in Node 1 covers Nodes 4 and 5. In this case, the synopsis comprises (1) the number of records, (2) the average value of the key attribute (`ORDERID`), (3) the average value of an non-key attribute (`PRICE`), and (4) the maximum value of an non-key attribute (`ORDERDATE`).

Algorithm 1. Synopsis-aware search (SAS)

```
1: SAS(n_r)                                         ▷ n_r: a reference to the root node
2: function SAS(n)
3:     N ← Fetch n
4:     if N is an internal node then
5:         for each e ∈ N do
6:             if e satisfies query constraint then
7:                 if e contains synopsis sufficient for a disjoint part of query then
8:                     R ← R ∪ e                    ▷ R: result buffer (initially R = ∅)
9:                 else
10:                    SAS(e's child reference)
11:    else
12:        for each e ∈ N do
13:            if e satisfies query constraint then
14:                R ← R ∪ e
```

Exact Search Exploiting Embedded Synopsis. Here, we present a search algorithm (synopsis-aware search; SAS) for exact query processing that utilizes the synopsis embedded in B+Tree to improve the processing efficiency.

Algorithm 1 presents the pseudo code of SAS, which extends a conventional depth-first search on B+Tree. The conventional algorithm examines internal nodes of B+Tree until it reaches the leaf nodes, where the matched records are retrieved to the result buffer to form the query answer. In contrast, if the embedded synopsis is mathematically sufficient to form a part of the query answer, SAS merely retrieves the synopsis information to the result buffer, but it no longer examines its child nodes (at lines 6 to 8). Hence, SAS can reduce the search cost even though it can still obtain an exact answer to the given query.

Approximate Search Exploiting Embedded Synopsis. SAS is useful to reduce the search cost, but it works only if the embedded synopsis is mathematically sufficient for a given query. Here, we presents the improved version (SAS+) that extends SAS to approximate processing. Even if the embedded synopsis may not be perfectly sufficient to a given query, it may be able to provide the approximate answer; this solution potentially offers the performance improvement to more queries, and may be acceptable for use cases where query responsiveness matters more than query exactness, such as rough statistics surveys.

This paper presents an *inter-attribute result composition* technique that composes separate results obtained from multiple synopsis-embedded B+Tree to offer an approximate query answer. Figure 2 illustrates its example. Assume that a separate synopsis-embedded B+Tree is organized for each attribute that is likely to appear in query constraints. The figure exemplifies two B+Tree structures, one by `ORDERID` and another by `SHIPDATE`. The search process is two-fold. First, SAS is applied to each of the B+Tree structures; specifically speaking, the B+Tree structure by `ORDERID` offers a partial result that ignores a query constraint on `SHIPDATE`, and vice versa. Second, the obtained partial results are mathematically composed to an approximate answer. How to compose partial

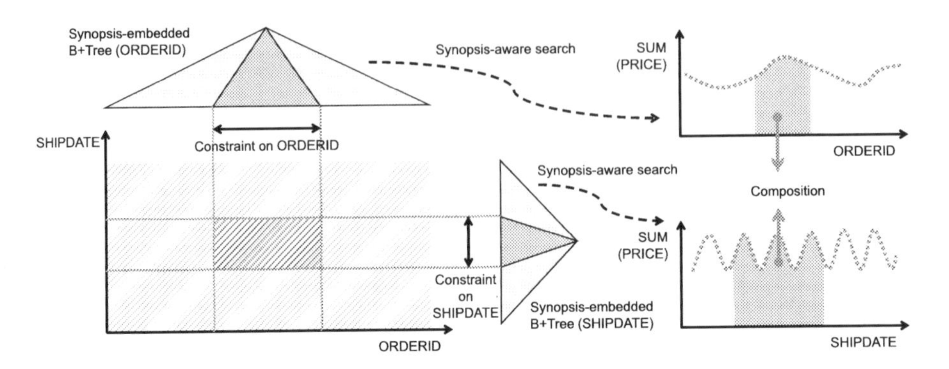

Fig. 2. Inter-attribute result composition. For a query having a composed constraint, separate synopsis-aware search offers a partial result and the inter-attribute result composition technique composes the results to obtain an approximate answer.

results depends on the query design. The example query is supposed to report a sum value; thus, arithmetic multiplication works for the composition. This mathematical composition assumes that the concerned attributes are statistically independent. If the attributes are correlated, the obtained result possibly contain non-negligible errors. Hence, the improved algorithm (SAS+) additionally decides whether the search process should utilize the synopsis (to be composed into an approximate answer), or obtain the answer by traversing to the leaf nodes, based on the correlation value between the concerned attributes.

3 Experiment

This section presents the experiments that we conducted to clarify how effectively the proposed synopsis-aware approaches worked. We implemented an experimental query engine in C; the engine incorporated the synopsis-aware search algorithms (SAS and SAS+) based on the well-known B+Tree design [2]. We tested the query performance by using this query engine and the TPC-H datasets. The datasets were generated by the standard *dbgen* tool (generating a uniform dataset) [6] and the revised *dbgen* tool (generating a skewed dataset) [7], both with the scale factor of 10, and they were loaded into the data structures managed by the experimental query engine. The zipf value was set to 1.0 for the skewed dataset. Four different test queries, Q1 to Q4, were tested on the datasets. All the experiments were performed on the two-socket server with dual Intel Xeon 6132 processors (each having 14 processing cores at 2.60 GHz), 96 GB main memory, 28 TB storage (composed of 24 hard disks by RAID6) running CentOS Linux 7.7. All the data structures were stored in the ext4 file system constructed in the storage space. The node size of B+Tree was set to 4096 bytes. We performed five trials of each test; this paper reports their average value.

For each dataset, we prepared two relation files, `LINEITEM (ORDERKEY)` and `LINEITEM (SHIPDATE)`, and embedded the following synopsis information: the number of records; the sum, maximum and minimum values of the key attribute (L_ORDERKEY and L_SHIPDATE); and the sum, maximum and minimum values of an additional attribute (L_EXTENDEDPRICE). This synopsis configuration yielded only 0.80% space overhead for the uniform and skewed datasets.

Exact Query Processing of Search Queries with Sngle Constraints. First, the paper presents that the synopsis-aware search (SAS) performs significantly faster when the embedded synopsis information is sufficient.

We executed the following test queries (Q1, Q2 and Q3) on LINEITEM (ORDERKEY).

```
Q1: SELECT SUM(L_EXTENDEDPRICE) FROM LINEITEM
    WHERE L_ORDERKEY BETWEEN 54000001, 60000001
Q2: SELECT MAX(L_EXTENDEDPRICE) FROM LINEITEM
    WHERE L_ORDERKEY BETWEEN 54000001, 60000001
Q3: SELECT MAX(L_SHIPDATE) FROM LINEITEM
    WHERE L_ORDERKEY BETWEEN 54000001, 60000001
```

Compared with the baseline case (having no synopsis embedded), SAS successfully reduced the necessary execution time for Q1 and Q2; the baseline took 0.549–0.553 s and 0.533–0.535 s respectively, whereas SAS only took 0.020–0.021 s for both, achieving speedups by factors of 26.3 to 27.5. In contrast, SAS could not speed up Q3, but worked comparably with the baseline case; the baseline took 0.649–0.650 s, whereas SAS also took 0.636–0.638 s. This difference was clearly because the embedded synopsis satisfied the constraint and derivative attributes of Q1 and Q2, but it did not for Q3.

Approximate Query Processing of Search Queries with Composite Constraints. Second, the paper presents that the synopsis-aware search plus approximate processing (SAS+) performs faster with moderate error rates.

As additional synopsis, we further embedded the correlation value between the key attribute and another attributes (L_ORDERKEY and L_SHIPDATE) for each file. This synopsis configuration yielded only 1.62% space overhead for the uniform and skewed datasets. Then, we executed the following test query (Q4).

```
Q4: SELECT SUM(L_EXTENDEDPRICE) FROM LINEITEM
    WHERE L_ORDERKEY BETWEEN 36000000, 42000000
    AND L_SHIPDATE BETWEEN 1992-01-01, 1992-12-31
```

Note that Q4 limits the search space along with the L_ORDERKEY and L_SHIPDATE attributes. LINEITEM (ORDERKEY) or LINEITEM (SHIPDATE) do not hold the synopsis information that perfectly offers an exact query result, but they offer an opportunity of approximate processing.

Compared with the baseline case (13.280 s and 13.280 s), SAS significantly reduced the necessary execution time (0.041 s and 0.040 s) for the uniform and skewed datasets. However, note that the query processing was approximate for Q4. SAS induced only negligible (0.2%) error for the uniform dataset, but it involved substantial (14.3%) error for the skewed dataset. This was seemingly because SAS assumed the uniformity of the attribute value distribution and applied the synopsis-based approximation to the entire space. By contrast, SAS+ utilized the inter-attribute correlation information to limit the search space to which the synopsis-based approximation was applied. SAS+ also performed significantly faster (0.521 s) than the baseline method with a small (3.0%) error for the uniform dataset. Interestingly, SAS+ offered the balanced property for the skewed dataset. Specifically speaking, SAS+ offered a much smaller (6.6%) error than SAS (14.3%) with much smaller execution time (6.684 s) than the baseline case (13.280 s).

4 Related Work and Conclusion

Related Work. Utilizing a precomputed synopsis has been actively studied for conventional exact query processing and approximate query processing [3,5]. Typically utilized synopses can be roughly grouped into histograms, wavelets and sketches. Recently, Liang et al. proposed building a tree of partial aggregates in advance [4]. This idea is close to our study. However, our paper further presents an idea of embedding a synopsis into an existing data structure such as B+Tree and an technique, inter-attribute result composition, to apply the approximation to queries having composite constraints.

Conclusion. This paper has proposed *synopsis embedment*, an idea of incorporating synopsis into an existing data structure. The embedded synopsis allows the query processing on the data structure to reduce the search cost. This paper has presented efficient search algorithms on the synopsis-embedded B+Tree structure. The presented experiments have clarified their performance superiority in scenarios of exact and approximate query processing. There remain open problems in the synopsis embedment; we would like to explore how to manage the freshness of embedded synopsis and the accuracy of query answers.

Acknowledgements. This work was in part supported by JSPS Grant-in-Aid for Scientific Research (B) JP20H04191.

References

1. Comer, D.: The Ubiquitous B-Tree. ACM Comput. Surv. **11**(2), 121–137 (1979)
2. Elmasri, R., Navathe, S.B.: Fundamentals of Database Systems, 3rd edn. Addison-Wesley-Longman, Prentice Hall,(2000)
3. Li, K., Li, G.: Approximate query processing: what is new and where to go? - A survey on approximate query processing. Data Sci. Eng. **3**(4), 379–397 (2018)
4. Liang, X., Sintos, S., Shang, Z., Krishnan, S.: Combining aggregation and sampling (nearly) optimally for approximate query processing. In: Proceedings of the ACM SIGMOD, pp. 1129–1141 (2021)
5. Liu Q.: Approximate query processing. In: Liu, L., Ozsu, M.T. (eds.) Encyclopedia of Database Systems. Springer, Boston (2009). https://doi.org/10.1007/978-0-387-39940-9_534
6. Transaction Processing Performance Council: TPC-H benchmark specification. http://www.tpc.org/tpch/. Accessed 25 Mar 2022
7. YSU Data Lab: TPC-H-Skew Public: TPC-H benchmark with skew factor enabled. https://github.com/YSU-Data-Lab/TPC-H-Skew. Accessed 25 Mar 2022

Advanced Analytics Methodologies and Methods

Diversity-Oriented Route Planning for Tourists

Wei Kun Kong[1,2](✉), Shuyuan Zheng[1], Minh Le Nguyen[2], and Qiang Ma[1]

[1] Department of Social Informatics, Kyoto University, Yoshida-Honmachi, Sakyo-ku, Kyoto 606-8501, Japan
kong.diison@gmail.com

[2] School of Information Science, Japan Advanced Institute of Science and Technology, 1-1 Asahidai, Nomi, Ishikawa 923-1292, Japan

Abstract. Touring route planning is an essential part of e-tourism, and significantly aids the development of the tourism industry. Several models have been proposed to formalize the touring route-planning problem with promising results. Most existing models consider only one tourist and return the same or similar results to tourists if they issue the same or similar queries when multiple users use the system simultaneously, which may cause congestion problems. In this study, we introduce a novel diversity-oriented touring route planning problem, and propose a multi-agent reinforcement learning approach with a dynamic reward mechanism. Experimental results show that our method significantly improves the diversity of the planned routes, reducing the bias of visiting locations and improving the total gain of all tourists.

Keywords: Diversity-oriented route planning · Multi-agent reinforcement learning

1 Introduction

Over the past few decades, tourism has experienced continued expansion and diversification to become one of the largest and fastest-growing economic sectors in the world.

Several touring route planning models have been proposed [8,11,13] to plan optimized routes for single tourists. The existing models can achieve promising performance for single tourists' requests, but may present tourists with prisoner's dilemma [15] in that the touring route planning model recommends homogenized routes to tourists when they have similar requests. When a large number of tourists with the same request are presented with homogenized planned routes from a model, it will inevitably bring great pressure to both the spots and the traffic, and cause tourists themselves to stack into congestion. This problem is represented in Fig. 1.

C. Strauss et al. (Eds.): DEXA 2022, LNCS 13427, pp. 243–255, 2022.
https://doi.org/10.1007/978-3-031-12426-6_20

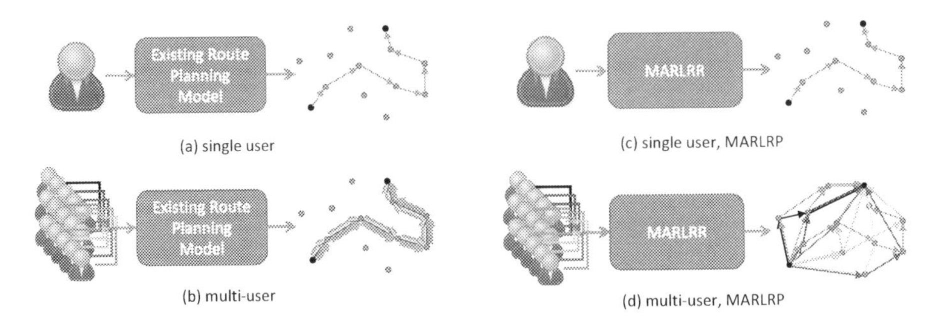

Fig. 1. An illustration of the existing route planning models and MARLRP.

As existing models [3,4,6,9,10,12,17] are designed for single tourist, they ignore the congestion problem caused by homogenized planned routes. According to [2], the congestion of spots in one period has a significant negative impact on the number of tourists in the next period, which may be owing to the decline in tourist experience caused by the congestion. A survey by the Kyoto government [18] in 2019 shows that people dissatisfied with traveling to Kyoto are the most dissatisfied owing to congested spots, accounting for 20.2%.

For tourists, the only concern is planning a route that meets their own conditions and can maximize gains. In most of the existing route planning task settings, only single tourist is considered with no capacity limitations for spots, which cannot meet real sightseeing circumstances. To address this issue, we introduced a novel diversity-oriented touring route planning task (DRP) in this paper. In DRP, the spots are limited in capacity, and spot congestion affects the reward of visiting the spots. The aim of DRP is to maximize the total reward for all tourists by recommending diverse routes to tourists.

As the first step in DRP, we propose a multi-agent reinforcement learning model for route planning (MARLRP). We evaluated the proposed model in terms of the diversity of the planned routes, total reward, total static reward of all tourists, and the balance between spots' attendance, and showed that MARLRP significantly outperformed the baseline. Figure 1 illustrates the effectiveness of MARLRP in improving the diversity of the planned routes.

Moreover, we proposed a nonlinear congestion-reward function according to a tourism economic model, which can reflect the function between tourists' gain and possible spot congestion well. We combined the simulated touring environment with real-world statistical data during training and testing, which significantly improved the applicability of the proposed method.

2 Related Works

The orienteering problem (OP) [10] is defined as follows. Given n nodes in an Euclidean plane each with a score $s(i) \geq 0$ (note that $s(1) = s(n) = 0$), find a route of maximum score through these nodes beginning at 1 and ending at n of length (or duration) no greater than the time budget T_{max}.

The team orienteering problem (TOP) [4] extends the single-competitor version of the OP to a multi-competitor version. A team of several competitors start at the same point, and each member of the team visits as many control points as possible within a prescribed time limit, and then ends at the finish point. Once a team member visits a point and receives a reward, other team members can no longer be awarded to visit this point. Each member of a team must select a subset of control points to visit, such that there is minimal overlap in the points visited by each member of the team, the time limit is not violated, and the total team score is maximized. However, TOP is not like an ordinary multiagent task. The relationship between a competitor and other competitors in TOP is similar to the relationship between a competitor and its avatars. There no competition among competitors, and they cooperate to complete the work of a single competitor in the OP.

The vehicle routing problem [17] is defined as follows: Let $G = (V, A)$ be a directed graph, where $V = \{0, \ldots, n\}$ is the vertex set and $A = \{(i, j) : i, j \in V, i \neq j\}$ is the arc set. Vertex 0 represents the depot, whereas the remaining vertices correspond to customers. A fleet of m identical vehicles with capacity Q is based on the depot. The fleet size is given a priori, or is a decision variable. Each customer i has a non-negative demand, q_i.

The diverse profit variants of the classic OP change the united fixed profits of the spots to different values. In OP with variable profits [7], the underlying assumption is that the collection of scores at a particular node requires either a number of discrete passes or a continuous amount of time at that node. The collected score of node i depends on the associated collection parameter $\alpha_i \in [0, 1]$. The proposed discrete and continuous models [7] were formulated as a linear integer programming model and a non-linear integer programming model, respectively. It was shown that the discrete model can be solved for instances with up to 200 nodes within 2 h of computational time. However, the continuous model requires more computation time for instances with 75 nodes.

The team orienteering problem with decreasing profits (DP-TOP) [1] extends TOP with decreasing profits. Each node's profit is a decreasing function of time, and owing to the complexity of the problem, the column generation approach (CG) is introduced to reformulate and calculate the lower and upper bounds of the initial DP-TOP integer programming model. The evolutionary local search (ELS) has also been proposed to solve this problem. TOP benchmark instances [4] were modified by adding variable profits to nodes, and almost all instances could be solved optimally by CG with the cost of computational time, whereas ELS was less competitive in terms of the quality of the solutions.

The multi-agent orienteering problem (MOP) is a multi-agent planning problem in which individual agents are self-interested and interact with each other when they arrive at the same nodes simultaneously. [5] studied MOP with time-dependent capacity constraints. Owing to capacity constraints, each node can only receive a limited number of agents simultaneously. If more agents are present, all agents will have to wait for some additional queuing time. Therefore, the main focus is to identify a Nash equilibrium in which individual agents cannot improve their current utilities through deviation. The problem was formulated as an integer programming model in a game-theoretic framework. Two solution approaches were proposed: a centralized approach with integer linear programming that computes the exact global solution, and a variant of the sampled fictitious play algorithm [14] that efficiently identifies equilibrium solutions. However, the first approach does not scale well, and can only solve for very small instances. Computational experiments can obtain equilibrium solutions in randomly generated instances.

The selfish OP (SeOP) [16] models the problem of crowd congestion at certain venues as a variant of the OP, the main issue of which is to provide route guidance to multiple selfish users (with budget constraints) moving through a venue simultaneously. SeOP combines OP with selfish routing, which is a game between selfish agents looking for minimum latency paths from the source to the destination along the edges of a network available to all agents. Thus, SeOP is a variant of MOP, in which agents have selfish interests and individual budget constraints. Similar to selfish routing, the Nash equilibrium is employed as the solution concept in solving SeOP. [16] proposed DIRECT, an incremental and iterative master-slave decomposition approach to compute an approximate equilibrium solution. Similar to existing flow-based approaches, DIRECT is scale invariant in terms of the number of agents. A theoretical discussion of the approximation quality and experimental result clearly shows that the non-pairwise formulation achieves the same solution quality as the pairwise formulation using a fraction of the number of constraints, and the master-slave decomposition achieves solutions with an adjustable approximation gap using a fraction of the full-path set.

There is no existing model that considers the impact of spot congestion on tourists' experience while assuming that the profits of tourists visiting spots are related to the degree of spot congestion, which can better conform to reality. Our proposed diversity-oriented tour route planning task fills this gap.

3 Methodology

3.1 Preliminary

The diversity-oriented touring route planning problem is defined as follows. Given a set of spots $\mathcal{S}$ and a set of tourists $\mathcal{T}$, the reward of a tourist visiting a spot varies with the congestion of the spot. Diversity-oriented touring route planning plans a route for every tourist to maximize the total reward of all tourists $\mathcal{R}$. The definitions of $\mathcal{S}$, $\mathcal{T}$, and $\mathcal{R}$ are provided in Definitions 1 to 7.

Definition 1 (Spot Set $\mathcal{S}$). *$\mathcal{S}$ is a set of spots $\mathcal{S} = \{s_1, \ldots, s_N\}$. Every spot s_n is represented by the location l_n, touring time $t_n^{(t)}$, capacity c_n, maximum reward $r_n^{(max)}$, minimum reward $r_n^{(min)}$, number of tourists in the spot num_n, and real touring reward r_n.*

$$s_n = \{l_n,\ t_n^{(t)},\ c_n,\ r_n^{(max)},\ r_n^{(min)},\ num_n, and\ r_n\}.$$

Definition 2 (Tourist set $\mathcal{T}$). *$\mathcal{T}$ is a set of tourists. Every tourist t_m is represented by the time budget $t_m^{(b)}$, start location $l_{start}^{(m)}$, speed v_m, start spot $s_{start}^{(m)} \in S$, and end spot $s_{end}^{(m)} \in S$.*

$$\mathcal{T} = \{t_1, \ldots, t_M\}$$

$$t_m = \{t_m^{(b)},\ l_{start}^{(m)},\ v_m,\ s_{start}^{(m)} \in S,\ s_{end}^{(m)} \in S\}$$

Definition 3 (Tourist t_m's route $\mathcal{S}^{(m)}$). *Tourist t_m's route $\mathcal{S}^{(m)}$ is an ordered sequence of spots $\mathcal{S}^{(m)} = \left\langle s_1^{(m)}, \ldots, s_k^{(m)} \right\rangle$. Tourists were prohibited from visiting the same spot twice: $\left\| \left\{ s_1^{(m)}, \ldots, s_k^{(m)} \right\} \right\| = k$. Tourists must visit their starting spot at the first and ending spots at the last: $s_1^{(m)} \equiv s_{start}^{(m)},\ s_k^{(m)} \equiv s_{end}^{(m)}$.*

$$\mathcal{S}^{(m)} = \left\langle s_1^{(m)}, \ldots, s_k^{(m)} \right\rangle$$

where, $\forall s_i^{(m)} \in \mathcal{S},\ \left\| \left\{ s_1^{(m)}, \ldots, s_k^{(m)} \right\} \right\| = k,\ s_1^{(m)} \equiv s_{start}^{(m)},\ s_k^{(m)} \equiv s_{end}^{(m)}$.

Definition 4 (Reward r_n). *The reward r_n earned by tourists visiting spot s_n is based on num_n, c_n, $r_n^{(min)}$, and $r_n^{(max)}$.*

$$r_n = f(num_n,\ c_n,\ r_n^{(min)},\ r_n^{(max)})$$

Definition 5 (Commuting time $t_{i,j,m}^{(c)}$). *The commuting time $t_{i,j,m}^{(c)}$ represents the time it takes tourist t_m to travel from spot $s_i^{(m)}$ to spot $s_j^{(m)}$.*

$$t_{i,j,m}^{(c)} \equiv t^{(c)}(s_i^{(m)}, s_j^{(m)}, v_m) \equiv t^{(c)}(l_i^{(m)}, l_j^{(m)}, v_m) = \frac{d\left(l_i^{(m)}, l_j^{(m)}\right)}{v_m}$$

Definition 6 (Time Constraint for t_m). *For any tourist, the total time cost of commuting and touring spots should not exceed the time budget of that tourist.*

$$t^{(c)}(l_{start}^{(m)}, l_1^{(m)}, v_m) + \sum_{i=1}^{k-1} t_{i,i+1,m}^{(c)} + \sum_{i=1}^{k-1} t_i^{(t)} \leq t_m^{(b)}$$

where, $s_i^{(m)} \in \mathcal{S}^{(m)},\ s_{i+1}^{(m)} \in \mathcal{S}^{(m)}$.

Definition 7 (Total reward $\mathcal{R}$). *The total reward $\mathcal{R}$ is the sum of all rewards earned by all tourists sequentially visiting all spots on their routes.*

$$\mathcal{R} = \sum_{m=1}^{M} \sum_{n=1}^{N} r_n \cdot \mathbf{1}_{\mathcal{S}^{(m)}}(s_n)$$

where, $\mathbf{1}_{\mathcal{S}^{(m)}}(s_n) = \begin{cases} 1, & if \quad s_n \in \mathcal{S}^{(m)} \\ 0, & if \quad s_n \notin \mathcal{S}^{(m)} \end{cases}$.

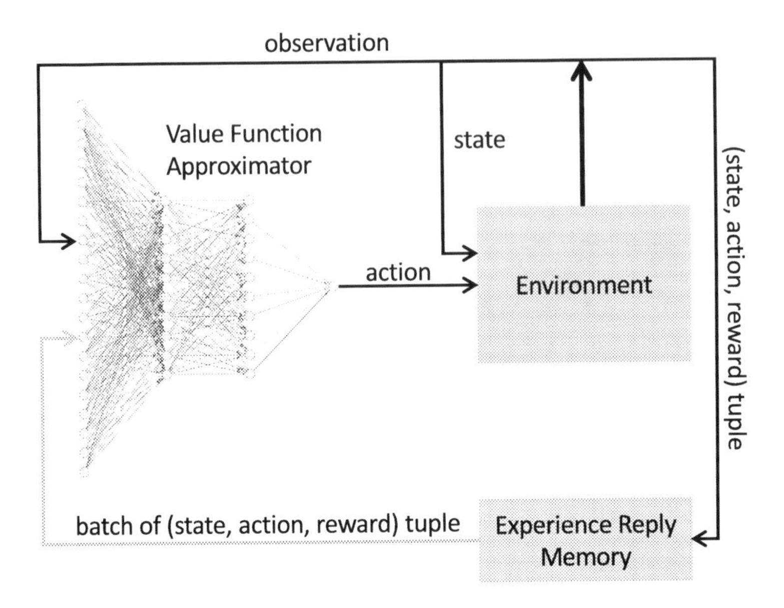

Fig. 2. Overview of MARLRP. In DRP, the state represents the current state of the tourist, including the information about the tourist such as its remaining time, and the information about the tourist's tour such as the spot the tourist just finished visiting. The action represents the action taken by the tourist, including going to the next spot and waiting. The reward represents the rewards a tourist receives after visiting a spot.

3.2 Model

In this section, we describe our proposed model and its use in planning optimal routes for multiple tourists to maximize the total gain. Our model can be roughly divided into two components: the value function approximator and environment. The first module is a 3-layer feed-forward neural network that learns to generate an action according to observations. The second module is a simulator of the real touring environment, which contains spots, tourists, constraints, and rules of how tourists receive rewards. Figure 2 and Algorithm 1 show the overall architecture of our model.

$$r_n = max\left(r,\ r_n^{(min)}\right) \tag{1}$$

where, $r = \begin{cases} r_n^{(max)} \cdot \cos\left(\frac{num_n}{2c_n}\pi\right), & 0 \leq num_n < 2c_n \\ r_n^{(min)}, & 2c_n \leq num_n \end{cases}.$

The environment contains information on 72 spots in Kyoto, a group of randomly initialized tourists, and incentive strategies for tourists visiting these spots. We adopt Eq. 1 as the basic reward function, which goes like Fig. 3. The basic assumption is that the greater the number of tourists in a spot, the greater

Algorithm 1: Overview of MARLRP Training

```
1  Initialize network Q;
2  Initialize target network Q̂;
3  Initialize experience replay buffer D;
4  Initialize Environment env;
5  Initialize Tourists T;
6  Initialize the frequency sync_freq of syncing model weights from Q to Q̂;
7  /* Sample phase
8  ε ← setting new epsilon with ε-decay;
9  Choose an action a from state s using policy ε-decay(Q);
10 Tourist in T takes action a, observed reward r, and next state s';
11 Save transition (s, r, a, s', done) to D;
12 /* Train phase
13 Sample a random minibatch of N transitions from D;
14 while i in range(N) do
15     if done_i then
16         y_i = r_i;
17     else
18         y_i = r_i + γ max_{a'∈A} Q̂(s'_i, a');
19     end
20 end
21 Calculate the loss L = 1/N Σ_{i=0}^{N-1} (Q(s_i, a_i) − y_i)^2;
22 Update Q using the SGD algorithm by minimizing the loss L;
23 Every sync_freq steps, copy weights from Q to Q̂;
```

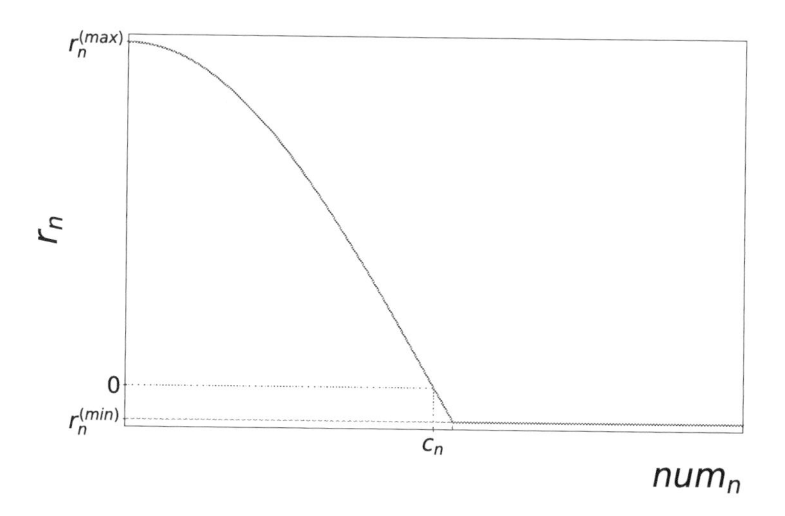

Fig. 3. Illustration of the reward function.

the expectation of congestion in the spot, and the rate at which the expectation of congestion will also increase as the number of tourists increases [2].

3.3 Real-Time Statistics of the External Environment

Real-time statistical information of the external environment was adopted from [19], including geographic location, number of tourists in spots at different time periods, and score of spots at different time periods. The scores of spots were calculated from the average scores of photos uploaded by Flickr users, and the number of people at different spots in different time periods was counted by the number of Flickr users who posted photos at the spots. In some time periods, the number of photos posted by Flickr users in many spots was zero, and therefore, we excluded the data in these time periods. For the case where the number of people in the spot was zero, we adopt Laplace smoothing, and add the minimum non-zero number of people to all spots' tourist numbers. We update the spot statistics after tourists perform a certain number of actions. As we only obtained available data in seven time periods, we cyclically updated the spot data randomly.

4 Experiment and Result

4.1 Experiment Setting

We conducted experiments based on three settings: identical, similar, and random settings. The details of each setting are as follows.

- Identical: The tourists have the same time budget, start location, start spot, and end spot with each other.
- Similar: The start spots and end spots for all tourists are selected from two sets containing spots close to each other. Tourists have the same time budget and start location.
- Random: The time budget, start location, start spot, and end spot of all tourists are generated randomly.

For experiments with similar and random settings, we tested the model thrice to eliminate the effects of randomness. $r_n^{(min)}$ was set to half the average of $r_n^{(max)}$ for all spots, and the capacity of the spot was set to two times of the number of people in the spot at this time.

We also trained a baseline model that shares the same settings as our proposed MARLRP model, except that the observation and reward function of the baseline is not based on r_n but on $r_n^{(max)}$ during training and evaluation. During evaluation, we recorded both the total reward and total static reward of all tourists.

4.2 Evaluation Metrics

We adopted the total reward ($\mathcal{R}$) to measure the total actual rewards gained by all tourists, total static reward ($\mathcal{R}_s$) to measure the total maximum rewards all tourists can obtain in an ideal situation, and the Gini coefficient of the percentage of spot attendance (G_s) to measure the balance between spots' attendance, and average edit distance of the planned routes (ED_r) to measure the diversity among the planned routes, as shown in Eq. 2, Eq. 3, Eq. 4 and Eq. 5, respectively.

$$\mathcal{R} = \sum_{m=1}^{M} \sum_{n=1}^{N} r_n \cdot \mathbf{1}_{\mathcal{S}^{(m)}}(s_n), \tag{2}$$

$$\mathcal{R}_s = \sum_{m=1}^{M} \sum_{n=1}^{N} r_n^{(max)} \cdot \mathbf{1}_{\mathcal{S}^{(m)}}(s_n) \tag{3}$$

$$G_s = \frac{\sum_{k=1}^{N} \sum_{l=1}^{N} \left| \frac{num_k}{c_k} - \frac{num_l}{c_l} \right|}{2 \sum_{k=1}^{N} \sum_{l=1}^{N} \frac{num_l}{c_l}} \tag{4}$$

$$ED_r = \frac{\sum_{k=l}^{N} \sum_{l=1}^{N} \text{lev}(\mathcal{S}^{(k)}, \mathcal{S}^{(l)})}{|\mathcal{T}|} \tag{5}$$

$$\text{where, } \text{lev}(\mathcal{S}^{(k)}, \mathcal{S}^{(l)}) = \begin{cases} |\mathcal{S}^{(k)}| & \text{if } |\mathcal{S}^{(l)}| = 0 \\ |\mathcal{S}^{(l)}| & \text{if } |\mathcal{S}^{(k)}| = 0 \\ \text{lev}(\text{tail}(\mathcal{S}^{(k)}), \text{tail}(\mathcal{S}^{(l)})) & \text{if } \mathcal{S}^{(k)}[0] = \mathcal{S}^{(l)}[0] \\ 1 + \min \begin{cases} \text{lev}(\text{tail}(\mathcal{S}^{(k)}), \mathcal{S}^{(l)}) \\ \text{lev}(\mathcal{S}^{(k)}, \text{tail}(\mathcal{S}^{(l)})) \\ \text{lev}(\text{tail}(\mathcal{S}^{(k)}), \text{tail}(\mathcal{S}^{(l)})) \end{cases} & \text{otherwise} \end{cases}$$

4.3 Experimental Result

Table 1 lists the results of the experiments with identical and similar settings. The planned routes of MARLRP and baseline are shown in Fig. 4. In both experimental settings, our proposed model can plan routes for tourists with a higher diversity than the baseline model. The visit distributions of MARLRP and baseline are shown in Fig. 6. When tourists issue the same query, the baseline model recommends the same path for tourists. Even when the users' queries are similar, the spots visited by tourists remain highly homogeneous. The high robustness of the baseline makes it difficult to avoid congestion in popular spots. In contrast, our proposed model enables tourists to visit spots in a more decentralized

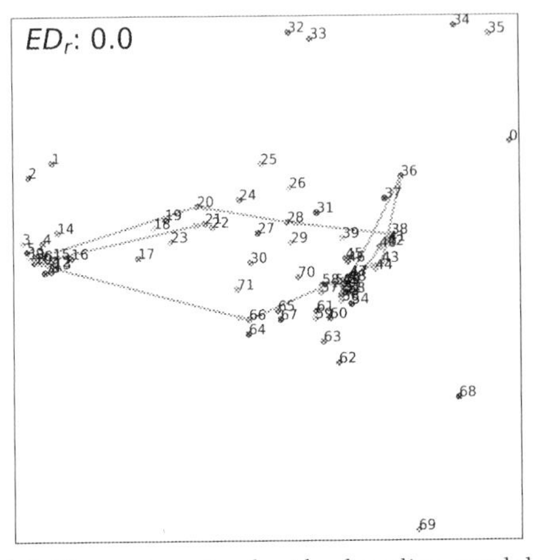

(a) Planned routes by the baseline model with identical setting and 100 tourists.

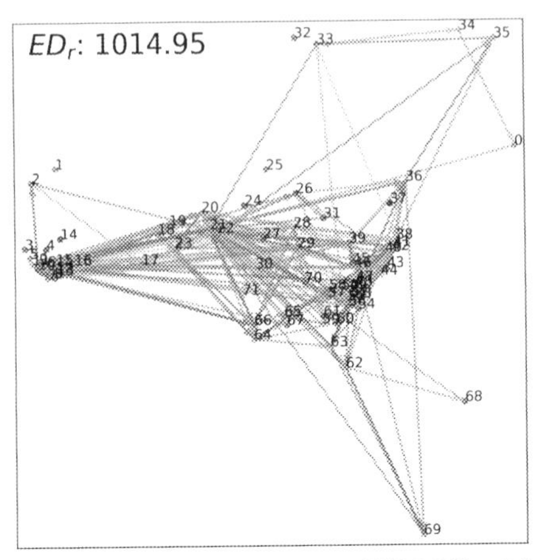

(b) the routes planned by MARLRP with identical setting and 100 tourists.

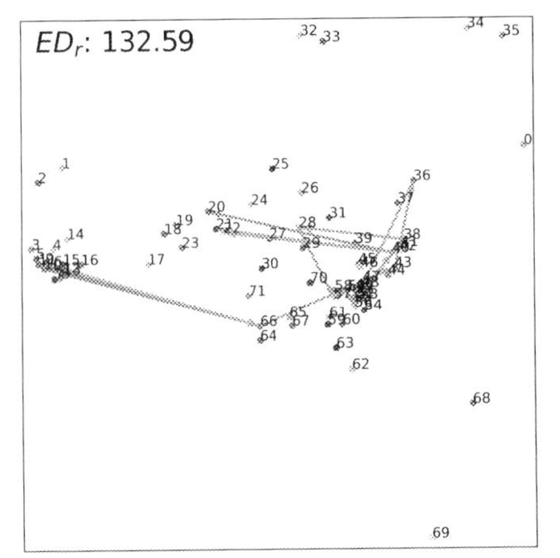

(c) Planned routes by the baseline model with similar setting and 200 tourists.

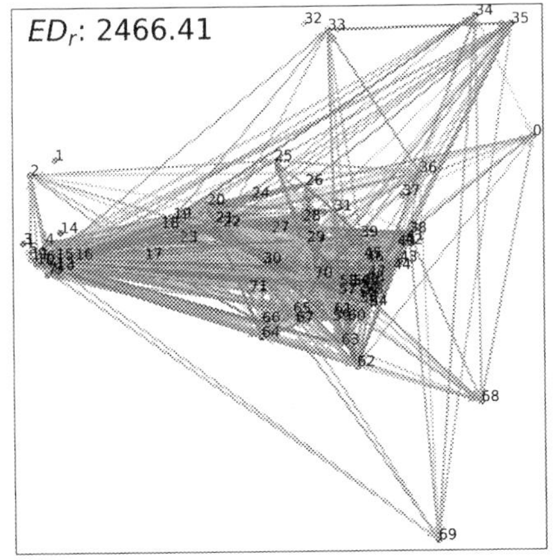

(d) Routes planned by MARLRP with similar setting and 200 tourists.

Fig. 4. Visualization of the planned routes.

manner in both experimental settings. We found that MARLRP enables tourists to not only receive more rewards, but also to diversify the planned routes and balance visits between spots.

The trend of the rewards with the number of tourists in the experiments using a random setting is shown in Fig. 5. As the planned spots are concentrated in a small number of spots, it is difficult for the baseline model to improve the overall reward of tourists, even when the number of tourists increases. The proposed model generates routes for tourists based on the congestion degree of

Table 1. Results of MARLRP and the baseline

		$\lvert\mathcal{T}\rvert$	$\mathcal{R}$	$\mathcal{R}_s$	G_s	ED_r
Identical	MARLRP	50	**2553.39**	4012	**0.73**	**449.26**
		100	**2618.04**	4149.69	**0.74**	**1014.95**
		150	**2618.04**	4149.69	**0.74**	**1216.63**
	Baseline	50	−468.32	**6136.69**	0.89	0
		100	−1941.29	**12273.39**	0.89	0
		150	−3416.81	**18410.09**	0.89	0
Similar	MARLRP	100	**5884.12**	9407.91	**0.46**	**1020.58**
		200	**7336.88**	12225.18	**0.46**	**2412.95**
		300	**7803.54**	13165.69	**0.48**	**3532.64**
	Baseline	100	−2183.09	**12279.02**	0.87	66.04
		200	−5123.23	**24558.20**	0.87	132.48
		300	−8067.51	**36832.81**	0.87	199.69

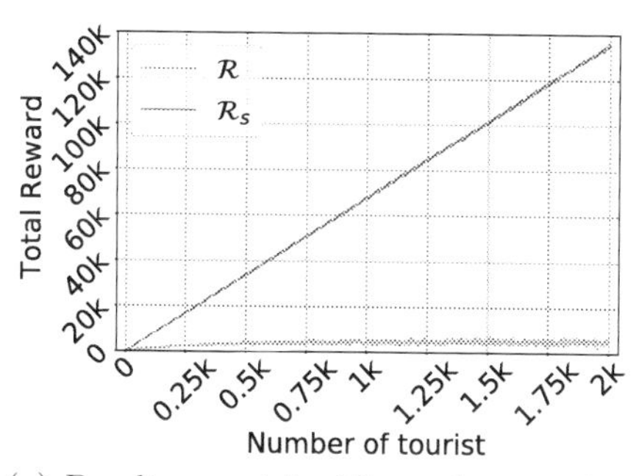

(a) Baseline model with random setting, number of tourist from 1 to 2000.

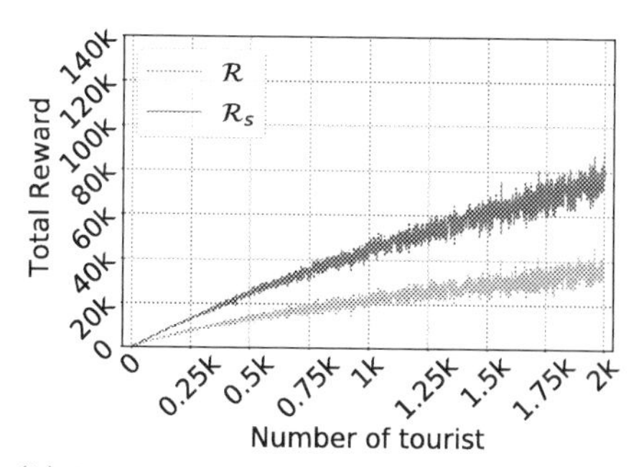

(b) MARLRP with random setting, number of tourist from 1 to 2000.

Fig. 5. Reward trend with number of tourists.

spots, to deal with a larger number of queries from tourists and increase the overall reward for tourists. MARLRP performed better than the baseline under the pressure of a large number of tourists.

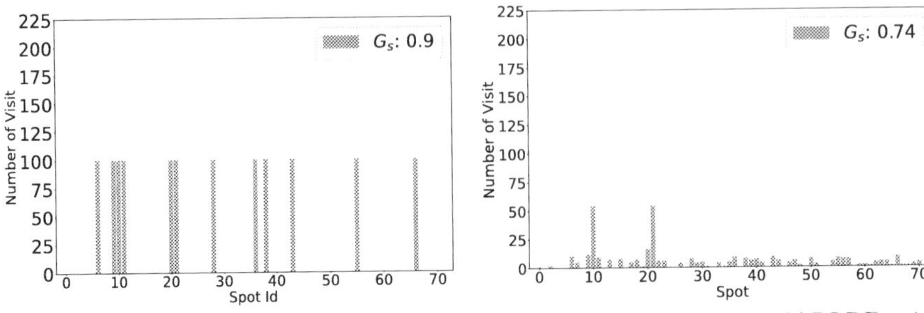

(a) Visit distribution by baseline model with Identical setting and 100 tourists.

(b) Visit distribution by MARLRP with identical settings and 100 tourists.

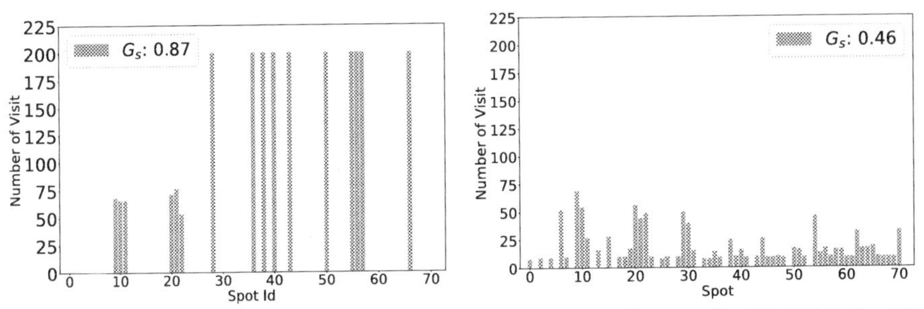

(c) Visit distribution by the baseline model with Similar setting and 200 tourists.

(d) Visit distribution by MARLRP with Similar setting and 200 tourists.

Fig. 6. Visualization of the visit distribution.

5 Conclusion

In this study, we introduced a novel diversity-oriented tour route planning task. We proposed a multi-agent reinforcement learning approach for solving a diversity-oriented touring route planning task, and proposed a reward-congestion function according to a tourism economic model. We combined a simulated environment with real-world statistics during training and testing to improve the applicability of our proposed model. According to our experimental result, the proposed model can significantly improve the total rewards for all tourists, reduce the stress on popular spots, balance the number of visits to different spots, and present tourists with more diverse plans.

Acknowledgements. This work was supported by JST SPRING, Grant Number JPMJSP2102. This work was partly supported by MIC SCOPE (201607008) and the joint work of Kyoto University and Yahoo!, Japan.

References

1. Afsar, H.M., Labadie, N.: Team orienteering problem with decreasing profits. Electron. Notes Discrete Math. **41**, 285–293 (2013)

2. Albaladejo, I.P., González-Martínez, M.: Congestion affecting the dynamic of tourism demand: evidence from the most popular destinations in Spain. Curr. Issue Tour. **22**(13), 1638–1652 (2019)
3. Archetti, C., Feillet, D., Hertz, A., Speranza, M.G.: The capacitated team orienteering and profitable tour problems. J. Oper. Res. Soc. **60**, 831–842 (2009). https://doi.org/10.1057/palgrave.jors.2602603
4. Chao, I.M., Golden, B.L., Wasil, E.A.: The team orienteering problem. Eur. J. Oper. Res. **88**, 464–474 (1996)
5. Chen, C., Cheng, S.F., Lau, H.C.: Multi-agent orienteering problem with time-dependent capacity constraints. Web Intell. Agent Syst. Int. J. **12**(4), 347–358 (2014)
6. Christofides, N.: The vehicle routing problem. Revue française d'automatique, informatique, recherche opérationnelle. Recherche opérationnelle **10**(V1), 55–70 (1976)
7. Erdoğan, G., Laporte, G.: The orienteering problem with variable profits. Networks **61**, 104–116 (2013)
8. Gavalas, D., Konstantopoulos, C., Mastakas, K., Pantziou, G.: A survey on algorithmic approaches for solving tourist trip design problems. J. Heuristics **20**(3), 291–328 (2014). https://doi.org/10.1007/s10732-014-9242-5
9. Geem, Z.W., Tseng, C.-L., Park, Y.: Harmony search for generalized orienteering problem: best touring in China. In: Wang, L., Chen, K., Ong, Y.S. (eds.) ICNC 2005. LNCS, vol. 3612, pp. 741–750. Springer, Heidelberg (2005). https://doi.org/10.1007/11539902_91
10. Golden, B.L., Levy, L., Vohra, R.V.: The orienteering problem. Nav. Res. Logist. **34**, 307–318 (1987)
11. Gunawan, A., Lau, H.C., Vansteenwegen, P.: Orienteering problem: a survey of recent variants, solution approaches and applications. Eur. J. Oper. Res. **255**, 315–332 (2016)
12. Ilhan, T., Iravani, S.M.R., Daskin, M.S.: The orienteering problem with stochastic profits. IIE Trans. **40**, 406–421 (2008)
13. Kumar, S., Panneerselvam, R.: A survey on the vehicle routing problem and its variants. Intell. Inf. Manag. **4**, 66–74 (2012)
14. Lambert, T.J., Epelman, M.A., Smith, R.L.: A fictitious play approach to large-scale optimization. Oper. Res. **53**, 477–489 (2005)
15. Rapoport, A., Chammah, A.M., Orwant, C.J.: Prisoner's Dilemma: A Study in Conflict and Cooperation, vol. 165. University of Michigan Press, Ann Arbor (1965)
16. Varakantham, P., Mostafa, H., Fu, N., Lau, H.C.: DIRECT: a scalable approach for route guidance in selfish orienteering problems. In: AAMAS (2015)
17. Verbeeck, C., Sörensen, K., Aghezzaf, E.H., Vansteenwegen, P.: A fast solution method for the time-dependent orienteering problem. Eur. J. Oper. Res. **236**, 419–432 (2014)
18. Kyoto City Official Website: Kyoto Tourism Comprehensive Survey (2019). https://www.city.kyoto.lg.jp/sankan/page/0000271655.html
19. Xu, J., Sun, J., Li, T., Ma, Q.: Kyoto sightseeing map 2.0 for user-experience oriented tourism. In: 2021 IEEE 4th International Conference on Multimedia Information Processing and Retrieval (MIPR), pp. 239–242 (2021)

Optimizing the Post-disaster Resource Allocation with Q-Learning: Demonstration of 2021 China Flood

Linhao Dong[1], Yanbing Bai[1(✉)], Qingsong Xu[2], and Erick Mas[3]

[1] Center for Applied Statistics, School of Statistics, Renmin University of China, Beijing, China
ybbai@ruc.edu.cn

[2] Data Science in Earth Observation, Technical University of Munich, 80333 Munich, Germany

[3] International Research Institute of Disaster Science, Tohoku University, Sendai 980-8572, Japan

Abstract. Reasonable and efficient allocation plan of emergency resources is key to successful disaster response. The Q-learning algorithm may provide a new approach to the resource allocation problem, which can take a variety of factors into consideration and timely respond to subsequent changes. In this paper, we propose a reinforcement learning Q-learning model in which the subsequent changes of disasters are included. Furthermore, in order to obtain a convincing result, three penalty functions are presented to represent three key factors, including efficiency, effectiveness and fairness. We test our proposed method under the background of actual flood disaster and compare results with real-time disaster development. Our empirical results show that the proposed model is in line with the development of disaster, and extremely sensitive to information on subsequent changes in the disaster situations, verifying its effectiveness in allocation of emergency resources.

Keywords: Disaster relief · Resource allocation · Reinforcement learning · Q-learning

1 Introduction

When devastating disasters (e.g., earthquakes, hurricanes, and floods) occur, the resources we have in a short period of time are limited. Under such urgent circumstances, the emergency center must formulate a reasonable resource allocation plan. The issue of resource allocation is key to humanitarian logistics and is directly related to the well-being of survivors and the effectiveness of disaster relief operations.

Due to a large amount of computation, high complexity, and few relevant historical samples of such problems, traditional optimal planning methods are

L. Dong and Y. Bai—Equal contribution.

C. Strauss et al. (Eds.): DEXA 2022, LNCS 13427, pp. 256–262, 2022.
https://doi.org/10.1007/978-3-031-12426-6_21

difficult to handle such problems. Some researchers summarized the emergency resource allocation algorithms [1] and mentioned that a completely accurate algorithm can only be applied to small-scale situations. In recent years, many scholars have begun to apply reinforcement learning algorithms to decision-making problems in the control field, and it turns out that they have excellent application prospects [2,3]. However, the existing reinforcement learning (RL) model is difficult to make timely adjustments according to the follow-up changes of the disasters. Moreover, no empirical analysis is carried out in the actual scene.

With these problems provided, we built a *reinforcement learning Q-learning model* in which the subsequent changes of disasters are added. Three types of losses are added to represent three key factors, so we can obtain a practical and reference value configuration scheme. Then, an empirical evaluation is conducted to discuss its performance.

2 Related Work

RL has been applied to resource allocation problems in various fields, such as communication [4], computational resource allocation [5], and smart cities [6]. Notably, the Q-learning algorithm in reinforcement learning has effectively been applied to solve many problems [7].

In terms of post-disaster resource allocation, some researchers [8] have proposed a theoretical method for applying Q-learning to disaster scenarios [9]. In addition, the Q-learning method [10] is applied to road network restoration problems after disasters. A Markov decision process is used as a multi-agent assessment and response system [11], with reinforcement learning designed to ensure the integration of the emergency response team. The Q-learning algorithm is yet to be developed in this regard.

3 Methodology

3.1 Model Architecture

Assuming a disaster area with only one emergency response center, all resources are planned and allocated by the rescue center. We need to formulate a resource allocation plan within T time units after the disaster as soon as possible and allocate it to N different areas to achieve the best rescue performance.

We define the regional state as $S_t(S_t = (S_{1,t}, S_{2,t}, ..., S_{N,t}))$, which represents the degree of resource demand in the region. The local response center's decision $Y_t(Y_t = (Y_{1,t}, Y_{2,t}..., Y_{N,t}), Y_t \leq C)$ is the allocation plan for the amount of resources allocated to each affected area for each time period t. A complete action set Y_C is available in each step. Due to poor post-disaster conditions, it is assumed that the local emergency center supplies C units of rescue resources in each time period.

In order to select an excellent program, three aspects are taken into consideration, namely efficiency, effectiveness and fairness. The objective function is proposed based on these standards.

3.2 Optimization Function

Based on the above analysis, three penalty functions are proposed to define the objective value.

Specifically, the effectiveness penalty $\phi(S_{i,t})$ is based on the demand state, and the efficiency penalty $\chi(d_i, Y_{i,t})$ is based on distance d ($d = (d_1, d_2, ..., d_N)$) and the fairness penalty $\Theta(S_{i,T+1})$ is based on the final state. The objective function is defined by weighting the three penalty functions. Our goal is to find the best state-action pair which maximizes the objective function.

The model will calculate three performance indicators at each time unit, accumulate and weight them to obtain the final objective function, the objective function is as follows:

$$min \quad \xi_1 \sum_{i=1}^{N} \sum_{t=1}^{T} \chi(c_i, Y_{i,t}) + \xi_2 \sum_{i=1}^{N} \sum_{t=1}^{T} \phi(S_{i,t}) + \xi_3 \sum_{i=1}^{N} \Theta(S_{i,T+1}), \tag{1}$$

where $\{\xi_1, \xi_2, \xi_3\}$ are weights of the three types of indicators to balance the influence of each factor on the final solution.

Algorithm 1. Update of Q-matrix

Input: The current state S_t^k and current action Y_t^k

Generate subsequent changes of disasters D_i^t

Calculate the next state S_{t+1}^k and reward R_t^k based on the reward function

$$S_{t+1}^k = S_t^k - Y_t^k + D_t^k$$

$$R_t^k = -\chi(Y_t^k) - \phi(S_{t+1}^k) \quad if \quad t = 2, 3, ..., T-1$$

Update $Q(S_t^k, Y_t^k) \leftarrow Q(S_t^k, Y_t^k) + \alpha(R_t^k + \gamma \max_{Y_C} Q(S_{t+1}^k, Y_t^k) - Q(S_t^k, Y_t^k))$

Update $S_t^k \leftarrow S_{t+1}^k$

Repeat this process and obtain the final Q-matrix

Use the final Q-matrix to calculate the final path and action

---

3.3 Reward Update (R) and State Update (S)

The reward function will also be defined around these three loss functions. In this paper, the state of the planning period is divided into an initial time, an intermediate time and a final time period. The reward functions of the 3 time stages are determined in Algorithm 1.

The time of the first action choice is called the initial state of the planning period. The reward is not only related to the state when the action is chosen but also to the state it will reach. Therefore, the reward function for the state at time period $t = 1$ is shown in Eq. (2). Second, the intermediate state of the reward from time period 2 to time period $T - 1$ corresponds to a demand state-based effectiveness penalty, a distance-based efficiency penalty, which is shown in Eq. (3). The reward of the final state corresponds to the distance-based efficiency penalty and the final state-based fairness penalty at the final time T, as shown in Eq. (4). In addition, the reward values used to update the Q-matrix in this paper are all negative values, so the Q-matrix selects the reward with the smallest absolute value.

$$R = -\chi(Y_1) - \phi(S_{i,1}) - \phi(S_{i,2}) \quad if \quad t = 1, \tag{2}$$

$$R = -\chi(Y_t) - \phi(S_{i,t+1}) \quad if \quad t = 2, 3, ..., T-1, \tag{3}$$

$$R = -\chi(Y_T) - \Theta(S_{T+1}) \quad if \quad t = T, \tag{4}$$

4 Experiment

In this section, to prove the effectiveness of the proposed RL method, the model is applied to the actual background of the flood disaster in Henan in July 2021 to generate the optimal allocation of resources.

4.1 Data Preparation

Initial Environment. This paper adopts the classical k-means method for data clustering modeling. We evaluate the state of the region as 4 levels. The higher the level, the higher the demand for resources in the region and the more severe the disaster. The clustering results are put into subsequent models as the initial definition of the Q-learning environment.

4.2 Numerical Experiment

Q-learning algorithm is affected by three parameters, namely the exploration rate ϵ, the learning rate α and the discount factor γ. The initial parameters are set to $\epsilon = 0.5$, $\alpha = 0.8$, $\gamma = 0.2$, and the maximum number of training sets is set to $K = 50000$. The initial weights are set to $\{0.3, 0.6, 0.1\}$.

This numerical experiment is to conduct rescue work in 15 counties ($N = 15$) in Henan province under the background of flood disasters. According to the above clustering methods, we used Henan geological data and meteorological data on July 16 to obtain the initial state of each county, and then we used this state value to define the initial environment. After that, we set the time interval of the scheme to 14 days ($T = 14$), and used the meteorological forecast information within 14 days to calculate the subsequent changes of the disaster. We input this information into the model as subsequent changes to the environment. Assuming that we have $C = 10$ units of relief materials in each time period (2 units each time for large counties, and 1 for small counties), the final model will generate the resource allocation plan for the next 14 days.

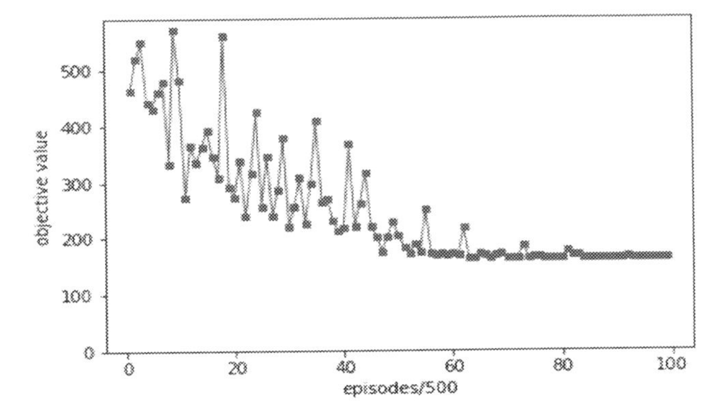

Fig. 1. Objective value

4.3 Analysis of Results

According to Fig. 1, after 50,000 steps, the objective value has fully converged to 162.84. In the first 10,000 steps, the model explores rapidly with a high frequency, and the objective function decreases rapidly. After 20,000 steps, the descending speed becomes significantly slower. Finally, after 40,000 steps, the model can generate the optimal allocation plan stably.

In order to verify the rationality of the scheme, we checked the official survey report on this disaster and counted the precipitation data from July 16 to 30.

Table 1. The model's resource allocation plan

County	T = 1	T = 2	T = 3	T = 4	T = 5	T = 6	T = 7	...	T = 14
Nanyang	○	○	○	○	•	•	•	...	○
Xuchang	•	○	•	•	○	•	○	...	○
Luoyang	○	•	•	○	•	•	○	...	○
Jiaozuo	•	•	○	•	•	○	•	...	•
Xinxiang	•	•	•	•	○	○	•	...	○
...	...	...	...	...	...	...	...	...	
Kaifeng	○	○	○	•	○	•	•	...	○

We compared the resource allocation plan generated by the model; the rescue strategy of the plan is basically in line with the development of the disaster. As shown in Table 1, in the early stage of the disaster ($T = 1, 2, 3$), the plan focused on the northern part of Henan province, all of which have high precipitation. In the mid-disaster, resources were continuously provided to areas around Zhengzhou, and priority was given to rescuing the counties which save transportation costs. After July 23, the pressure imposed by the disaster became relieved and the allocation of resources has become less burdened. In addition, although the allocation of resources is biased, the plan still takes into account all the affected counties. It is also quite fair and reasonable when evaluating from a humanitarian perspective. In general, in the case of limited resources, the model can make an exceptional judgment on such a long-term disaster.

5 Conclusion

We propose a model based on the reinforcement learning Q-learning method, which focuses on the post-disaster emergency resource allocation. This model is applied to the actual background of flood disasters in Henan province in China. For future development, we plan to apply this set of construction methods to more scenarios to improve its generalized application.

Acknowledgment. This research was supported by Public Health & Disease Control and Prevention, Major Innovation & Planning Interdisciplinary Platform for the "Double-First Class" Initiative, Renmin University of China (No. 2022PDPC), fund for building world-class universities (disciplines) of Renmin University of China. Project No. KYGJA2022001, fund for building world-class universities (disciplines) of Renmin University of China. Project No. KYGJF2021001, Beijing Golden Bridge Project seed fund (No. ZZ21021), National Natural Science Foundation of China (Grant No. 72004226). This research was supported by Public Computing Cloud, Renmin University of China.

References

1. Yu, L., Zhang, C., Yang, H., Miao, L.: Novel methods for resource allocation in humanitarian logistics considering human suffering (2018)
2. Mao, M.A., Menache, I., Kandula, S.: Resource management with deep reinforcement learning. In: ACM, pp. 50–56 (2016)
3. Mnih, V., Kavukcuoglu, K., Silver, D., Graves, A.: Playing Atari with deep reinforcement learning. arXiv preprint arXiv:1312.5602 (2013)
4. Aihara, N., Adachi, K., Takyu, O.: Q-learning aided resource allocation and environment recognition in LoRaWAN with CSMA/CA. IEEE Access **7**, 152126–152137 (2019)
5. Kim, Y., Kim, S., Lim, H.: Reinforcement learning based resource management for network slicing. Appl. Sci. **9**(11), 2361 (2019)
6. Semrov, M., Todorovski, L., Srdic, A.: Reinforcement learning approach for train rescheduling on a single-track railway. Transp. Res. Part B: Methodol. **86**, 250–267 (2016)
7. Xu, Y., Xia, J., Wu, H., Fan, L.: Q-learning based physical-layer secure game against multiagent attacks. IEEE Access **7**, 49212–49222 (2019)
8. Yu, L., Zhang, C., Jiang, J., et al.: Reinforcement learning approach for resource allocation in humanitarian logistics. Expert Syst. Appl. **173**, 114663 (2021)
9. Chowdhury, M.U., Erden, F., Guvenc, I.: RSS-based Q-learning for indoor UAV navigation. In: IEEE Military Communications Conference, pp. 121–126 (2019)

10. Su, P.Z., Lianqi, D., et al.: Multicrew scheduling and routing based on deep Q-learning. In: AAAI-22 Workshop on Machine Learning (2021)
11. Nadi, Y., Edrisi, A.: Adaptive multi-agent relief assessment and emergency response. Int. J. Disaster Risk Reduct. **24**, 12–23 (2017)

ARDBS: Efficient Processing of Provenance Queries Over Annotated Relations

Sareh Mohammadi(✉) and Nematollaah Shiri

Concordia University, Montreal, Canada
sareh.mohammadi@mail.concordia.ca, shiri@cse.concordia.ca

Abstract. We investigate efficient methods for processing provenance queries over annotated relations. Using provenance semirings as the base theory for evaluating SPJU (select-project-join-union) queries over annotated relations and its extension that incorporates difference operation, we propose a rewriting approach that transforms SQL queries containing SPJUD operations into semantic-aware queries. This approach considers each semantics as a first-class citizen which results in simplified structures and reduced processing costs. The idea is implemented in the lightweight middle ARDBS (Annotated Relational Database System), and the results of our experiments using different datasets indicate improved efficiency of ARDBS compared to the existing solutions.

Keywords: Annotated relations · Provenance semirings · Performance

1 Introduction

Annotations provide explanation of some sort in different semantics about the tuples in a query result. To this end, several solutions extended relational algebra for specific annotation types [3, 6–8, 18], and a number of prototype systems [1, 5, 9, 12–16] have developed the capabilities of standard relational database technologies to support annotation-aware computation.

Green et al. [11] have proposed provenance semirings as a theoretical foundation for positive queries over annotated relations. It assumes that the database relations have an additional column for annotation of some domain K defined based on the underlying semantics. In addition, two generic operators $\oplus$ and $\otimes$ are defined that operate over the annotations to formulate annotation expressions based on the corresponding relational algebra expression of the query. Intuitively, $\oplus$ is used for annotations of tuples obtained through projection and union, while $\otimes$ is used on those obtained through cross product and join. The work in [11] also introduces polynomial semiring as a general provenance that can be mapped to other commutative semirings. To support queries with negation, Geerts et al. [8] define the operator $\ominus$ and identify conditions under which they extend semirings with negation.

Existing annotation-aware query processing solutions [1, 2, 5, 9, 12, 13, 16] mainly target a specific provenance semantic type when processing queries. In some frameworks [1, 9, 13, 16], the base semantics provide a generality from which other semantics of

C. Strauss et al. (Eds.): DEXA 2022, LNCS 13427, pp. 263–269, 2022.
https://doi.org/10.1007/978-3-031-12426-6_22

annotations can be derived, but often at extra cost due to the complex structures of the base annotations and additional annotation conversion steps. In this paper, we promote an approach for supporting multiple semantic types when each semantic is handled directly through query processing. This approach avoids the extra time and space requirements of the previous solutions. Our proposed solution rewrites SPJUD queries based on query patterns we identified that are compatible with the theory of provenance semirings [11]. We developed a running prototype system called annotated relational database system (ARDBS, pronounced as RDBS) which runs on top of a standard SQL engine as a middleware and performs a lightweight preprocessing to rewrite input SQL queries into semantic-aware queries. The results of our experiments indicate promise for ARDBS to yield an efficient and easy to use engine to support applications with provenance aware queries.

The rest of this paper is organized as follows. In Sect. 2, we discuss the strategies to support annotation-aware queries and present the solution approach. Section 3 introduces ARDBS and reports the experiments and performance evaluation results. Concluding remarks and future work are provided in Sect. 4.

2 Semantic-Aware Query Processing

In this section, we propose an *all-primary* approach for supporting different types of annotation semantics. Figure 1 compares this approach with the *primary/secondary* approach followed by the existing frameworks such as Perm [9], GProM [1], and ProvSQL [15, 16]. While the *primary/secondary* approach processes all queries in a default semantic and converts annotations in the postprocessing step to the desired semantics, in *all-primary* approach each input query q is rewritten based on the input semantic S into a *semantic-aware* query Q_S. The purpose of semantic-aware queries is to perform annotation calculations related to each semantic as close as possible to the corresponding RA operations to directly produce outputs in the desired semantics (Rs). Ideally, when processing such queries, annotations should also be accessed and processed once the data is being accessed, rather than process annotations later. Bringing forward the annotation computations and eliminating annotation conversions cause improvements in query processing time compared with the primary/secondary approach. In the following, we discuss the requirements for transforming standard queries into semantic-aware ones.

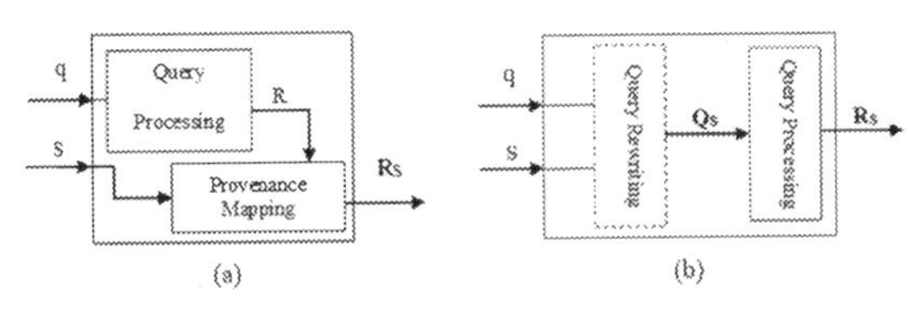

Fig. 1. (a) Primary/Secondary (b) All-Primary approaches

To transform an input query q to a semantic-aware query Q_S, we introduce three rewriting patterns (RP) corresponding to the SPJ types of queries and queries with union and difference operations, shown in Table 1. In these RPs we use a set of annotation functions (AFs), inspired by the operations in the semiring framework, that implement each semantics' calculations. $\mathbf{H}_S$ is the Horizontal function that operates on the annotation columns appearing in each row in the result of the cross products. It is defined based on the associative operation $\otimes$ in provenance semirings and aggregates the input annotations in a single annotation column, e.g., *joinAnn* column in RP1. Vertical function $\mathbf{V}_S$ corresponds to the operation $\oplus$ and comes with a *Group By* command to aggregate the annotation column of duplicated tuples. This early aggregation reduces the size of intermediate results and helps to prevent inconsistencies when a query contains difference operation. $\mathbf{M}_S$ is a function that implements difference operation $\ominus$. Note that the subscript "S" indicates that the functionality of H, V, and M depends on the selected semantics. For more details of these functions, interested readers are referred to [10, 11].

Table 1. Rewriting patterns for RA expressions with select-project-join-union-difference.

Standard RA Exp.	Rewriting Patterns (RP's)
$\pi_L(\sigma_C(R_1 \times \ldots \times R_k)$	**RP1:** $\gamma_{L,\, V_s(\text{joinAnn})}(\pi_{L,\, H_s(Ann_1,\ldots,Ann_k) \rightarrow joinAnn}(\sigma_C(R_1 \times \ldots \times R_k))$
$R_i \cup R_j$	**RP2:** $\gamma_{L, V_S(Ann)}(R_i \cup_B R_j)$
$R_i - R_j$	**RP3:** $\pi_{L_i, M_s(Lann, Rann) \rightarrow Ann}\left(\left(\gamma_{L_i, V_s(Ann) \rightarrow Lann} R_i\right) ⟕_L \left(\gamma_{L_j, V_s(Ann) \rightarrow Rann} R_j\right)\right)$

In RP2 and RP3, L refers to the list of attributes in source tables R_i and R_j, assuming that they are compatible. In RP2, U_B denotes the bag union operation that preserves duplicate tuples so that their annotations can be aggregated. In RP3, $⟕_L$ is left-outer-join over list L that is used to identify "identical" tuples appearing in both R_i and R_j. In annotation-aware negation, since the presence of a tuple in the result depends on both the relational algebra (RA) difference and the annotation difference, the operator $⟕_L$ provides the necessary arguments for $\mathbf{M}_S$ to operate on. The following example shows how the rewriting patterns work.

Example. This example illustrates the rewriting steps for an SPJU query q_1, which is the union of two SPJ subqueries over the input relations shown in Fig. 2. These relations have an annotation column *Ann* that contains unique identifiers assigned to tuples.

$$q_1 : \pi_{A,C}(\sigma_{B=D}(R_1 \times R_2)) \bigcup \pi_{A,C}(R_1)$$

If we show the left and right subqueries in q_1 as q_{11} and q_{12}, i.e., $q_1 = q_{11} \cup q_{12}$, and show their rewritings as Q_{11} and Q_{12}, respectively, then the rewriting of q_1 is:

$$Q_1 : \gamma_{A.C.V_{Why}(Ann) \rightarrow Ann}(Q_{11} \cup_B Q_{12}), \text{ where}$$

Q_{11}: $\gamma_{A.C.V_{Why}(joinAnn) \rightarrow Ann}\left(\pi_{A.C.H_{Why}(Ann_1.Ann_2) \rightarrow joinAnn}(\sigma_{B=D}(R_1 \times R_2))\right)$ and
Q_{12}: $\gamma_{A.C.V_{Why}(Ann)}(R_1)$

Suppose the semantics considered in this rewriting is *why-provenance* [3, 4], which describes each set of source tuples that witnesses the existence of an output tuple. Then the corresponding horizontal function H_{Why} returns the pairwise union over the sets of annotations, and the vertical function V_{Why} returns the union of the annotations involved. Figure 3 depicts the query expression tree for Q_1 together with the intermediate results. The left branch in this tree shows the steps to process Q_{11} and also the way the horizontal function works on the result of the cross product in order to determine the annotation of the joined tuples. The result of Q_{11} is shown as R_5, obtained by applying V_{Why} to R_4. After performing the bag union on R_5 and R_6, the resulting tuples shown as R_7 are aggregated by applying V_{why} over the annotation column that yields relation R_8 as the output of query Q_1

R_1

A	B	C	Ann_1
g	b	c	p
h	a	d	q

R_2

D	E	Ann_2
a	g	r
d	c	s
a	h	t
b	g	u

Fig. 2. Example annotated relations

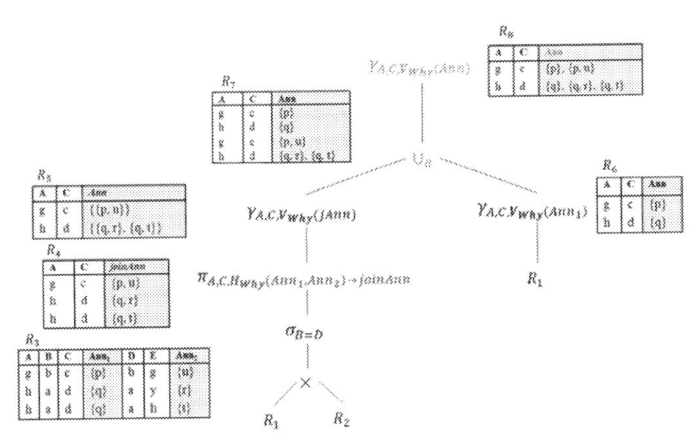

Fig. 3. The processing steps of Q_1

3 ARDBS Framework

We have provided a lightweight algorithm that implements the all-primary approach. ARDBS is the development of this algorithm working as an intermediate layer between users and the standard relational DBMS. To evaluate the performance of ARDBS, we performed several experiments using the benchmark datasets induced by TPC-H suits [17] in which relations were extended with annotations. We also created and used other annotated datasets in our experiments. We measured the performance of ARDBS mainly in terms of execution time under different situations. Due to the lack of space, here we just report a set of experiments that compare the query execution times in ARDBS,

ProvSQL [15, 16] and GProM [1]. ProvSQL is similar to ARDBS for using provenance semirings as the base theory, and GProM is similar to ARDBS for using a middleware approach as its architecture. Since ProvSQL is implemented on a Linux-based version of PostgreSQL, we have performed all experiments on the same environment and using a PC with 3.2 GHZ, 3 Cores, and 16 GB RAM. In addition, since DBMS execution times for a specific query may vary, to obtain more accurate times, we restarted the server and cleared the memory before each run and found the average execution time of three runs for each query.

The experiments are performed using an annotated database with nine relations generated over two schemas that record information about *products* and *ordered items*. While we consider the *products* relation to be fixed (with 1887 tuples) throughout the tests, we consider different varying instances of *order-item* relation, ranging from 10^2 to 10^7 tuples. Figure 4 shows the execution times when a simple SPJ query pattern is used over *products* and different sizes of *order-items*. For ARDBS and ProvSQL, we considered the bag semantics, while for GProM, we consider its default provenance type. In our experiments, we noted that both ProvSQL and GProM failed to complete query processing when the number of tuples in relation *order-item* exceeded 2×10^5 and 10^6 tuples, respectively. Our results indicate that for database sizes that ProvSQL could complete its query processing, ARDBS was significantly faster. In such cases ARDBS and ProvSQL produced the same results. We conclude that the performance advantage of ARDBS over ProvSQL is mainly due to the *all-primary* approach used in ARDBS and *primary-secondary* in ProvSQL.

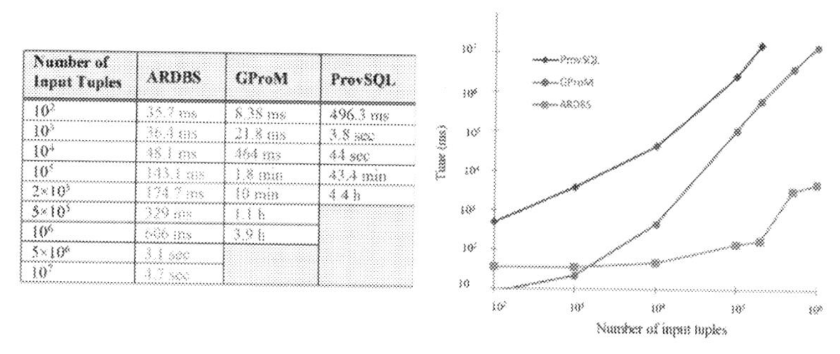

Number of Input Tuples	ARDBS	GProM	ProvSQL
10^2	35.7 ms	8.38 ms	496.3 ms
10^3	36.4 ms	21.8 ms	3.8 sec
10^4	48.1 ms	464 ms	44 sec
10^5	143.1 ms	1.8 min	43.4 min
2×10^5	174.7 ms	10 min	4.4 h
5×10^5	329 ms	1.1 h	
10^6	606 ms	3.9 h	
5×10^6	3.1 sec		
10^7	3.7 sec		

Fig. 4. Comparing query execution times by ProvSQL, GProM, and ARDBS

4 Conclusion and Future Work

We introduced a lightweight, middleware approach for extending the capabilities of relational DBMSs for processing annotated data in different semantics. Our proposed solution approach as implemented in ARDBS reduces the workload of annotated query processing and improves the query execution time by directly translating the input queries according to the desired semantic. The extended relational algebra implemented in query rewriting phase makes the solution least dependent on the underlying engine and hence more extensible to support user-defined provenances. Besides, users can easily express queries over annotated data. The results of numerous experiments using a wide range of relation sizes, show effectiveness of ARDBS. The focus of this paper was on SPJUD provenance queries. We plan to explore the remaining types of queries, such as aggregate queries, as our future work.

References

1. Arab, B.S., Feng, S., Glavic, B., Lee, S., Niu, X., Zeng, Q.: GProM - a swiss army knife for your provenance needs. Quar. Bull. Comput. Soc. IEEE Tech. Committee Data Eng. **41**(1), 51–62 (2018)
2. Bhagwat, D., Chiticariu, L., Tan, W.-C., Vijayvargiya, G.: An annotation management system for relational databases. VLDB J. **14**(4), 373–396 (2005). https://doi.org/10.1007/s00778-005-0156-6
3. Buneman, P., Khanna, S., Wang-Chiew, T.: Why and where: a characterization of data provenance. In: Van den Bussche, J., Vianu, V. (eds.) ICDT 2001. LNCS, vol. 1973, pp. 316–330. Springer, Heidelberg (2001). https://doi.org/10.1007/3-540-44503-X_20
4. Cheney, J., Chiticariu, l., Tan, W.-C.: Provenance in Databases: Why, How, and Where. Now Publishers Inc (2009)
5. Chiticariu, L., Tan, W.-C., Vijayvargiya, G.: DBNotes: a post-it system for relational databases based on provenance. In: Proceedings of the 2005 ACM SIGMOD International Conference on Management of Data (SIGMOD 2005), pp. 942–944. Association for Computing Machinery, New York, NY, USA (2005)
6. Cui, Y.: Lineage tracing in data warehouses. Ph.D. Dissertation. Stanford University, Stanford, CA, USA (2002)
7. Fuhr, N., Rölleke, T.: A probabilistic relational algebra for the integration of information retrieval and database systems. ACM Trans. Inf. Syst. **15**(1), 32–66 (1997)
8. Geerts, F., Poggi, A.: On database query languages for K-relations. J. Appl. Log. **8**(2), 173–185 (2010)
9. Glavic, B., Miller, R.J., Alonso, G.: Using SQL for efficient generation and querying of provenance information. In: Tannen, V., Wong, L., Libkin, L., Fan, W., Tan, W.-C., Fourman, M. (eds.) In Search of Elegance in the Theory and Practice of Computation. LNCS, vol. 8000, pp. 291–320. Springer, Heidelberg (2013). https://doi.org/10.1007/978-3-642-41660-6_16
10. Green, T.J.: Containment of conjunctive queries on annotated relations. Theory Comput. Syst. **49**, 429–459 (2011)
11. Green, T.J., Karvounarakis, G., Tannen, V.: Provenance semirings. In: Proceedings of the Twenty-Sixth ACM SIGMOD-SIGACT-SIGART Symposium on Principles of Database Systems, pp. 31–40. Association for Computing Machinery (2007)
12. Green, T.J., Karvounarakis, G., Ives, Z.G., Tannen, V.: Provenance in ORCHESTRA. IEEE Data Eng. Bull **33**(3), 9–16 (2010)
13. Lee, S., Ludäscher, B., Glavic, B.: PUG: a framework and practical implementation for why and why-not provenance. VLDB J. **28**(1), 47–71 (2018). https://doi.org/10.1007/s00778-018-0518-5
14. Mutsuzaki, M., et al.: Trio-One: Layering uncertainty and lineage on a conventional DBMS. In: Proceeding of Conference on Innovative Data Systems Research (CIDR), Pacific Grove, California (2007)
15. Senellart, P.: Provenance and probabilities in relational databases. SIGMOD Rec. **46**(4), 5–15 (2017)
16. Senellart, P., Jachiet, L., Maniu, S., Ramusat, Y.: ProvSQL: Provenance and Probability Management in PostgreSQL. In: Proceedings of the VLDB Endowment (PVLDB), vol. 11, issue 12, pp. 2034–2037 (2018)

17. TPC. TPC benchmarks. http://www.tpc.org/
18. Zimanyi, E.: Query evaluation in probabilistic relational databases. Theoret. Comput. Sci. **171**(1997), 179–219 (1997)

Analysis of Extracellular Potential Recordings by High-Density Micro-electrode Arrays of Pancreatic Islets

Jan David Hüwel[1(✉)], Anne Gresch[2], Tim Berger[2], Martina Düfer[2], and Christian Beecks[1]

[1] Department of Mathematics and Computer Science, University of Hagen, Hagen, Germany
{jan.huewel,christian.beecks}@fernuni-hagen.de

[2] Institute of Pharmaceutical and Medicinal Chemistry, University of Münster, Münster, Germany
{agresch,t_berg21,martina.duefer}@uni-muenster.de

Abstract. Pancreatic β-cells form highly connected networks within the islets of Langerhans ensuring an adequate insulin secretion. Extracellular potential recordings are frequently used to investigate these networks based on their electrical activity. High-density micro-electrode arrays provide an efficient digital technology to record simultaneous signals of multiple individual cells within these networks of pancreatic islets, where electrical peaks caused by glucose stimulation are a well established indicator for activity patterns or regions within an islet. In this short paper, we propose a simple yet effective method for the analysis of extracellular potential recordings of pancreatic islets. The preliminary results of our proposal indicate that activity patterns differ across cells and that our approach is fundamental for further cross-domain research.

Keywords: Spike detection · Time series analysis · Islet of Langerhans

1 Introduction

Type 2 diabetes mellitus is one of the most common diseases which affects more than 500 million adults worldwide in 2021 according to the International Diabetes Federation [1]. Research on this disease thus strongly affects general health and has a potentially life-saving impact on a significant percentage of the global population. One important research direction is concerned with the functionality of pancreatic islets. These islets are mainly composed of β-cells which are responsible for the production of insulin as a response to rising blood glucose levels in human bodies [8].

This research was supported by the research training group "Dataninja" (Trustworthy AI for Seamless Problem Solving: Next Generation Intelligence Joins Robust Data Analysis) funded by the German federal state of North Rhine-Westphalia.

C. Strauss et al. (Eds.): DEXA 2022, LNCS 13427, pp. 270–276, 2022.
https://doi.org/10.1007/978-3-031-12426-6_23

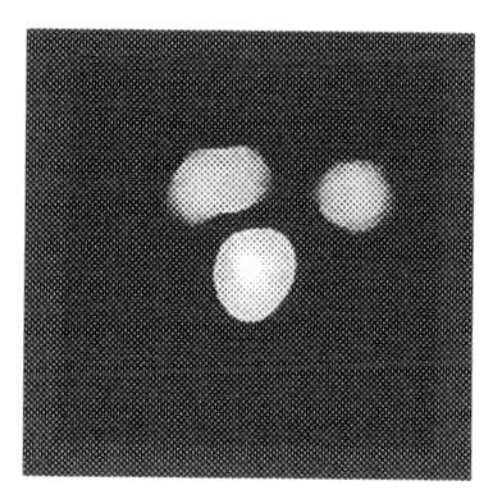

Fig. 1. Islets attached to the CMOS-MEA Chip. The black square is the sensor chip. The upper left islet was used for data analysis of experiment 1 and both upper islets were used for data analysis of experiment 2.

The β-cells of the pancreas are located in highly connected networks that ensure adequate insulin secretion through inter-cellular communications. If this network is disrupted, it can lead to a decrease of insulin secretion [5], which in turn can lead to negative health effects. One major research goal is thus to understand the communication behavior of individual cells within the cellular network of a pancreatic islet. To this end, we focus our investigation on the electrical activity of β-cells, since this activity is a widely accepted indicator of functionality [8] that is coupled to insulin secretion, and aim to develop a peak-based method to detect and analyse electrical activity patterns.

In our experimental setup, the electrical activity was recorded by placing isolated islets from mice on a CMOS-MEA chip [2] (cf. Fig. 1), which generates a large amount of time series data with a frequency of 1 kHz showing waves and peaks (cf. Fig. 2). In this short paper, we focus our investigation on the peak signals and implemented a spike-based activity analysis on the extracellular potential recordings of the pancreatic islets. Spikes were extracted as an electrical signature for analyzing active regions of the islet. The aim was to show that this restriction yields enough information to make conclusive observations on the data.

2 Physiological Background

The endocrine part of the pancreas comprises the islets of Langerhans. The term islet refers to a collection of different cell types. In humans, around 60–75% of these islets consists of so-called β-cells, which are responsible for the secretion of insulin, while the remaining 15–30% of the islets contain α-cells secreting glucagon. The rest is mixed with δ-cells, γ-cells, and ϵ-cells [8]. In comparison to humans, rodents show a similar cell composition. While for humans the aforementioned cells are spread across the islet, rodents tend to form a core of β-cells that are surrounded by other cell types [8].

The β-cells are electrically excitable cells. The conversion of a glucose stimulus into an electrical signal, i.e. the depolarization of the membrane potential, ultimately triggers the secretion of insulin [8]. The resulting changes of the membrane potential are picked up by our sensors on the level of one or two neighbouring single islet cells at most.

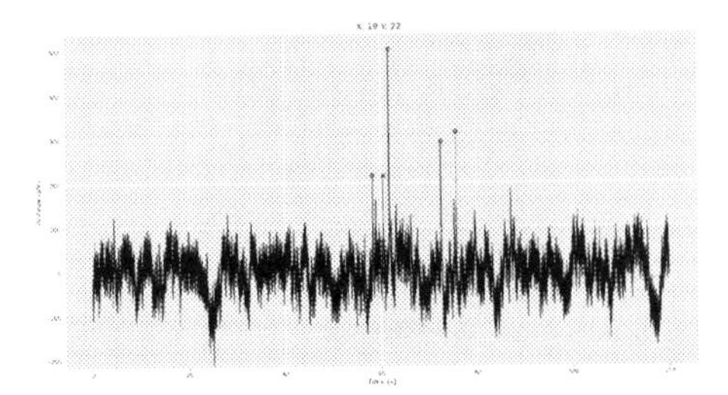

Fig. 2. Exemplary trace of one sensor covered by one or two cells within a pancreatic islet stimulated with 10 mM glucose. Red circles mark detected peaks crossing a peak score value of 400.

To be able to record the extracellular voltage changes, pancreatic islets were isolated from mice and placed on the CMOS-MEA chips from Multichannel Systems (Fig. 1). The resulting time series data was filtered with a Butterworth 2nd order filter between 0.1 500 Hz and processed at a recording frequency of 1 kHz over a span of two minutes. Electrical activity was induced by application of a stimulatory glucose concentration. By changing from a high (10 mM) to a low (3 mM) glucose concentration, the electrical activity disappears. β-cells have a diameter of about 15 µm [8]. Since the distance between the CMOS-MEA5000 sensors is 16 µm [2], each sensor will be covered approximately by a single cell.[1]

3 Spike Detection in Micro-electrode Arrays

The data investigated in the scope of this work is recorded by means of a 65 $\times$ 65 sensor array producing time series over a span of two minutes with a frequency of 1 kHz. To detect spikes in the electrical activity, each sensor is treated separately. Our goal is therefore to label peaks in univariate time series as spikes.

While there are several methods for solving this task [3,6,10], we decided to utilize a simple yet efficient method in our first investigation. Some common methods rely on the derivative of the time series [3] or require the data to follow certain patterns [10] and were therefore not optimally applicable in our rather unsupervised scenario.

To assess whether a certain point in a time series is a spike or not, we make use of the scoring system introduced by Palshikar [6]. The so-called *peak score* PS can be used as a means to determine how strong of a peak a given point is. There are multiple versions of the peak score, but the one used throughout this short paper is shown in Eq. (1). Given a certain point x_i of a time series x, the peak score $PS(k, x, i)$ determines the average difference between the point x_i and all neighboring points up to distance k. For negative peaks, this formula can simply be negated.

$$PS(k, x, i) = \frac{\Sigma_{j=i-k}^{i+k} (x_i - x_j)}{2k} \tag{1}$$

[1] The 3D characteristics of an islet, i.e. the multiple layers of cells above the sensors, have not been considered in this analysis.

The computation of the peak score is accelerated by filtering the data and testing only local extreme points. This also assures that a peak that stretches over multiple points is only classified as a singular spike. Additionally, we include a postfiltering that sorts out minor peaks as noise [6]. For this purpose, we take the average of all calculated scores μ and the standard deviation σ as well as a deviation factor T_{std}. A peak x_i is considered noise if $PS(k, x, i) \leq \mu + T_{std} \cdot \sigma$. Finally, we use a threshold on the peak score to limit the classification as spikes to high peaks. While the effect of thresholding is examined in the following section, Algorithm 1 shows the pseudocode of the peak detection algorithm.

Algorithm 1: Peak detection algorithm for time series data

Data: Time series $(t, x) = (t_i, x_i)_{i=1,..,N}$, range k, deviation factor T_{std}, peak score threshold T_{ps}

Result: List of peak positions and peak scores $L = (t_j, s_j)_{j=1,...,n}$

1 $L \leftarrow ()$
2 $C \leftarrow ()$
3 **for** $i \leftarrow k$ **to** $N - k$ **do**
4   **if** $x_{i-1} < x_i > x_{i+1}$ **or** $x_{i-1} > x_i < x_{i+1}$ **then**
5         $C \leftarrow C \cup (t_i, PS(k, x, i))$
6 $\mu \leftarrow \frac{1}{|C|} \sum_{(t,s) \in C} s$
7 $\sigma \leftarrow \left(\frac{1}{|C|} \sum_{(t,s) \in C} (s - \mu)^2 \right)^{-2}$
8 **for** $(t, s) \in C$ **do**
9   **if** $s > \mu + T_{std}\sigma$ **and** $s > T_{ps}$ **then**
10         $L \leftarrow L \cup (t, s)$

The peak detection algorithm is applied to each individual time series and can thus be computed for the complete sensor array in parallel. For a given time series, this algorithm returns a list of peak positions and peak scores.

An example of the application to a small segment is depicted in Fig. 2. This figure also shows how effective the postfiltering and thresholding sort out the regular noise from actual spikes.

4 Preliminary Experiments

In the first series of experiments, we aim to show that spikes occur with a particularly high amplitude dependent on the glucose level. Furthermore, we examine the effect of different thresholds T_{ps}.

To detect only spikes with high amplitudes (cf. Fig. 2), we empirically set the peak score threshold $T_{ps} = 400$, the range $k = 100$ and the deviation factor $T_{std} = 3$. Stimulation with 3 mM glucose yields a very small number of spikes per sensor, exceeding a peak value of 400 (Fig. 3 on the left). The number of spikes increases with a glucose concentration of 10 mM (Fig. 3, on the right). As can be seen in the figure, the border matches the shape of the islet and implies more spikes in its outer area.

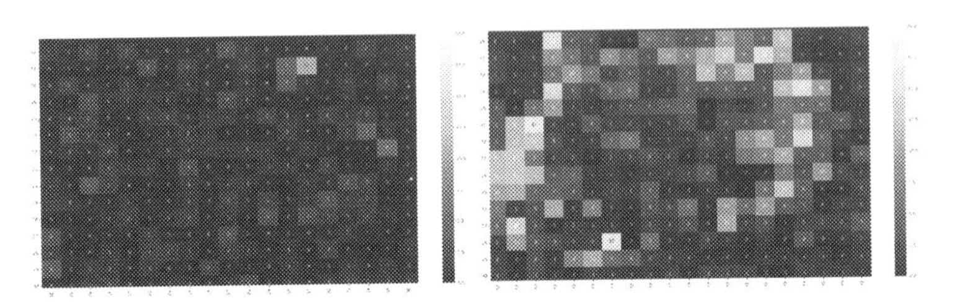

Fig. 3. Number of spikes detected within 2 min for each electrode of an islet stimulated with 3 mM (left) and 10 min after switching to 10 mM glucose (right)

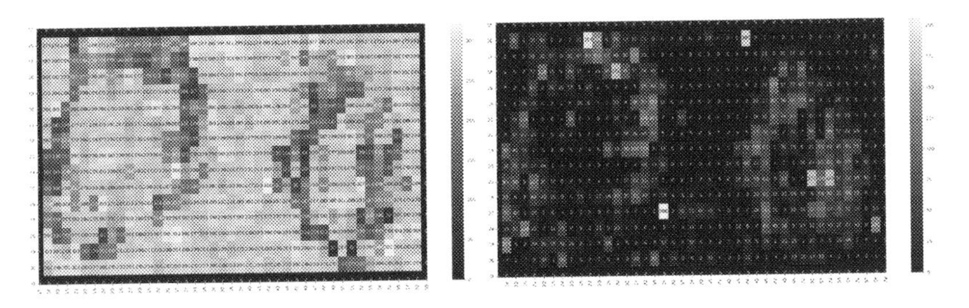

Fig. 4. Heatmaps of peak occurrences with different peak score thresholds. Left: All peaks with a positive peak score. Right: All peaks with a peak score over 100.

The second experiment examines the effect of the peak score threshold for spike classification. Here, we use a 10 mM stimulant and a deviation factor of $T_{std} = 2$. In Fig. 4 we show the cumulative spikes per sensor in a recording of two pancreatic islets. Between the two heatmaps, we changed the minimum peak score that an extreme point needs to be classified as a spike. In the left heatmap, this value is 0, and in the right one it is 100. While both plots indicate the borders of the islets, it is notable that the outer areas seem to have less activity spikes overall, but the existing spikes have higher peak scores than those in other areas. This complies with the observation in the previous experiment where a peak score of 400 was mainly present on the edge of the islet as well.

5 Discussion

The CMOS-MEA technology enables - due to its high electrode density - the measurement of the extracellular voltage changes of the individual islet cells within the islet cell association. The observed glucose-dependency of the electrical activity of the islet in the first experiment has been shown in different studies as well [7] and can be understood as a proof of concept for our method. The fact that the islet in question can be well perceived in the cumulative spike heatmap indicates the potential of our approach.

The second experiment has shown that the outer areas of the pancreatic islets had overall less spikes in electrical activity than the rest of the cells, but the measured spikes seem to be more extreme. On the contrary, the inner areas exhibit more minor spikes, to a point where it is comparable to the noise in empty areas. One reason for this could be a regional variation in activity within the islet, as it is currently postulated in the literature [5,9]. It is also possible that the detected spikes can be assigned to specific cell types, causing unevenly distributed activity (cf. Sect. 2). However, this observation requires further experiments to gain definite explanations.

6 Conclusions and Future Work

Our experiments have shown that the proposed method allows for the observation of known phenomena, proving the consistency of our results. In future work, we aim to further analyse activity within and across β-cells based on peak extraction. Specifically, we intend to adapt the signature matching distance [4] for searching similar patterns of activity between neighboring cells.

Overall, we conclude that the presented approach is promising for further cross-domain research and that further examinations in this field can lead to interesting insights in the functionality of β-cells and the network between different cells within the islet of Langerhans.

References

1. International diabetes federation. https://diabetesatlas.org. Accessed 24 Feb 2022
2. Multichannel systems. https://www.multichannelsystems.com. Accessed 28 Feb 2022
3. Azzini, I., et al.: Simple methods for peak detection in time series microarray data. In: Proceedings of CAMDA 2004 (Critical Assessment of Microarray Data) (2004)
4. Beecks, C., Kirchhoff, S., Seidl, T.: Signature matching distance for content-based image retrieval. In: Proceedings of the 3rd ACM Conference on International Conference on Multimedia Retrieval, pp. 41–48 (2013)
5. Johnston, N.R., et al.: Beta cell hubs dictate pancreatic islet responses to glucose. Cell Metab. **24**(3), 389–401 (2016)
6. Palshikar, G.: Simple algorithms for peak detection in time-series. In: Proceedings of 1st International Conference on Advanced Data Analysis, Business Analytics and Intelligence, vol. 122 (2009)
7. Pfeiffer, T., et al.: Rapid functional evaluation of beta-cells by extracellular recording of membrane potential oscillations with microelectrode arrays. Pflügers Archiv-Eur. J. Physiol. **462**(6), 835–840 (2011)
8. Rorsman, P., Ashcroft, F.M.: Pancreatic β-cell electrical activity and insulin secretion: of mice and men. Physiol. Rev. **98**(1), 117–214 (2018)

9. Satin, L.S., Zhang, Q., Rorsman, P.: "take me to your leader": an electrophysiological appraisal of the role of hub cells in pancreatic islets. Diabetes **69**(5), 830–836 (2020)
10. Scholkmann, F., Boss, J., Wolf, M.: An efficient algorithm for automatic peak detection in noisy periodic and quasi-periodic signals. Algorithms **5**(4), 588–603 (2012)

Intelligent Air Traffic Management System Based on Knowledge Graph

Jiadong Chen, Xueyan Li, Xiaofeng Gao(✉), and Guihai Chen

MoE Key Lab of Artificial Intelligence, Department of Computer Science and Engineering, Shanghai Jiao Tong University, Shanghai, China
{chenjiadong998,karroyan}@sjtu.edu.cn, {gao-xf,gchen}@cs.sjtu.edu.cn

Abstract. In recent years, there have been many studies on knowledge graphs but few studies in air management systems. Nowadays, the problem of insufficient deployment capacity like flight delay, occurs in existing civil air traffic control systems in China. The authors aim to construct the knowledge graph of air traffic control system and design related intelligent applications. This research supports some application scenarios such as flight delay analysis and so on. The authors propose a knowledge embedding model Translating-spatio-temporal Embedding (Trans-ST), supporting spatiotemporal information, which is evolved from the Translating Embedding (TransE) model after incorporating the embedding of information on time and space. The knowledge triples are mapped onto the hyperplane determined by the spatiotemporal information. The experiment is based on the real dataset and an encyclopedic dataset from YAGO. It shows that knowledge inference effect of Trans-ST is better than that of traditional model and state-of-art model.

Keywords: Knowledge graph · Spatio-temporal knowledge embedding · Aerial traffic management

1 Introduction

The goal of this study is to construct a knowledge graph for air management system and support some applications in air management system. The construction of knowledge graph is dived into four core technologies: knowledge extraction, knowledge representation, knowledge fusion and knowledge reasoning. Knowledge extraction is responsible for extracting triples of entities, attributes and relationships from various types of data sources. However, most of the data obtained is still very rough and requires subsequent cleaning, integrating and validation which will be done in the knowledge fusion step. The data will also be standardized in this step. Knowledge reasoning begins when the data has all been added into knowledge graph. This step will further expand the data volume of the knowledge graph. Some hidden relational triples can be inferred based on the data and structure of the current knowledge graph.

C. Strauss et al. (Eds.): DEXA 2022, LNCS 13427, pp. 277–283, 2022.
https://doi.org/10.1007/978-3-031-12426-6_24

Our contributions can be summarized as follows:

1. Data collection and information extraction: This study uses crawlers to crawl various websites, collects a large number of encyclopedic data sets and processes the semi-structured tabular data into structured data. Then, we extracts unstructured data such as flow control messages for subsequent use.
2. Knowledge embedding: This study proposed a knowledge embedding model supporting temporal and spatial information embedding. A comparison experiment prove that knowledge inference ability of our model is better.
3. Real-world scenario application: This study also supports the application of flight delay analysis based on the aviation knowledge graph.

2 Related Work

Knowledge graphs are mainly divided into general knowledge graphs and industry ones. General knowledge graph was first proposed by Google in 2012. Wikidata, DBpedia and YAGO, covering many semi-structured and structured data, lay the foundation for the construction of general knowledge graphs. There are also many applications including very classic internet searches and queries such as question answering system [13]. Information retrieval task [3,8,10] and knowledge inference [2,6,7] are also important tasks. General knowledge graph focuses on the breadth of knowledge rather than the depth of knowledge. Lately, the increasing scale of encyclopedic knowledge graphs (KGs) calls for personalized knowledge graph summarization [9,11].

In the existing general knowledge graph project, there is a linguistic project called wordnet. It is an English vocabulary database and maintained by Princeton University professors. Unlike ordinary dictionaries, in wordnet, all words are blocked into a synonym network between which there are various connections. General knowledge graph projects which are based on encyclopedic data include traditional Wikipedia, DBpedia.

3 Trans-ST Knowledge Embedding Model

3.1 Fact Representation

In Trans-ST model, a fact that is temporally and spatially aware takes the form of $(h, r, t, \tau_s, \tau_e, l)$. h and t represent head and tail entity, respectively. Notation r denotes the relation between h and t. τ denote the time when this fact is established. l is used to denote the geographical information of the fact. There may be some kinds of instant facts. The relation does not continue for a long time. For example, the relations such as wasBornIn and arrivedAt are all in this type. In those cases, two time are set to be the same. That can be explained as $\tau_s = \tau_e$.

3.2 Spatial-Temporal Projection of Facts

Hyte [4] represents each timestamp as a hyperplane characterized by its normal vector. When adding space information into the model, two different ways to construct the hyperplane are proposed. One is mapping the time data and space data into a new hyperplane. Another way is to construct both time hyperplane and space hyperplane.

One Hyperplane. This method divides the space into a lot of space blocks of the city according to longitude and latitude. A very straightforward approach to model construction is for each triple to combine the time and space inside it as a new piece of data. The product of the time block and the space block will be the number of hyperplanes corresponding to the spatiotemporal data. Time and space are used as the index of the rows and columns of this matrix of time-space slices. In actual operation, this matrix is flattened into a long vector and sent to the embedded layer in the neural network to train and learn. In this kind of model, the authors represent each time-space slice by vector $e_{\tau l}$. The head vector is represented by vector e_h. The relation vector is represented by vector e_r. The tail vector is represented by vector e_t. They will be translated over the hyperplane determined by time-space information $e_{\tau l}$. The projection of e_h, e_r and e_t can thus be represented as $P_{\tau l}(e_h) = e_h - (e_{\tau l})^\top e_h(e_{\tau l})$, $P_{\tau l}(e_r) = e_r - (e_{\tau l})^\top e_r(e_{\tau l})$,$P_{\tau l}(e_t) = e_t - (e_{\tau l})^\top e_t(e_{\tau l})$ The time-space vector $e_{\tau l}$ should be normalized and let $||e_{\tau l}||^2 = 1$.

Two Hyperplanes. Another model is to map time and space into two planes. For some relational triples that only have time information or only have spatial information, using the time and space index to do the embedding model will make this data wasted. The second model of this study attempts to make another attempt. This is to map time and space to a plane separately. It is the same in the first model to divide time and space into time slices and space slices. In this kind of model, the authors represent each time slice by vector e_τ. The authors represent each space slice by vector e_l. The head vector, relation vector and tail vector are still represented as e_h, e_r, and e_t. They will also be translated over the two kinds of hyperplanes. The projection on time-related hyperplane of e_h, e_r, and e_t can thus be represented as $P_\tau(e_h) = e_h - (e_\tau)^\top e_h(e_{\tau l})$, $P_\tau(e_r) = e_r - (e_\tau)^\top e_r(e_{\tau l})$,$P_\tau(e_t) = e_t - (e_\tau)^\top e_t(e_{\tau l})$, The projection on space-related hyperplane of e_h, e_r and e_t can be represented as $P_l(e_h) = e_h - (e_l)^\top e_h(e_l)$, $P_l(e_r) = e_r - (e_l)^\top e_r(e_l)$, $P_l(e_t) = e_t - (e_l)^\top e_t(e_l)$

The time-related vector e_τ should be normalized and let $||e_\tau||^2 = 1$.

The space-related vector e_l should be normalized and let $||e_l||^2 = 1$.

3.3 Translation and Loss Function

One Hyperplane. First the authors will consider a positive tuple which is a true fact in the dataset. The positive tuple and negative tuple will be discussed later in next section. A positive tuple that remains valid at the moment

τ at location l : (h, r, t, τ, τ, l). The authors expect that the mapping $P_{\tau l}(e_h) + P_{\tau l}(e_r) \approx P_{\tau l}(e_t)$ holds true. The process of making this establish is also called translation. h, r and t holds true on the spatial-temporal hyperplane determined by τ and l. In this way, the authors define the scoring function $f_{\tau l}(h, r, t) = \|P_{\tau l}(e_h) + P_{\tau l}(e_r) - P_{\tau l}(e_t)\|_{l_1/l_2}$. For an arbitrary timestamp τ and location l, the authors construct a subset of tuples in which all the facts are valid at the moment and place. Denote the subset by $\mathcal{D}^+_{\tau l}$. The authors parallel construct $\mathcal{D}^-_{\tau l}$ by negative sampling. For the moment τ and space l, the authors develop the loss function $\mathcal{L}_{\tau l} = \sum_{x \in \mathcal{D}^+_{\tau l}} \sum_{y \in \mathcal{D}^-_{\tau l}} [f_{\tau l}(x) - f_{\tau l}(y) + \gamma]_+$. $\gamma > 0$ is a margin hyperparameter. The overall loss will be calculated by $\mathcal{L} = \sum_{\tau \in |T|, l \in |L|} \sum_{x \in \mathcal{D}^+_{\tau l}} \sum_{y \in \mathcal{D}^-_{\tau l}} \left[f_{\tau l_x}(x) - f_{\tau l_y}(y) + \gamma\right]_+$.

Two Hyperplanes. The loss function of this model is in fact very similar with the previous one. The authors still consider the positive tuple only. The score of the negative tuple will be the same with the positive tuple. The positive tuple here is still: (h, r, t, τ, τ, l). One of the optimization goals will be that mapping $P_\tau(e_h) + P_\tau(e_r) \approx P_\tau(e_t)$ holds true. Another optimization goal will be that mapping $P_l(e_h) + P_l(e_r) \approx P_l(e_t)$ holds true. h, r and t should meet the basic equation $h + r = t$ on both hyperplanes determined by space l and time τ. In this model, the score function is defined as $f_\tau(h, r, t) = \|P_\tau(e_h) + P_\tau(e_r) - P_\tau(e_t)\|_{l_1/l_2}$, $f_l(h, r, t) = \|P_l(e_h) + P_l(e_r) - P_l(e_t)\|_{l_1/l_2}$.

The way to construct the positive and negative samples is the same with the previous model. They are denoted as D^+ and D^-. The loss function in this model is $\mathcal{L}_{\tau l} = \sum_{x \in \mathcal{D}^+_{\tau l}} \sum_{y \in \mathcal{D}^-_{\tau l}} [f_\tau(x) - f_\tau(y) + f_l(x) - f_l(y) + \gamma]_+$.

The overall loss of this model will be calculated by $\mathcal{L} = \sum_{\tau \in |T|, l \in |L|} \sum_{x \in \mathcal{D}^+_\tau} \sum_{y \in \mathcal{D}^-_{\tau l}} [f_\tau(x) - f_\tau(y) + f_l(x) - f_l(y) + \gamma]_+$.

4 Experiments

4.1 Baseline Models

In this experiment, to evaluate the performance of our model, several methods are used to compare. TransE [1]: This is the basic model in Trans family. All other variant models are based on it. The basic idea is also used in out Trans-ST model. **TransH** [12]**:** This is the first model which uses hyperplanes to do projection. The hyperplanes in TransH model is determined by relations. Our model Trans-ST also uses hyperplanes to embed the space and time information. **HyTE** [4]**:** This is the latest model which include time information into the process of embedding. Our model Trans-ST is partly based on the idea of HyTE, but it includes not only time information but space information.

4.2 Performance and Analysis

Entity Prediction. The results of entity prediction is shown in Table 1. It demonstrates that the performance of our first model with one hyperplane is the best in both head and tail prediction. The result of the second model with two hyperplanes is less than one hyperplane. This shows the attempt in embedding time and space information into one vector is better. Our model Trans-ST outperforms the state-of-art link prediction [5] model HyTE in both datasets. Compared with traditional TransE and TransH model, Trans-ST has a very significant process. This is very reasonable and validates our hypothesis that when time and space information are both known, the entity is easier to be confirmed.

Table 1. Mean Rank and Hits@10 for entity prediction on YAGO10K

Datasets	Metrics	TransE	*TransH*	*Hyte*	*Trans-ST*(m_1)	*Trans-ST*(m_2)
YAGO10K	Mean Rank (head)	1984	1786	1043	765	884
	Mean Rank (tail)	404	348	107	78	99
	Hits@10 (head)	1.0	1.8	15.0	23.5	20.4
	Hits@10 (tail)	2.5	4.9	36.4	46.6	40.0
ATC11K	Mean Rank (head)	2246	1952	1287	978	1023
	Mean Rank (tail)	524	451	123	98	105
	Hits@10 (head)	0.8	1.5	13.6	21.3	16.7
	Hits@10 (tail)	1.9	4.0	32.3	40.8	38.7

Relation Prediction. The results of relation prediction is shown in Table 2. It shows that the performance of our model Trans-ST(m1) is still the best in both datasets. But it does not make a big progress compared with HyTE. The authors analyze this problem. It may be the space information is determinative in entity prediction. But it cannot help that much in confirming an entity.

Table 2. Mean Rank and Hits@10 for relation prediction

Datasets	Metrics	*TransE*	*TransH*	*Hyte*	*Trans-ST*(m_1)	*Trans-ST*(m_2)
YAGO10K	Mean Rank	1.92	1.63	1.33	1.21	1.35
	Hits@10 (head)	70.6	72.8	79.2	82.2	77.9
ATC11K	Mean Rank	2.30	1.87	1.54	1.33	1.49
	Hits@10 (head)	67.3	69.8	76.4	80.7	78.0

The attempts of embedding space and time information is still not very mature. It can help in link prediction [5] task, but there still remains improvement.

5 Conclusion

This paper successfully constructs the prototype of the knowledge graph in the field of air traffic control. In addition to semi-structured data and structured data, this study also deals with non-structural data. This paper uses the knowledge embedding model to map the relational triples in the graph database into the vector space. This paper first proposes a model Trans-ST that embeds time and space information into the knowledge graph. The model maps time and space-related information into a hyperplane, allowing the traditional triplet optimization function to be implemented on that hyperplane. The experimental results show that the proposed model is more applicable to the aviation-related data with time and space dimensions.

Acknowledgement. This work was supported by the National Key R&D Program of China [2018YFB1004700]; the National Natural Science Foundation of China [61872238, 61972254]; and the State Key Laboratory of Air Traffic Management System and Technology [SKLATM20180X].

References

1. Bordes, A., Usunier, N., Garcia-Duran, A., Weston, J., Yakhnenko, O.: Translating embeddings for modeling multi-relational data. In: Advances in Neural Information Processing Systems (NeurIPS), pp. 2787–2795 (2013)
2. Chakrabarti, S.: Knowledge extraction and inference from text: shallow, deep, and everything in between. In: ACM International ACM SIGIR Conference on Research & Development in Information Retrieval (SIGIR), pp. 1399–1402 (2018)
3. Corcoglioniti, F., Dragoni, M., Rospocher, M., Aprosio, A.P.: Knowledge extraction for information retrieval. In: European Semantic Web Conference, pp. 317–333 (2016)
4. Dasgupta, S.S., Ray, S.N., Talukdar, P.: Hyte: Hyperplane-based temporally aware knowledge graph embedding. In: Conference on Empirical Methods in Natural Language Processing (EMNLP). pp. 2001–2011 (2018)
5. Dong, Y., et al.: Link prediction and recommendation across heterogeneous social networks. In: IEEE International Conference on Data Mining (ICDM), pp. 181–190 (2012)
6. Guo, S., Wang, Q., Wang, L., Wang, B., Guo, L.: Knowledge graph embedding with iterative guidance from soft rules. In: AAAI Conference on Artificial Intelligence (AAAI) (2018)
7. Hamilton, W., Bajaj, P., Zitnik, M., Jurafsky, D., Leskovec, J.: Embedding logical queries on knowledge graphs. In: Advances in Neural Information Processing Systems (NeurIPS), pp. 2026–2037 (2018)
8. Liu, Y., Xu, B., Yang, Y., Chung, T., Zhang, P.: Constructing a hybrid automatic q&a system integrating knowledge graph and information retrieval technologies. In: Foundations and Trends in Smart Learning, pp. 67–76 (2019)
9. Muhammad, I., Kearney, A., Gamble, C., Coenen, F., Williamson, P.: Open information extraction for knowledge graph construction. In: Kotsis, G., Tjoa, A.M., Khalil, I., Fischer, L., Moser, B., Mashkoor, A., Sametinger, J., Fensel, A., Martinez-Gil, J. (eds.) DEXA 2020. CCIS, vol. 1285, pp. 103–113. Springer, Cham (2020). https://doi.org/10.1007/978-3-030-59028-4_10
10. Sun, H., Bedrax-Weiss, T., Cohen, W.W.: PullNet: open domain question answering with iterative retrieval on knowledge bases and text. arXiv:1904.09537 (2019)
11. Safavi, T., et al.: Personalized knowledge graph summarization: from the cloud to your pocket. In: IEEE International Conference on Data Mining (ICDM), pp. 528–537 (2019)

12. Wang, Z., Zhang, J., Feng, J., Chen, Z.: Knowledge graph embedding by translating on hyperplanes. In: AAAI Conference on Artificial Intelligence (AAAI) (2014)
13. Lan, Y., Wang, S., Jiang, J.: Multi-hop knowledge base question answering with an iterative sequence matching model. In: IEEE International Conference on Data Mining (ICDM), pp. 359–368 (2019)

Analytical Algebra: Extension of Relational Algebra

Jakub Peschel(✉), Michal Batko(✉), and Pavel Zezula(✉)

Masaryk University, Brno, Czechia
{jpeschel,batko,pzezula}@mail.muni.cz
https://disa.fi.muni.cz/

Abstract. In the context of contemporary data, the processing of information is crucial. This paper proposes an extension to the traditional database relational algebra, which enriches the data model and provides additional complex-data operations. Specifically, we focus on analytical operators from the areas of data mining and similarity search, such as frequent pattern mining or similarity search queries. The proposed approach can be easily extended by additional algebraic operators. To demonstrate the capabilities of our analytical algebra, we show three practical use cases with different levels of the expression complexity.

Keywords: Analytical algebra · Relational algebra · Analytical operator

1 Introduction and Related Work

Relational database management systems (RDBMS) provide the traditional way of managing structured data. Relational algebra is their internal representation of the queries that are used to retrieve the stored data. This representation is useful because the RDBMS query optimizer can use transformation rules to lower the costs of the query execution.

However, relational algebra was primarily designed for primitive data types, such as strings and numbers. Even though extensions for various complex data types exist (e.g. geo-spatial extensions [13]), they are mainly used for specialized tasks and therefore are of limited use in the general data processing. Moreover, the equality-based predicates, typically used for filtering the data in relational algebra, are hardly usable for various modern data types such as images, where the equality needs to be relaxed to similarity.

Another set of operations not covered by the traditional relational algebra is the data mining techniques that allow for discovering new knowledge hidden in the data. Examples of interesting knowledge discovery techniques are frequent

This research was supported by ERDF "CyberSecurity, CyberCrime and Critical Information Infrastructures Center of Excellence" (No. CZ.02.1.01/0.0/0.0/16_019/0000822).

C. Strauss et al. (Eds.): DEXA 2022, LNCS 13427, pp. 284–290, 2022.
https://doi.org/10.1007/978-3-031-12426-6_25

item-set mining, sub-sequence mining and graph mining for the discovery of association rules.

Thus, we propose an extension of the relational data model with a data type for collections and a set of new operators that allow using similarity-based queries and frequent pattern mining. Because there are many different techniques for data processing and new ones emerge over time, we design our approach to be further extensible.

The benefit of such a model stems from the possibility of storing and processing new types of data and the possibility of using new techniques inside the RDBMS previously available only in third-party tools. Other benefits are the ability to abstract from algorithmic details of individual techniques and describe a process of extracting information from the data in a way that allows the application of query optimization techniques.

Related Work. There have been several attempts to extend the relational data model on different levels over the years that focus on specific aspects of the problem. In the following section, several of the approaches are described.

In the area of similarity search, an extension of relational algebra was proposed in [1]. This algebra provides a general abstraction of the objects and similarity measures and provides algebraic operations incorporated in relational algebra. However, the paper does not tackle the operator extensibility, so it does not support the data mining operations.

Similarly, the extension of SQL called SimSeQL [5] provides a similarity search through the engine MESSIF in the form of "blade". These approaches process queries outside of the systems and prevents the utilisation of RDBMS tools like query optimisation.

In the area of data mining, algebra for a description of data mining queries with the possibility of incorporation into a relational model was proposed in [6]. This algebra is based on 3W-model [10] and provides a closed set of operators for data mining queries. The closeness of the algebra makes it impossible to adapt it to emerging topics in data mining as well as incorporate other areas of data analysis.

2 Analytical Algebra

As anticipated above, relational algebra provides a suitable abstraction for data processing but lacks complex-data operators that are needed for various data analytical tasks. In the work [12], the formalisation of several analytical tasks has been proposed. However, a special data model was used there, and the approach lacked some necessary methods (e.g. for filtering the data) to express complex use-cases such as [11]. Therefore, we decided to extend the traditional relational algebra into *analytical algebra*: we add an array type usable in the relational schema, and we define several operators for effective use of similarity search and data mining.

2.1 Array Type

Multiple data analytical tasks work with complex data structures, which often contain sequential, or temporal elements, describe the multi-dimensionality of its objects or consist of multiple parts. These can be effectively modelled using the array approach known from programming languages. In fact, the support for an array data type is already present in some of the modern RDBMS, so the analytical algebra operators can be easily implemented.

The array is declared by a statement of the inner domain followed by square brackets and may contain a number of objects; for example, `NUMBER[5]` represents an array containing five numbers, i.e. a five-dimensional vector, or `STRING[]` is an array of an unlimited number of strings, e.g. a list of words of a document.

For the manipulation of the arrays, following functions are applicable: Index Access ([]), Length ($len()$), Contains ($contains()$), Subset ($subset()$), Concatenation ($cat()$) Deduplication ($distinct()$), Ordering ($order()$), Equals ($=$) and Unnest ($unnest()$). To create an array, an aggregate function $nest()$ can be used.

2.2 Analytical Relation and Analytical Operators

Relational algebra operates over data stored in the relations. The relation is a Cartesian product of data domains, and atomic data types are considered for the purpose of relational algebra. We will denote these data types as standard types. We define *analytical relation* as a Cartesian product of standard types and array-type for analytical algebra. Thus, the standard relation is a special case of the analytical relation, which only contains the standard type attributes.

Analytical relation can be described with analytical relation's schema. The relational schema is usually denoted with capital letters and contains names of the individual attributes followed by types of the attributes, which can be omitted for readability reasons if the type can be deduced. Example of relational schema for relation a is:

$$A(name : STRING, age : NUMBER, classes : STRING[])$$

The main building blocks of analytical algebra are the *analytical operators.* These are data analytical functions that are applied to analytical relations to provide new relations as a result. We divide these operations into mining operators, similarity-based operators, and relational operators.

Due to the enormous size of the data analytical field, we restrained our focus on an area of frequent pattern mining and similarity search. However, other areas can be added to the analytical algebra in the same manner due to the extensible nature of the analytical algebra.

Mining Operators. In mining operators, we present a collection of frequent pattern mining tasks. In these tasks, the user defines a frequency threshold determining a minimal number of pattern occurrences in an analytical relation needed to include the pattern in the result.

Frequent Item-Set Mining. The task analyses the collection of sets for the frequently occurring subset. The task was introduced as a problem of obtaining items bought frequently in supermarkets, sometimes referred to as market basket analysis [2].

The frequent item-set mining operator is denoted by IM and, as an argument, takes an array-type attribute and minimal frequency threshold. The operator treats the array-type attribute values as a collection of sets. Duplicity and orders are forgotten for individual arrays, resulting in the collection of sets.

The IM operator follows a formula:

$$IM_{r.a,\theta}(r) = \{(p, f) | \exists t_a \in r.a, p \subseteq t_a; f = count(p, r.a), f \geq \theta\}$$

where $r.a$ is a column of array-type, t_a is one array, θ is a minimal frequency threshold defined by the user, p is a frequently occurring subset and f and the function $count(p, r.a)$ counts the number of occurrences of the frequent pattern p.

The output relation's schema is:

$$IM(pattern : A[], frequency : NUMBER)$$

where A is a domain of the attribute $r.a$.

Due to the statistical differences in datasets, there is a possibility to select different families of algorithms to optimise the time [9].

Frequent Sequence Mining. For the analysis of frequent subsequences, the task of frequent sequence mining was proposed in [2]. The array type is suitable for this task, and the input relation can be viewed without modification.

Frequent sequence mining is denoted as SM and, similarly to frequent item-set mining, accept array-type attribute and minimal frequency threshold as an input. The SM operator follows a formula:

$$SM_{r.a,\theta}(r) = \{(p, f) | \exists t_a \in r.a, p \subseteq^* t_a; f = count(p, r.a), f \geq \theta\}$$

where $r.a$ is a column of array-type, t_a is one array, θ is a minimal frequency threshold defined by the user, p is a frequently occurring subsequence and f and the function $count(p, r.a)$ counts the number of occurrences of the frequent pattern p. In the context of this formula, $\subseteq^*$ symbol is used to denote the subsequence or the equal sequence, similarly to usage in the case of sets.

The output relation's schema is the same as for the frequent item-set mining:

$$SM(pattern : A[], frequency : NUMBER)$$

where A is domain of the attribute $r.a$.

There have been proposed several different approaches to frequent sequence mining that can be similar to frequent item-sets mining, divided into three main families Apriori-based [3], vertically-based [4] and tree based [8].

Due to the bigger number of characteristics of the data, it is expected that the differences in the different algorithm's performances may be even bigger than in the frequent item-set mining, but more research is needed.

Similarity-Based Operators. The second group of operators is similarity-based operators. These operations use distance computations to estimate the similarity or dissimilarity of objects. In the following paragraphs, we denote the distance function with *dist* and is expected to return NUMBER.

There exist numerous distance functions; one notable group are metric functions. The distance function is classified as a metric function if it meets three metric postulates: the identity of indiscernibles, symmetry, triangle inequality [14].

This paper describes just two join operators: range join and k-nn join. The reason for just these two operators is that typical k-nn search and range search are just special cases of these two join operators as described at the end of each operator.

Range Join. The first described similarity join operator is range join [7]. This operator combines two relations based on the distance between compared attributes. Computed distance is compared against a user-defined threshold, and the pair are combined if the distance is smaller than the defined threshold.

We denote this operation with RJ, and it follows a formula:

$$\begin{aligned} RJ_{dist(q.a,r.b),d}(q,r) = \{&t \in q \times r \times \mathcal{R}, t = (a,\ldots,b,\ldots,distance)| \\ &t.distance = dist(t.a,t.b), t.distance \leq d\} \end{aligned}$$

where $q.a$ is an attribute of the relation q, $r.b$ is an attribute of the relation r, $dist$ is a used distance function able to compare a similarity between domains of $q.a$ and $r.a$, $t.a$ and $t.b$ are corresponding attributes in a composition of tuples from r and q and d is a maximal distance threshold.

The output relation's schema for the resulting relation of this operator is:

$$RJ = Q + R + (distance : NUMBER)$$

where Q is relation's schema of relation q of R is relation's schema of relation q.

K-NN Join. The second similarity-based operator is k nearest neighbours join. It is a binary operator that retrieves a fixed amount of objects from data relation for each object from query relation. The benefit of this operator is the knowledge of the maximal size of the result compared to the unpredictability of the result of the range join.

The operator is denoted $KNNJ$ and follows a formula:

$$\begin{aligned} KNNJ_{dist(q.b,r.a),k}(q,r) = \{&t \in q \times r \times \mathcal{R}, t = (a,\ldots,b,\ldots,distance)| \\ &t.distance = dist(t.a,t.b)\} = X, |X| = k \wedge \forall x \in X, \\ &\forall y \in q \times r \times \mathcal{R} - X : x.distance \leq y.distance \end{aligned}$$

where a is attribute, q is object with same domain as a, $dist$ is a used distance function and k is the number of retrieved objects.

The output relation's schema is:

$$KNNJ = Q + R + (distance : NUMBER)$$

where Q is relation's schema of q, and R is relation's schema of q.

3 Examples of Usage

In this section, we focus on the usage of the proposed algebra. We prepared two examples with different levels of complexity.

Community Mining. With the growth of social networks, the task of detecting communities grew in importance. Problems with communities stem from a lack of formal definitions. As such, *communities* are vaguely defined as more densely interconnected nodes of networks. One of the definitions proposed in [11] defines communities as two-part structures consisting of dense cores and interconnected surroundings. The following algebraic expression can describe the process of obtaining such communities:

$$\begin{aligned}
&let\ graph &&= (source, nbrs) && (1)\\
&let\ mgraph &&= \rho_{graph(source,nbrs)}(\Pi_{source,cat(nbrs,source)}(graph)) && (2)\\
&let\ cores &&= \rho_{cores(members,frequency)}\sigma_{frequency>len(pattern)\wedge} && \\
& && \quad {}_{\wedge len(pattern)>\theta}(IM_{mgraph.nbrs,\theta}(mgraph)) && (3)\\
&let\ assign &&= RJ_{ComDist(cores.members,mgraph.nbrs),d}(cores, mgraph) && (4)\\
&let\ aggregate &&= \rho_{aggregate(core,members)}\ {}_{query}\mathcal{G}_{nest(source)}(assign) && (5)\\
&let\ c_check &&= aggregate \bowtie_{subset(aggregate.members,cores.pattern)} cores && (6)\\
&let\ comms &&= {}_{members}\mathcal{G}(c_check) && (7)
\end{aligned}$$

Public Space Analysis. These days, most public space is being captured by surveillance cameras. The analysis of such recordings could improve several areas, such as ad targeting for groups and improvements in investigations of security forces. The process of frequently co-occurring people in public space can be described with the following algebraic expression:

$$\begin{aligned}
&let\ occurences &&= (face, time) && (1)\\
&let\ o_1 &&= \rho_{o_1}(occurences) && (2)\\
&let\ o_2 &&= \rho_{o_2}(occurences) && (3)\\
&let\ face_pairs &&= \rho_{f(f_1,t_1,f_2,t_2,dist)}(RJ_{FaceDist(o_1.face,o_2.face)}(o_1, o_2)) && (4)\\
&let\ fgroups &&= \rho_{g(id,faces,times)}({}_{f_1}\mathcal{G}_{nest(f_2),nest(t_2)}(face_pairs)) && (5)\\
&let\ time_bins &&= {}_{time}\mathcal{G}_{nest(id)}(\rho_{t(time,face)}(\Pi_{unnest(times),id}(fgroups))) && (6)\\
&let\ groups &&= IM_{time_bins.bins,\theta}(\rho_{time_bins(time,bins)}(time_bins)) && (7)
\end{aligned}$$

4 Conclusion

In this paper, we described the concept of analytical algebra as an extension of relational algebra; we described array data type and provided a basic set of functions for this type. We proposed the formalisation of several analytical operators that can be used to create complex data analytical queries. In the use-case section, we showed the relevance of this approach for the task of community mining and proposed other use-cases for the analysis of public space and market transactions.

References

1. Adali, S., Bonatti, P., Sapino, M.L., Subrahmanian, V.: A multi-similarity algebra. In: Proceedings of the 1998 ACM SIGMOD International Conference on Management of Data, pp. 402–413 (1998)
2. Agrawal, R., Imieliński, T., Swami, A.: Mining association rules between sets of items in large databases. In: Proceedings of the 1993 ACM SIGMOD International Conference on Management of Data, pp. 207–216 (1993)
3. Agrawal, R., Srikant, R.: Mining sequential patterns. In: Proceedings of the Eleventh International Conference on Data Engineering, pp. 3–14. IEEE (1995)
4. Ayres, J., Flannick, J., Gehrke, J., Yiu, T.: Sequential pattern mining using a bitmap representation. In: Proceedings of the Eighth ACM SIGKDD International Conference on Knowledge Discovery and Data Mining, pp. 429–435 (2002)
5. Budikova, P., Batko, M., Zezula, P.: Query language for complex similarity queries. In: Morzy, T., Härder, T., Wrembel, R. (eds.) ADBIS 2012. LNCS, vol. 7503, pp. 85–98. Springer, Heidelberg (2012). https://doi.org/10.1007/978-3-642-33074-2_7
6. Calders, T., Lakshmanan, L.V., Ng, R.T., Paredaens, J.: Expressive power of an algebra for data mining. ACM Trans. Database Syst. (TODS) **31**(4), 1169–1214 (2006)
7. Dohnal, V., Gennaro, C., Zezula, P.: Similarity join in metric spaces using eD-index. In: Mařík, V., Retschitzegger, W., Štěpánková, O. (eds.) DEXA 2003. LNCS, vol. 2736, pp. 484–493. Springer, Heidelberg (2003). https://doi.org/10.1007/978-3-540-45227-0_48
8. Han, J., et al.: PrefixSpan: mining sequential patterns efficiently by prefix-projected pattern growth. In: Proceedings of the 17th International Conference on Data Engineering, pp. 215–224. Citeseer (2001)
9. Heaton, J.: Comparing dataset characteristics that favor the Apriori, Eclat or FP-growth frequent itemset mining algorithms. In: SoutheastCon 2016, pp. 1–7. IEEE (2016)
10. Johnson, T., Lakshmanan, L.V., Ng, R.T.: The 3W model and algebra for unified data mining. In: VLDB, pp. 21–32 (2000)
11. Peschel, J., Batko, M., Valcik, J., Sedmidubsky, J., Zezula, P.: FIMSIM: discovering communities by frequent item-set mining and similarity search. In: Reyes, N., et al. (eds.) SISAP 2021. LNCS, vol. 13058, pp. 372–383. Springer, Cham (2021). https://doi.org/10.1007/978-3-030-89657-7_28
12. Peschel, J., Batko, M., Zezula, P.: Algebra for complex analysis of data. In: Hartmann, S., Küng, J., Kotsis, G., Tjoa, A.M., Khalil, I. (eds.) DEXA 2020, Part I. LNCS, vol. 12391, pp. 177–187. Springer, Cham (2020). https://doi.org/10.1007/978-3-030-59003-1_12
13. Strobl, C.: PostGIS. In: Shekhar, S., Xiong, H. (eds.) Encyclopedia of GIS, pp. 891–898. Springer, Boston (2008). https://doi.org/10.1007/978-0-387-35973-1_1012
14. Zezula, P., Amato, G., Dohnal, V., Batko, M.: Similarity Search: The Metric Space Approach, vol. 32. Springer Science, New York (2006). https://doi.org/10.1007/0-387-29151-2

BLOCK-OPTICS: An Efficient Density-Based Clustering Based on OPTICS

Kota Yukawa[1(✉)] and Toshiyuki Amagasa[2]

[1] Graduate School of Science and Technology, University of Tsukuba, Tsukuba, Japan
yukawa@kde.cs.tsukuba.ac.jp

[2] Center for Computational Sciences, University of Tsukuba, Tsukuba, Japan
amagasa@cs.tsukuba.ac.jp

Abstract. This paper proposes an efficient density-based clustering method based on OPTICS. Clustering is an important class of unsupervised learning methods that group data points based on similarity, and density-based clustering detects dense regions of data points as clusters. The ordering points to identify the clustering structure (OPTICS) is one of such methods that has been widely used due to its advantages over the predecessor DBSCAN. The computational cost of OPTICS is not too high ($O(n \log n$ where n is the number of data points), but it still suffers from a long execution time when the input size is large and/or the number of dimensions is high. The reason is that the algorithm requires massive distance calculation for each data point against other points to calculate the density around it. To make it more efficient, we propose BLOCK-OPTICS, which allows us to perform OPTICS more efficiently by eliminating unnecessary distance calculations for high-density areas. The experimental results using a synthetic dataset and several real-world datasets show that BLOCK-OPTICS is much faster than the original OPTICS while maintaining the same results.

Keywords: Clustering · DBSCAN · OPTICS

1 Introduction

Clustering is one of the most important data analysis methods by which we can cluster data points according to their similarity (or distance). Since it does not require any training dataset, it can be categorized as one of the unsupervised methods and are used for different objectives, such as data preprocessing, data summarization, etc.

There are many clustering methods, but density-based clustering has been widely used in many applications among other methods. The idea of density-based clustering is to detect the areas where data points exist at high density as clusters, and it offers many advantages, such as it can detect clusters of arbitrary

C. Strauss et al. (Eds.): DEXA 2022, LNCS 13427, pp. 291–296, 2022.
https://doi.org/10.1007/978-3-031-12426-6_26

shapes and different sizes and it is less sensitive to noisy data. DBSCAN [5] is the pioneering work of density-based clustering, and there are many successors. OPTICS [1] is an extension of DBSCAN by addressing the problem of detecting meaningful clustering with varying densities. More precisely, it orders the data points according to their spatial proximity and records for each point some density parameters, making it possible to detect hierarchical clusters without performing clustering by varying parameters.

Besides, Chen et al. have proposed an improvement of DBSCAN by eliminating unnecessary distance calculations for such data points that reside in dense areas [3]. As a result, they successfully reduced the execution time of DBSCAN without affecting the output. Therefore, we thought we could apply this idea to OPTICS to improve performance. However, none of the existing works has tried it.

This paper proposes BLOCK-OPTICS to improve the performance of OPTICS by employing the similar idea proposed by BLOCK-DBSCAN [3]. Specifically, it eliminates the distance calculations for data points that do not require density calculations, improving efficiency. The experimental evaluation using synthetic and real-world datasets shows that BLOCK-OPTICS successfully reduces the execution time without affecting the output.

2 Proposed Method

2.1 Approach

This section proposes the proposed scheme, *BLOCK-OPTICS*, which allows us to perform OPTICS clustering quicker. The idea is to eliminate unnecessary distance computations when the point is in a dense area and is a core point. This idea is inspired by *BLOCK-DBSCAN* [3], and the key theorem is as follows:

Theorem 1. *Let $N_r(p)$ denote the number of points from point p within radius r. If $|N_{\frac{\epsilon}{2}}(p_i)| \geq MinPts$, $\forall q \in N_{\frac{\epsilon}{2}}(p_i)$ is a core point, i.e., there are at least $MinPts$ neighbors within distance ϵ.*

Proof. When $\forall p_j, p_k \in N_{\frac{\epsilon}{2}}(p_i)$, $d_{i,j} \leq \frac{\epsilon}{2}$, $d_{i,k} \leq \frac{\epsilon}{2}$. From the trigonometric theorem on distance, we get $d_{j,k} \leq d_{i,j} + d_{i,k} \leq \epsilon$. This means that the distance between any two points within p_i distance $\frac{\epsilon}{2}$ is less than or equal to ϵ. Therefore, for q satisfying $\forall q \in N_{\frac{\epsilon}{2}}(p_i)$, $N_{\frac{\epsilon}{2}}(p_i) \subset N_{\frac{\epsilon}{2}}(q)$. For this reason, if $|N_{\frac{\epsilon}{2}}(p_i)| \geq MinPts$, $|N_{\frac{\epsilon}{2}}(p)| \geq MinPts$ is also satisfied. Consequently, $\forall q \in N_{\frac{\epsilon}{2}}(p_i)$ is a core point. □

In the proposed algorithm, each *inner-core point* $\boldsymbol{q}$ in the *inner-core block*, satisfying $|N_{\frac{\epsilon}{2}}(p_i)| \geq MinPts$, covers all points in the inner-core block within the range ϵ. Consequently, once we detect an inner-core block, we can immediately mark all inner-core points as core points without further investigation.

2.2 BLOCK-OPTICS Algorithm

The ExpandClusterOrder algorithm in BLOCK-OPTICS is shown in the Algorithm 1.

Algorithm 1. BLOCK-OPTICS:ExpandClusterOrder

```
Input: X // Vector set
    x // Vector
    ε // Distance threshold
    MinPts // Density threshold
    OrderdList
1: x.processed() // x to be processed
2: neighbors ← rangeQuery(x, ε)
3: x.core_distance ← CoreDistance(x, neighbors, MinPts)
4: OrderdList.append(x)
5: if x.core_distance! = UNDEFINED then
6:    seeds = PtiorityQueue()
7:    if p.core_distance ≤ ε/2 // p is a touched-inner core point then
8:       inner_core_block, not_inner_core_block
         ←devide_block(neighbors)
9:       process(inner_core_block)
10:      update(x, not_inner_core_block, seeds)
11:   else
12:      update(x, neighbors, seeds)
13:   while not seeds.empty() do
14:      y ← seeds.pop()
15:      if not y.processed then
16:         y.process() // y to be processed.
17:         neighbors ← rangeQuery(y, ε)
18:         y.core_distance ← CoreDistance(y, neighbors, MinPts)
19:         OrderdList.append(y)
20:         if y.core_distance! = UNDEFINED then
21:            if p.core_distance ≤ ε/2 // q is a Touched Inner Core Point then
22:               inner_core_block, not_inner_core_block
                  ←devide_block(neighbors)
23:               process(inner_core_block)
24:               update(y, not_inner_core_block, seeds)
25:            else
26:               update(y, neighbors, seeds)
```

The computing of the core distance in order is similar to that of the ExpandClusterOrder algorithm in OPTICS. The ϵ-neighborhood points are calculated by rangeQuery sequentially from the extracted points. Expanding the cluster is also performed for the ϵ-neighborhood points if the points form a cluster.

The conditional branches in Lines 7 and 21 indicate that the current point is a *touched-inner core point*. In this case, the ϵ-neighbor points are divided into *inner_core_block* or *not_inner_core_block* in Lines 8 and 22. *inner_core_block* is already a core point like a core point in DBSCAN without calculating core distance based on the theorem. Since these points are already marked as core points forming a cluster, all inner-core points are processed in Lines 9 and 23; i.e., process(*inner_core_block*) is the function that processes the points *inner_core_block*. Besides, those points *not_inner_core_block* being an ϵ- neighborhood but not in *inner_core_block* are processed using the update algorithm to updates the priority queue *seeds* as ordinary OPTICS in Lines 10 and 24.

In the meantime, in Lines 12 and 26, the current point is not a *touched-inner core point* but an outer-core point. In this case, the *seeds* are updated using all ϵ-neighbors as in the original OPTICS. Note again that, as described above, when the current point is a touched-inner core point, we can eliminate the calculations of core distance and reachability distance for the inner-core points in the current point's inner-core block, thereby making the process more efficient than the original OPTICS.

2.3 Merging Inner-Core Blocks

Similar to the original OPTICS, the output of the BLOCK-OPTICS algorithm is an ordered list of points according to the reachability distance, except that no reachability distance is calculated for those points in inner-core blocks. For this reason, we cannot judge only by the reachability distance whether the subsequent block is reachable or not.

To address this problem, in BLOCK-DBSCAN, we apply merging process for inner-core blocks. Let *icbs* be a set of extracted inner-core blocks. For $\exists(icb_1, icb_2) \in icbs$, if $core\_distance(icb_1) \leq \epsilon, core\_distance(icb_2) \leq \epsilon$ holds, then we need to consider the following three cases depending on the minimum distance between the two blocks. Let tic_1 and tic_2 be touched-inner core points of the corresponding blocks.

1. $dist(tic_1, tic_2) \leq \epsilon$: the two blocks are always reachable, and we merge the two blocks as a single cluser.
2. $\epsilon < dist(tic_1, tic_2) < 2\epsilon$: we cannot determine whether the two blocks are reachable, and we compute reachability distances as in original OPTICS.
3. $2\epsilon \leq dist(tic_1, tic_2)$: two blocks are not reachable, and hence we do not merge them.

The concrete algorithm for merging clusters is similar to that in BLOCK-DBSCAN [3].

3 Experiments

To experimentally evaluate the effectiveness of the proposed method, BLOCK-OPTICS, we have conducted a set of experiments using synthetic and real-world datasets. This section presents the details of the experimental study.

3.1 Experimental Settings

In the following experiments, we set the number of dimensions is 2 and the number of data 1,000 as a synthetic dataset. We used the MoCap dataset [4] and the APS dataset [2] as the real-world datasets. To make the input data, we have applied the preprocessing: first, we completed the missing values as zero;

and we removed *User* and *Class* attributes for the MoCap dataset and *class* attribute for the APS dataset, respectively. Note that the datasets' size and dimensionality were 78,095 records/36 dimensions for the MoCap dataset and 76,000 records/170 dimensions for the APS dataset.

Both OPTICS and BLOCK-OPTICS need hyperparameters: distance threshold ϵ and density threshold $MinPts$. For the synthetic dataset, we used $(\epsilon, MinPts) = (0.2, 10)$. For the real-world datasets, we used the following sets of parameters: $(\epsilon, MinPts) = \{(70000, 5), (70000, 10)\}$ for the MOCAP dataset and $(\epsilon, MinPts) = \{(75000, 5), (75000, 10)\}$ for the APS dataset.

3.2 Clustering Experiments on Synthetic Dataset

The clustering results of OPTICS and BLOCK-OPTICS on the synthetic dataset are shown in Fig. 1. The two scatter plots show that the two algorithms produce the same clustering results.

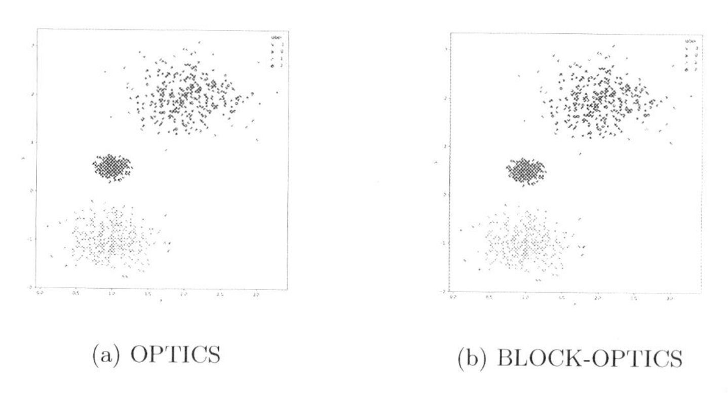

(a) OPTICS (b) BLOCK-OPTICS

Fig. 1. Clustering results for synthetic dataset.

3.3 Experiments with Real-World Datasets

Table 1. Execution time for real-world datasets.

Dataset	Method	ϵ	$MinPts$	ExpandClusterOrder [s]	Merge [s]
MoCap [4]	OPTICS	70000	5	13401.65	–
	BLOCK-OPTICS	70000	5	3833.03	70.32
	OPTICS	70000	10	14972.03	–
	BLOCK-OPTICS	70000	10	3055.32	24.7528
APS [2]	OPTICS	75000	5	201547.55	–
	BLOCK-OPTICS	75000	5	312.46	3317.38
	OPTICS	75000	10	207633.93	–
	BLOCK-OPTICS	75000	10	363.45	2931.78

Table 1 summarizes the results of the runtime comparison with two real-world datasets. The proposed method, BLOCK-OPTICS, significantly outperformed the original OPTICS in execution time. In the MOCAP dataset, the inner-core point excluded from the calculation was about 60%, while more than 90% of the points were detected as inner-core points in the APS dataset, resulting in a better performance. This is because the APS dataset is very densely clustered. The BLOCK-OPTICS method can efficiently perform clustering on datasets with dense clusters by eliminating redundant distance computations. If we look into the time breakdown, the fraction of the merging process is smaller for the MOCAP dataset. This result is because fewer touched-inner core blocks are detected in the MOCAP dataset than in the APS dataset, resulting in fewer merging operations. As a result, the fraction of execution time of the merging process to total execution time became smaller. Nevertheless, the time for merging is far shorter than that for *ExpandClusterOrder*.

4 Conclusions

In this paper, we have proposed a method for speeding up the density-based clustering method OPTICS, called BLOCK-OPTICS. The proposed method employs the concept of inner-core points proposed by BLOCK-DBSCAN, eliminating redundant distance calculations for the regions where data points densely exist. More precisely, we can skip distance calculations for the points in inner-core blocks, which turn out to be core points. Experimental results on synthetic and real-world datasets have shown that the proposed method performs clustering more efficiently than the original OPTICS. In the future, we plan to speed up clustering for more advanced algorithms, such as hierarchical clustering using HDBSCAN [6].

Acknowledgement. This paper was supported by Japan Society for the Promotion of Science (JSPS) KAKENHI under Grant Number JP22H03694 and the New Energy and Industrial Technology Development Organization (NEDO) Grant Number JPNP20006.

References

1. Ankerst, M., Breunig, M.M., Kriegel, H.P., Sander, J.: Optics: ordering points to identify the clustering structure. ACM SIGMOD Rec. **28**(2), 49–60 (1999)
2. APS failure at Scania trucks data set. https://archive.ics.uci.edu/ml/datasets/APS+Failure+at+Scania+Trucks
3. Chen, Y., Zhou, L., Bouguila, N., Wang, C., Chen, Y., Du, J.: Block-DBSCAN: fast clustering for large scale data. Pattern Recogn. **109**, 107624 (2021)
4. Mocap hand postures data set. https://archive.ics.uci.edu/ml/datasets/MoCap+Hand+Postures
5. Ester, M., Kriegel, H.P., Sander, J., Xu, X., et al.: A density-based algorithm for discovering clusters in large spatial databases with noise. In: KDD, vol. 96, pp. 226–231 (1996)
6. McInnes, L., Healy, J., Astels, S.: hdbscan: hierarchical density based clustering. J. Open Source Softw. **2**(11), 205 (2017)

What's Next? Predicting Hamiltonian Dynamics from Discrete Observations of a Vector Field

Zi-Yu Khoo(✉), Delong Zhang, and Stéphane Bressan

National University of Singapore, 21 Lower Kent Ridge Road, 119077 Singapore, Singapore
{khoozy,zhangdel,steph}@comp.nus.edu.sg

Abstract. We present several methods for predicting the dynamics of Hamiltonian systems from discrete observation of their vector field. Each method is either informed or uninformed of the Hamiltonian property. We empirically and comparatively evaluate the methods and observe that information that the system is Hamiltonian can be effectively informed, and that different methods strike different trade-offs between efficiency and effectiveness for different dynamical systems.

Keywords: Trajectory prediction · Physics-informed neural network · Data analysis · Inductive bias · Learning bias

1 Introduction

The prediction of the dynamics of systems is a relevant and crucial task to many applications in domains ranging from the hard to social sciences. The dynamics of a system is captured by the vector field formed by the time derivatives of the variables describing the system's current state. The system evolves along flow lines in the state space. The flow map is the function that, given an initial state and a time interval, outputs the state of the system after the time interval.

A Hamiltonian system [10] is a dynamical system governed by Hamilton's equations. The Hamiltonian property indicates the conservation of some quantity, typically the energy in mechanical and physical systems.

We design, present, and evaluate physics-informed methods for the prediction of the dynamics of a system from the observation of its vector field at discrete locations of the state space. We want to understand and quantify the significance of informing regression and integration of a vector field with physics information. In the spirit of an ablation study, we compare variants of the general method and different devices, a multilayer perceptron and a Gaussian process, with and without the information that the system is Hamiltonian. In the first stage, this concerns whether the vector field is under the constraints of Hamilton's equations. In the second stage, this concerns whether the vector field is integrated

C. Strauss et al. (Eds.): DEXA 2022, LNCS 13427, pp. 297–302, 2022.
https://doi.org/10.1007/978-3-031-12426-6_27

with a non-symplectic or symplectic integrator. The empirical comparative performance evaluation is conducted with data from several physical systems of an oscillator, a pendulum, a Henon Heiles system, a Morse potential model of a dyatomic molecule, and several abstract systems with logarithmic, inverse trigonometric, exponential, radical and polynomial Hamiltonian functions.

2 Related Work

We are interested in the problem of learning the vector field and the flow map from samples of the vector field. There are multiple statistical methods to learn correlated vector-valued functions [9,14,16]. Learning vector fields using machine learning has been addressed as multiple output regression by Hastie et al. [8] while state of the art works use neural networks that model ordinary differential equations [12] and partial differential equations [13]. Learning vector fields using kernel methods with regularization was introduced by Micchelli et al. [11]. Similarly, Boyle and Frean [4] introduced Gaussian processes to learn vector-valued functions. Recent works regularize the vector field [1].

Our work is similar to that of Greydanus et al. [6], who showed that a physics-informed neural network, similar to Bertalan et al.'s [2], can faster learn the Hamiltonian of mechanical systems and better predict their dynamics from selected samples of their vector fields than a neural network. Additionally, following Chen et al. [5], a symplectic integrator [7,15] can integrate the learned Hamiltonian vector field to predict a flow line.

3 Methodology

The general method for predicting the dynamics of a system from the observation of its vector field at discrete locations of its phase space comprises two successive stages: the learning or regression of the vector field from the samples, and the integration of the vector field into the flow map image of a state in the phase space for a prescribed time interval. We consider four variants of the general methods. They result from the obliviousness or awareness of information that a system is Hamiltonian during the first and second stages of the general method.

We consider two non-linear regression devices for the learning of a surrogate of the vector field, a multilayer perceptron neural network and a Gaussian process. The two devices are chosen as the main representatives of parametric and non-parametric non-linear regression statistical machine learning devices. They can learn a surrogate of the vector field, or learn a surrogate of the Hamiltonian and compute a surrogate of the vector field with automatic differentiation.

We consider the supervised learning of a surrogate $\hat{F}$ of the vector field F with two physics oblivious devices: a multilayer perceptron and a Gaussian process. They regress the vector-valued function oblivious to physics information. A training data set Z comprises N samples, $\frac{\mathrm{d}x}{\mathrm{d}t}$ and $\frac{\mathrm{d}y}{\mathrm{d}t}$, of the vector field for N states $z = (x, y)$ in the phase space of the system studied. When the surrogate is a multilayer perceptron, the weights minimise the mean squared error between the approximated vector field and the ground truth vector field. When the surrogate is a Gaussian process, the conditional expectation of the Gaussian process for the approximated vector field and the ground truth vector field is maximised.

We consider learning a physics-informed surrogate $\hat{H}$ of the Hamiltonian H learned from the same training data set Z under the constraints that the system is Hamiltonian before deriving the vector field. We adapt the method proposed by [2]. We use the constraints of Eq. 1 to define the loss function of the device. f_0 is an arbitrary pinning term. f_1 and f_2 are Hamilton's equations.

$$f_0 = \left(\hat{H}(x_0, y_0) - H_0\right)^2 \quad f_1 = \left(\frac{\partial \hat{H}}{\partial y} - \frac{\mathrm{d}x}{\mathrm{d}t}\right)^2 \quad f_2 = \left(\frac{\partial \hat{H}}{\partial x} + \frac{\mathrm{d}y}{\mathrm{d}t}\right)^2 \tag{1}$$

When the surrogate for the Hamiltonian is a multilayer perceptron neural network, the loss function is a linear combination of f_0, f_1 and f_2. When the surrogate is a Gaussian process, constraining its loss function leads to solving Eq. 2. In both cases, the derivative of the Hamiltonian at any new state of the phase space can be obtained from the surrogate by automatic differentiation.

$$\begin{bmatrix} \frac{\partial}{\partial z_1} k(z_1, Z)^\top k(Z, Z')^{-1} \\ \vdots \\ \frac{\partial}{\partial z_3} k(z_N, Z)^\top k(Z, Z')^{-1} \\ k(z_0, Z)^\top k(Z, Z')^{-1} \end{bmatrix} [H(Z)] = \begin{bmatrix} g(Z) \\ H_0 \end{bmatrix} \tag{2}$$

The flow map for predicting the dynamics of the system is computed by integrating the surrogate vector fields. Ignoring that the system is Hamiltonian, one can use the first order explicit Euler integrator [3]. Knowledge of the Hamiltonian system allows use of the implicit symplectic Euler integrator [7].

4 Performance Evaluation

Two experiments are conducted. In the first, we empirically compare the performance of the two physics-oblivious methods learning the vector field directly and of their two physics-informed counterparts learning the Hamiltonian. We use a testing data set of 20^{2n} vectors at evenly spaced states in the phase space for each system. Effectiveness is measured by mean squared error between the ground truth vectors and approximated vectors, and efficiency by the time taken for early stopping of the multilayer perceptron, or the time taken to fit a Gaussian process. For the second experiment, we evaluate the prediction of the dynamics of each Hamiltonian dynamical system by computing the flow map over the interpolated vector field. The vector field are learned with physics-informed methods in the first experiment, and combined with the Euler or symplectic Euler integrator. We use a testing data set of 5^{2n} evenly spaced states in the phase space for each system. Flow lines are calculated from the differential equations of the system with a symplectic Euler integrator for 50 time steps, with step size $h = 0.1$. The mean squared error between the ground truth flow line and the predicted flow line for each method, and prediction time is computed.

For both experiments, all multilayer perceptrons and physics-informed multilayer perceptrons set aside 20% of the training data set for validation based early stopping. Other settings follow Bertalan et al. [2]. All surrogates are trained in Google Colaboratory[1]. Training data set of size between 64 and 1024 are sampled uniformly at random from the vector fields of the eight example systems, respectively. The experiments are repeated for 20 unique random seeds, and the mean values are reported and compared. Results are plotted as Pareto plots.

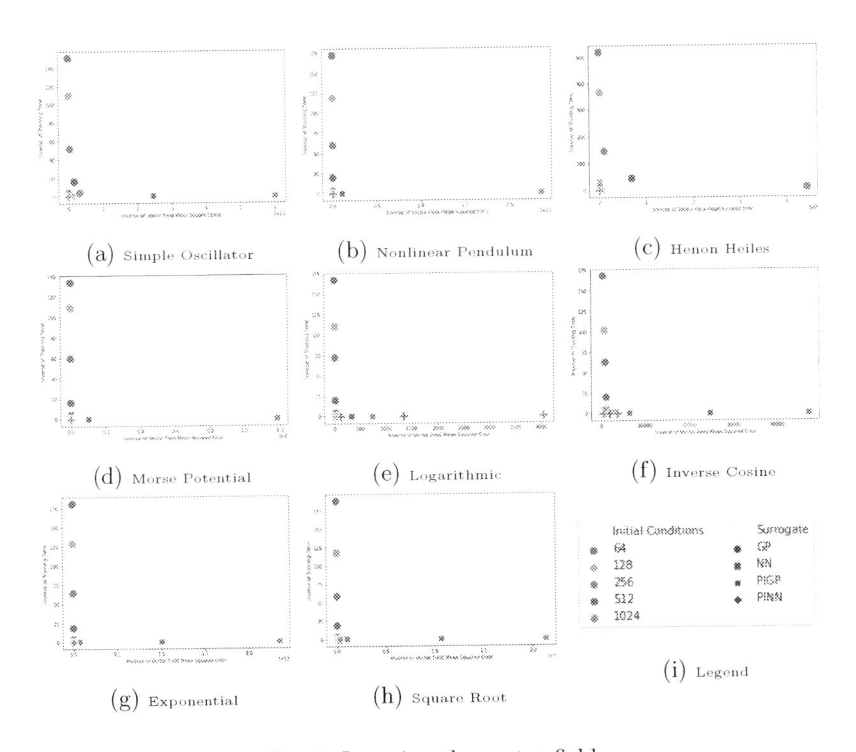

(a) Simple Oscillator (b) Nonlinear Pendulum (c) Henon Heiles (d) Morse Potential (e) Logarithmic (f) Inverse Cosine (g) Exponential (h) Square Root (i) Legend

Fig. 1. Learning the vector field

Figure 1 plots the Pareto plot for the inverse of the vector field approximation mean squared error (x axis) and the inverse of surrogate training time (y axis) for the different dynamical system indicated. The colors correspond to the different training data set of varying sizes. The circle, cross, square and plus symbols represent the Gaussian process (GP), multilayer perceptron (NN), physics-informed Gaussian process (PIGP) and physics-informed multilayer perceptrons (PINN) methods respectively. The most efficient method is the physics-informed Gaussian process, while multilayer perceptron can better extrapolate the vector field.

Figure 2 compares inverse prediction error (x axis) and inverse prediction time (y axis). The colors correspond to training data set of varying sizes. The

[1] Find code and results at github.com/zykhoo/predicting_hamiltonian_dynamics.

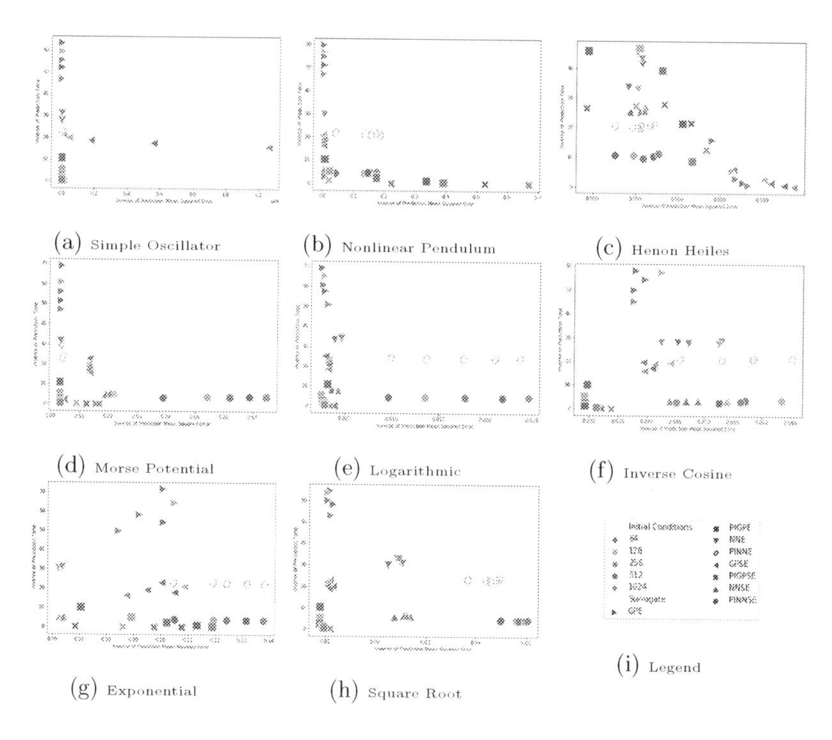

(a) Simple Oscillator (b) Nonlinear Pendulum (c) Henon Heiles (d) Morse Potential (e) Logarithmic (f) Inverse Cosine (g) Exponential (h) Square Root (i) Legend

Fig. 2. Predicting the flow map

right-pointing triangle, square, down-pointing triangle and empty circle symbols represent the Gaussian process (GPE), physics-informed Gaussian process (PIGPE), multilayer perceptron (NNE), and physics-informed multilayer perceptron (PINNE), all with the Euler integrator. The left-pointing triangle, cross, up-pointing triangle and filled circle symbols represent the same surrogates with symplectic Euler integrator. They learn and integrate the vector field. The most efficient method for the prediction of the dynamics is the physics-informed multilayer perceptron with symplectic integrator. This demonstrates the advantage of informing the methods of the Hamiltonian nature of the dynamical systems.

5 Conclusion

We design, present, and evaluate physics-informed methods for the prediction of the dynamics of a system from the observation of its vector field at discrete locations of the state space. We show that information that the system is Hamiltonian can be effectively informed in both the regression and integration of the vector field. The methods strike trade-offs between efficiency and effectiveness.

Acknowledgement. This research is partially supported by the Agency of Science, Technology and Research (A*STAR), by the National Research Foundation, Prime

Minister's Office, Singapore, under its Campus for Research Excellence and Technological Enterprise (CREATE) programme as part of the programme DesCartes, and by the Ministry of Education, Singapore, under its Academic Research Fund Tier 2 grant call (Award ref: MOE-T2EP50120-0019). Any opinions, findings and conclusions or recommendations expressed in this material are those of the authors and do not reflect the views of the National Research Foundation or of the Ministry of Education, Singapore.

References

1. Baldassarre, L., Rosasco, L., Barla, A., Verri, A.: Vector field learning via spectral filtering. In: Balcázar, J.L., Bonchi, F., Gionis, A., Sebag, M. (eds.) Machine Learning and Knowledge Discovery in Databases, pp. 56–71. Springer, Heidelberg (2010). https://doi.org/10.1007/978-1-4302-5990-9_2
2. Bertalan, T., Dietrich, F., Mezić, I., Kevrekidis, I.G.: On learning hamiltonian systems from data. Chaos: an Interdiscipl. J. Nonlinear Sci. **29**(12), 121–107 (2019)
3. Boyce, W.E., DiPrima, R.C., Meade, D.B.: Elementary Differential Equations and Boundary Value Problems. 11th edn. John Wiley & Sons, New York (2017)
4. Boyle, P., Frean, M.: Dependent Gaussian processes. In: Saul, L., Weiss, Y., Bottou, L. (eds.) Advances in Neural Information Processing Systems. vol. 17. MIT Press, London (2004)
5. Chen, Z., Zhang, J., Arjovsky, M., Bottou, L.: Symplectic recurrent neural networks. In: International Conference on Learning Representations (2020)
6. Greydanus, S., Dzamba, M., Yosinski, J.: Hamiltonian neural networks. In: Wallach, H., Larochelle, H., Beygelzimer, A., d'Alché-Buc, F., Fox, E., Garnett, R. (eds.) Advances in Neural Information Processing Systems. vol. 32. Curran Associates, Inc., Red Hook (2019)
7. Hairer, E., Wanner, G., Lubich, C.: Symplectic Integration of Hamiltonian Systems, pp. 179–263. Springer, Heidelberg (2002). https://doi.org/10.1007/3-540-30666-8_6
8. Hastie, T., Tibshirani, R., Friedman, J.: The Elements of Statistical Learning: Data Mining, Inference and Prediction. Springer, New York (2009). https://doi.org/10.1007/978-0-387-84858-7
9. Izenman, A.J.: Reduced-rank regression for the multivariate linear model. J. Multiv. Anal. **5**(2), 248–264 (1975)
10. Meyer, K.R., Hall, G.R.: Hamiltonian Differential Equations and the N-Body Problem, pp. 1–32. Springer, New York (1992). https://doi.org/10.1007/978-1-4757-4073-8_1
11. Micchelli, C.A., Pontil, M.: On learning vector-valued functions. Neural Comput **17**(1), 177–204 (2005)
12. Raissi, M.: Deep hidden physics models: deep learning of nonlinear partial differential equations. J. Mach. Learn. Res. **19**(1), 932–955 (2018)
13. Ruthotto, L., Haber, E.: Deep neural networks motivated by partial differential equations. J. Math. Imaging Vis. **62**(3), 352–364 (2020)
14. Van Der Merwe, A., Zidek, J.: Multivariate regression analysis and canonical variates. Can. J. Stat. **8**(1), 27–39 (1980)
15. de Vogelaere, R.: Methods of Integration which Preserve the Contact Transformation Property of the Hamilton Equations. University of Notre Dame (1956)
16. Wold, S., Ruhe, A., Wold, H., Dunn, III, W.J.: The collinearity problem in linear regression. The partial least squares (PLS) approach to generalized inverses. SIAM J. Sci. Stat. Comput. **5**(3), 735–743 (1984)

A Fuzzy Satisfaction Concept Based-Approach for Skyline Relaxation

Mohamed Haddache[1(✉)], Allel Hadjali[2], and Hamid Azzoune[3]

[1] DIF-FS/UMBB, University of Boumerdes, Boumerdes, Algeria
m.haddache@univ-boumerdes.dz
[2] LIAS/ENSMA, Poitiers, France
allel.hadjali@ensma.fr
[3] LRIA/USTHB, Algiers, Algeria
hazzoune@usthb.dz

Abstract. The study of the skyline queries has received considerable attention from several database researchers since the end of 2000's. Skyline queries are an appropriate tool that can help users to make intelligent decisions in the presence of multidimensional data. Based on the concept of Pareto dominance, the skyline operator returns the most interesting (not dominated in the sense of Pareto) objects of database. Skyline computation can gives a less number of objects witch are insufficient to serve the user needs. In this paper, we tackle this problem and we propose the solution that deals with it. The basic idea is to use fuzzy formal concept, we build the fuzzy formal concept lattice for dominated objects. Then, we retrieve the formal concept C^* that matches at best the ideal intent (the ideal intent represent the user query). The relaxed skyline is given by the set S_{relax}, formed by the union of skyline objects and the objects of the concept C^* and the size of S_{relax} is equal to k (where k is a user-defined parameter). Experimental study shows the efficiency and the effectiveness of our approach.

Keywords: Skyline queries · Relaxation · Fuzzy formal concept · Pareto dominance

1 Introduction

Skyline queries are one of the most multi-criteria methods that have gained a considerable interest in the last two decades. They are introduced by Borzsönyi in [2] to formulate multi-criteria searches. Since the end of 2000's, this concept has received much attention in the database community. It has been integrated in many database applications that require decision making (decision support [10,12], personalized recommendation like hotel recommender [12], crowdsourcing database [8]). Skyline process aims at identifying the most interesting (not dominated in sense of Pareto) objects from a set of data. They are then based on Pareto dominance relationship. This means that, given a set D of

C. Strauss et al. (Eds.): DEXA 2022, LNCS 13427, pp. 303–309, 2022.
https://doi.org/10.1007/978-3-031-12426-6_28

d-dimensional points (objects), a skyline query returns, the skyline S, i.e., set of points of D that are not dominated by any other point of D. A point p dominates (in Pareto sense) another point q iff p is better than or equal to q in all dimensions and strictly better than q in at least one dimension.

Many research studies have been conducted to develop efficient algorithms to skyline computation and introduce multiple variants of skyline queries in both complete and incomplete databases for more detail see the survey [7].

However, the skyline computation gives often a small number or an empty set of skyline objects which could be insufficient to serve the user needs. Several works have been developed for the purpose to relax the skyline and thus increasing its size, see [6] for the complete survey. In this paper, we address this problem but with another novel vision. In particular, the idea of the solution advocated is borrowed from the formal concept analysis field [11]. A formal concept is formed by an extent that contains a set of objects and intent that contains attributes describing these objects. In our context the set of objects is given by the set of dominated (no skyline) objects and the attributes correspond to the dimensions of skyline. The idea of our solution consists in building a formal concept lattice for dominated objects based on the satisfaction rate between each formal concept of lattice and the ideal intent. The relaxed skyline is given by the skyline objects union the objects of the concept that contains $k1$ objects and gives the highest satisfaction rate to the ideal intent (where $k1 + |S| = k$ and k is the objects user's need). Starting from this idea we develop a new algorithm to compute the relaxed skyline (S_{relax}). The rest of this paper is organized as follow. In Sect. 2, we define some necessary notions about the skyline queries, formal concept analysis, In Sect. 3, we detail the different steps of our approach. Section 4 is devoted to the experimental study that we have done and the Sect. 5 concludes the paper and outlines some perspectives for future work.

2 Background and Related Work

In this section, we recall some necessary notions about the skyline queries, formal concept analysis.

2.1 Skyline Queries

Skyline queries [2] are a popular example of preference queries. They are based on Pareto dominance principle which can be defined as follows:

Definition 1. *Let D be a set of d-dimensional data points (objects) and u_i and u_j two points of D. u_i is said to dominate, in Pareto sense, u_j (denoted $u_i \succ u_j$) iff u_i is better than or equal to u_j in all dimensions and strictly better than u_j in at least one dimension.*

Formally, we write:

$$u_i \succ u_j \Leftrightarrow \begin{cases} \forall k \in \{1,..,d\}, u_i[k] \geq u_j[k] \\ \wedge \\ \exists l \in \{1,..,d\}, u_i[l] > u_j[l] \end{cases} \tag{1}$$

where each tuple $u_i = (u_i[1], u_i[2], \cdots, u_i[d])$ with $u_i[k]$ stands for the value of the tuple u_i for the attribute A_k.

In Eq. (1), without loss of generality, we assume that the smaller value, the better.

Definition 2. *The skyline of D, denoted by S, is the set of points which are not dominated by any other point.*

Example 1. To illustrate the concept of the skyline, let us consider a database containing information on apartments as shown in Fig. 1(a). The list of apartments includes the following information: code apartment, distance between the home and the workplace, price. A company wants to rent apartments to their employees. The ideal candidate is an apartment that minimizes the distance between the apartment/work and the price. Applying the traditional skyline on the apartments list shown in Fig. 1(a), we have A_2 dominates in Pareto sense A_1, A_3, A_5, A_6 and A_7. Then, these apartments are discarded from the result set. However A_2 and A_4 are not dominated by any other apartment. They form the skyline S. See Fig. 1(b).

Code apartment	Distance /workplace(Km)	Price (euros)
A1	10	500
A2	3	400
A3	6	1300
A4	1	1600
A5	9	900
A6	13	700
A7	5	1400

(a) List of apartments

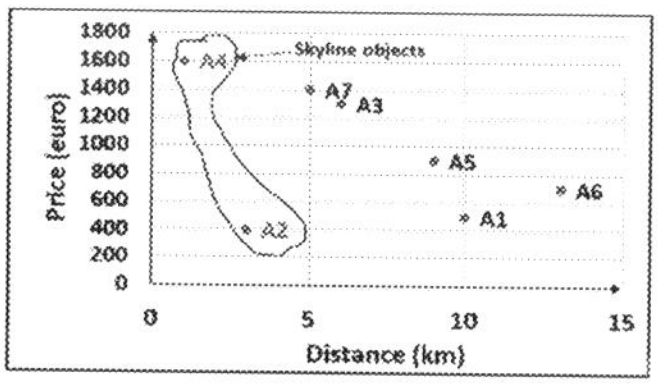

(b) Skyline apartments

Fig. 1. Skyline apartments.

2.2 Formal Concept Analysis

Formal concept analysis (FCA) is introduced by Wille in 1982 [11], as a method to represent knowledge and analysis the data. The AFC is based on formal context $\mathcal{K} = (\mathcal{O}, \mathcal{A}, \mathcal{R})$ which in the basic version represents a binary relation between the set of objects $\mathcal{O}$ and the set of attributes $\mathcal{A}$. The relation between the subset of objects and the subset of attributes is given by the Galois operator derivation Δ. This operator allows to define a formal concept.

A formal concept of formal context $\mathcal{K} = (\mathcal{O}, \mathcal{A}, \mathcal{R})$ is a pair $\langle O, A \rangle$ with $O \subseteq \mathcal{O}$, $A \subseteq \mathcal{A}$, $A^{\Delta} = O$ and $O^{\Delta} = A$. O is the extent of the formal concept and A its intent. The set of all formal concepts is equipped with a partial order denoted $\leq$, defined by $\langle O_1, A_1 \rangle \leq \langle O_2, A_2 \rangle$ iff $O_1 \subseteq O_2$ or $A_2 \subseteq A_1$. Ganter and Wille have proved in [4], that the set of all formal concept ordred by $\leq$ form a complete lattice of formal context $\mathcal{K}$.

In majority of application attributes are fuzzy than crisp; Burusco and Fuentes-Gonzales [3] and Belohlávek et al. [1] introduce a fuzzy formal concepts. A fuzzy formal concept is based on formal context given by the quadruplet $K = (L, \mathcal{O}, \mathcal{A}, \mathcal{R})$, where L is in general equal to $[0,1]$ and $\mathcal{R} \in L^{O \times \mathcal{A}}$ is a fuzzy relation defined from $\mathcal{O} \times \mathcal{A} \longrightarrow L$ which assigns to each object $o \in \mathcal{O}$ and for each property $a \in \mathcal{A}$ a degree $\mathcal{R}(o, a)$ for which the object o posses the attribute a. In [1], Belohlávek et al. proposed to use fuzzy implications that allow to generalize the Galois derivation operator in the fuzzy setting. This generalization is defined for a set $O \in L^{\mathcal{O}}$ and similarly defined for a set $A \in L^{\mathcal{A}}$ as follows:

$$O^{\Delta}(a) = \bigwedge_{o \in \mathcal{O}} (O(o) \rightarrow \mathcal{R}(o, a)) \tag{2}$$

$$A^{\Delta}(o) = \bigwedge_{a \in \mathcal{A}} (A(a) \rightarrow \mathcal{R}(o, a)) \tag{3}$$

$\bigwedge$ is the min conjunction operator and $\rightarrow$ is a fuzzy implication that verifies $(0 \rightarrow 0 = 0 \rightarrow 1 = 1 \rightarrow 1 = 1$ and $1 \rightarrow 0 = 0)$.

In our approach, we use the notion of satisfaction rate, for more detail, see [9].

3 Our Approach

In this section, we describe the different steps of our approach. First, we assume that we have: $\mathcal{O} = \{o_1, o_2, \ldots, o_n\}$ a database that contains n objects, $A = \{a_1, a_2, \ldots, a_d\}$ a set of d attributes (dimensions or properties), $\mathcal{R}(o_i, a_j)$: the function that gives the value of o_i w.r.t the attribute a_j, $degree(o_i, a_j)$: the degree for which the object o_i possesses the attribute a_j, $S = \{o_1, o_2, \ldots, o_t\}$: the set of skyline objects (t is the size of skyline $t \leq n$). D: the set of dominated objects, $D = \mathcal{O} - S$.

Our approach is based on the following steps

1. **Skyline and dominated objects computation:** First, we compute the skyline S and the set of the dominated objects $D = \mathcal{O} - S$ using the algorithm IBNL (Improvement BNL), see [6].
2. **Computing the degrees of dominated objects:** for each dominated object o_i from D, we compute the degree $degree(o_i, a_j)$ for which the object o_i passed the attribute a_j.
3. **Compute the ideal intent:** computes the ideal intent, i.e., the intent that maximizes the degree of objects attributes. This intent is given by: $inent\_ideal = \{a_1^1, a_2^1, \ldots, a_d^1\}$
4. **Building the fuzzy lattice for the objects of D:** we build the lattice for the dominated objects using Algorithm 1. Then, The goal of our approach is to retrieve the formal concept C^* of the lattice whose extent contains

k_1 objects ($k_1 + |S| = k$ and k is user-defined parameter) and whose intent matches at best the ideal intent, i.e., has the highest satisfaction rate w.r.t the ideal intent.

Algorithm 1: FSAASR

Input: A set of dominated objects D, a skyline S, k: the number of objects chosen by the user
Output: A relaxed skyline S_{relax}

1 $S_{relax} \leftarrow S$; $stop \leftarrow false$;
2 $Compute\_degre(D)$;
3 $Intent\_ideal = \{a_1^1, a_2^1, \ldots, a_d^1\}$; $Intent\_Min = Compute\_Intent\_Min()$; $Intent \leftarrow Intent\_Min$;
4 **while** $stop == false$ **do**
5     $Intent1 \leftarrow Next\_intent(Intent\_Min, 1)$; $extent1 \leftarrow Compute\_crisp\_Extent(Intent1)$; $Sat1 \leftarrow Compute\_sat(extent1, Intent\_ideal)$;
6   **for** $i := 2$ *to* d **do**
7         $Intent2 \leftarrow Next\_intent(Intent\_Min, i)$;/*compute the others intents of the intent */ $extent2 \leftarrow Compute\_crisp\_Extent(Intent2)$; $Sat2 \leftarrow Compute\_sat(extent2, Intent\_ideal)$;
8   **if** $Sat2 > Sat1$ **then**
9             $extent1 \leftarrow extent2$; $Sat1 \leftarrow Sat2$;
10            $Intent1 \leftarrow Intent2$;
11    $Intent \leftarrow Intent1$;
12   **if** $exten1.size() <= k$ **then**
13        $stop \leftarrow true$; $S_{relax} \leftarrow S_{relax} \cup extent1$;
14 **return** S_{relax};

4 Experimental Study

In this section, we present the experimental study that we have conducted. The goal of this study is to demonstrate the effectiveness of our approach and its ability to relax the less skyline. Also, we compare our approach to the approach proposed by Goncalves in [5]. All experiments were performed under Windows OS, on a machine with an Intel core i7 2,90 GHz processor, a main memory of 8 GB and 250 GB of disk. All algorithms were implemented with Java. Dataset benchmark is generated using the method described in [2].

4.1 Impact of Data Distribution

In this section, we study the impact of the variation of the distribution of dataset on size of relaxed skyline and the execution time. In this case, we use a dataset with $|O| = 5K$ $d = 2$. Figure 2(a) shows that the efficiency of the two approaches to relax the skyline is very high for all types of data. Figure 2(a) indicates that the execution time of the two approaches for anti-correlated data is low compared to the correlated or independent data. This is due to the size of the dominated objects set used to relax the skyline in the case of anti-correlated data compared to two others distributions. Figure 2(a) shows also that the execution time of our algorithm (FSAASR) is the best for all distributions.

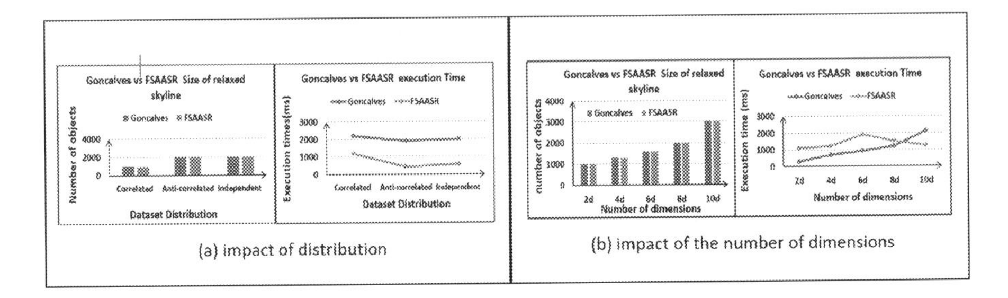

Fig. 2. Impact of distribution and the number of dimensions.

4.2 Impact of the Number of Dimensions

In this case, we use a correlated distribution data with $|O| = 100K$ and we vary the number of dimensions from 2 to 10. Figure 2(b) shows that the efficiency of the two approaches to relax the skyline is very high and it increases from 1000 Objects when $d = 2$ to 3000 for $d = 10$. The Fig. 2(b) shows also that the execution time increases with the increase of the number of dimensions for the approach developed by Goncalves. However, in this case of our algorithm (FSAASR) the execution time increases with the increases of number of dimensions, but when this number reaches 6, the execution time decreases as the number of dimensions increases. This is due to the size of $D = \mathcal{O} - S$ that decreases when the number of dimensions increases (The increase of the number of dimensions increases the size of S, in consequence decreases the size of D).

5 Conclusion and Perspectives

In this paper, we have addressed the problem of skyline relaxation, especially when the skyline contains a small number of points and we proposed a new approach to increasing its size. The basic idea of this approach is to build the fuzzy lattice of the dominated objects $D = \mathcal{O} - S$. Then, retrieve the formal concept C^* of the lattice whose extent contains k_1 objects ($k_1 + |S| = k$ and k is user-defined parameter) and whose intent matches at best the ideal intent, i.e., has the highest satisfaction rate w.r.t the ideal intent. An algorithm called **FSAASR** to compute the relaxed skyline is proposed. In addition, we implemented the approach proposed by Goncalves in [5] to compare its performances to that of our algorithm. The experimental study that we have done shows that our approach is good alternative to relax the skyline and have the best execution time compared to the approach proposed by Goncalves. As for future work, we will compare our approach to others approaches of relaxation.

References

1. Belohlávek, R.: Fuzzy galois connections. Math. Log. Q. **45**, 497–504 (1999)
2. Börzsönyi, S., Kossmann, D., Stocker, K.: The skyline operator. In: Proceedings of the 17th International Conference on Data Engineering, Heidelberg, Germany, 2–6 April 2001, pp. 421–430 (2001)
3. Burusco, A., Fuentes-González, R.: The study of the L-fuzzy concept lattice. Mathware Soft Comput. **1**(3), 209–218 (1994)
4. Ganter, B., Wille, R.: Formal Concept Analysis: Mathematical Foundations. Springer, Heidelberg (1999). https://doi.org/10.1007/978-3-642-59830-2
5. Goncalves, M., Tineo, L.: Fuzzy dominance skyline queries. In: Wagner, R., Revell, N., Pernul, G. (eds.) DEXA 2007. LNCS, vol. 4653, pp. 469–478. Springer, Heidelberg (2007). https://doi.org/10.1007/978-3-540-74469-6_46
6. Haddache, M., Hadjali, A., Azzoune, H.: Concept dissimilarity based approach for skyline relaxation. In: 4th International Conference on Advanced Aspects of Software Engineering, ICAASE 2020, Constantine, Algeria, 28–30 November 2020, pp. 1–8. IEEE (2020)
7. Kalyvas, C., Tzouramanis, T.: A survey of skyline query processing. ArXiv abs/1704.01788 (2017)
8. Lofi, C., El Maarry, K., Balke, W.-T.: Skyline queries over incomplete data - error models for focused crowd-sourcing. In: Ng, W., Storey, V.C., Trujillo, J.C. (eds.) ER 2013. LNCS, vol. 8217, pp. 298–312. Springer, Heidelberg (2013). https://doi.org/10.1007/978-3-642-41924-9_25
9. Raja, H., Djouadi, Y.: Projection extensionnelle pour la reduction d'un treillis de concepts formels flous. In: 22nd French Conferences on Fuzzy Logic and Its Applications (LFA 2013), Reims, France, pp. 103–110 (2013)
10. Tan, K., Eng, P., Ooi, B.C.: Efficient progressive skyline computation. In: Proceedings of 27th International Conference on Very Large Data Bases (VLDB), Roma, Italy, 11–14 September 2001, pp. 301–310 (2001)
11. Wille, R.: Restructuring lattice theory: an approach based on hierarchies of concepts. In: Rival, I. (ed.) Ordered Sets. ASIC, vol. 83, pp. 445–470. Springer, Dordrecht (1982). https://doi.org/10.1007/978-94-009-7798-3_15
12. Yiu, M.L., Mamoulis, N.: Efficient processing of top-k dominating queries on multi-dimensional data. In: Proceedings of the 33rd International Conference on Very Large Data Bases (VLDB), University of Vienna, Austria, 23–27 September 2007, pp. 483–494 (2007)

Quasi-Clique Mining for Graph Summarization

Antoine Castillon[1,2], Julien Baste[1], Hamida Seba[2], and Mohammed Haddad[2(✉)]

[1] Univ. Lille, CNRS, Centrale Lille, UMR 9189 CRIStAL, 59000 Lille, France
{antoine.castillon,julien.baste}@univ-lille.fr
[2] Univ Lyon, UCBL, CNRS, INSA Lyon, LIRIS, UMR5205, 69622 Villeurbanne, France
{hamida.seba,mohammed.haddad}@univ-lyon1.fr

Abstract. Several graph summarization approaches aggregate dense sub-graphs into super-nodes leading to a compact summary of the input graph. The main issue for these approaches is how to achieve a high compression rate while retaining as much information as possible about the original graph structure within the summary. These approaches necessarily involve an algorithm to mine dense structures in the graph such as quasi-clique enumeration algorithms. In this paper, we focus on improving these mining algorithms for the specific task of graph summarization. We first introduce a new pre-processing technique to speed up this mining step. Then, we extend existing quasi-clique enumeration algorithms with this filtering technique and apply them to graph summarization.

Keywords: Graph summarization · Quasi-clique mining · Pruning techniques

1 Background

Nowadays, several large-scale systems such as social networks or the link structure of the World Wide Web involve graphs with millions and even billions of nodes and edges [3]. The analysis of such graphs can prove to be very difficult given the huge amount of information contained in them. The goal of *graph summarization* (or *graph compression*) is to provide, given an input graph, a smaller summary [5]. Depending of the situation, this summary can assume different roles. Either it exists uniquely to save storage space and can be decompressed to obtain the original graph with or without loss of information, or it can be used directly to obtain information on the input graph such as paths, neighbourhoods and clusters. When summarizing a graph, one can focus on the optimisation of the size of the summary introducing a size threshold [1,5] or an information loss threshold [8]. Others adapt the summary so as to optimise some kind of queries [1,4].

This work is supported by Agence Nationale de la Recherche (ANR-20-CE23-0002).

C. Strauss et al. (Eds.): DEXA 2022, LNCS 13427, pp. 310–315, 2022.
https://doi.org/10.1007/978-3-031-12426-6_29

Even if these summarization methods perform well in term of execution time and compression rate, they do not retain all the information available in the input graph, especially the dense components which are mainly *cliques*, and *quasi-cliques* which are essential in the analysis of many real life networks. This motivated several graph summarization approaches based on dense subgraph mining [8,10]. These methods aggregate dense subgraphs into supernodes and keep track of all non-edges inside the supernode as correcting edges, to ensure lossless compression and allow the reconstruction of the original graph.

In this paper, we focus on enhancing the enumeration of dense subgraphs, and more precisely quasi-cliques, for the purpose of graph summarization. We first introduce a new pruning technique which allows an important speed up of quasi-clique enumeration algorithms. Then, we describe how to adapt the search for quasi-cliques to graph summarization by solving best suited relaxations of the Quasi-Clique Enumeration Problem.

The remainder of this document is organised as follows. Section 2 gives formal definitions of the concepts used in the paper. Section 3 introduces the new pruning technique and presents a relaxation of the quasi-clique enumeration problem. Section 4 compares the performances of the different quasi-clique mining schemes and their application to dense subgraph based summarization.

2 Preliminaries

In this paper all graphs are simple, undirected and consist of two sets: the vertex set V and the edge set E formed by pairs of V. Given $G = (V, E)$ a clique of G is a subset C of V such that the vertices in C are pairwise adjacent. We use the following definition for quasi-cliques which are a generalization of cliques to other dense subgraphs.

Definition 1 (γ-quasi-clique). *Given $G = (V, E)$, and $\gamma \in [0, 1]$, a subset of vertices $Q \subseteq V$ is called a γ-quasi-clique if for all $v \in V$, $d_Q(v) \geq \gamma(|Q| - 1)$, where $N_Q(v)$ represents the neighbours of v in Q and $d_Q(v) = |N_Q(v)|$.*

Definition 2 (Quasi-Clique Enumeration Problem). *Given $G = (V, E)$, a density threshold $\gamma \in [0, 1]$ and a size threshold m, the* Quasi-Clique Enumeration Problem (QCE) *consists of finding all maximal γ-quasi-cliques in G of size greater than m.*

QCE is well known to be difficult (the associated decision problem being NP-hard [2]), and have already been studied a lot trough several aspects. We focus on the exhaustive enumeration and its most classic method: the Quick algorithm [7] which performs a depth-first exploration of the solution space tree and it prunes the branches which cannot lead to a new solution. During the search, the algorithm keeps in memory X the current subset and a set `cand` formed by vertices outside X which are the next candidates to add to X. The pruning occurs according to these sets and several rules, and actually corresponds to the removal of vertices from `cand`. These rules are mostly based on the diameter and the degree threshold of quasi-cliques. For example, with $\gamma > 0.5$ the diameter of a γ-quasi-clique is at most 2.

3 Contributions

3.1 A New Pruning Technique

While the Quick algorithm is efficient in the general case, it is not hard to point out cases where it is not. One can, for instance, take the example depicted in Fig. 1. The pruning rules of the Quick algorithm are based only on the vertices and their degree, hence, in the case where each vertex is contained in a quasi-clique, the pruning techniques cannot be properly applied and the algorithm proceeds to the exploration of almost the whole solution space tree.

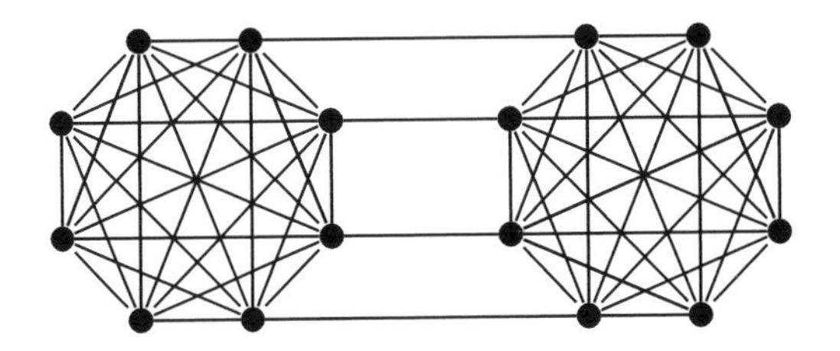

Fig. 1. Example of the inefficiency of the Quick algorithm.

In order to handle these cases, we introduce a new technique based on sets of vertices. According to Definition 1, each vertex v in a quasi-clique Y has at least $\lceil \gamma(|Y|-1) \rceil$ neighbours in Y. This property motivates the following pruning rule (see Algorithm 1): given the density and size threshold γ and m, if $\{u, v\} \in E$ is such that u and v share strictly less than $2\lceil \gamma(m-1) \rceil - m$ common neighbours then we can remove $\{u, v\}$ from E.

Algorithm 1. Pre_processing(G, γ, m)

repeat
  delete all $u \in V$ s.t. $d(u) < \lceil \gamma(m-1) \rceil$
  delete all $\{u, v\} \in E$ s.t. $|N(u) \cap N(v)| < 2\lceil \gamma(m-1) \rceil - m$
until G has not been modified during the last loop

3.2 A Redundancy-Free Graph Summarization Approach

While the Quick algorithm is an effective method to tackle the QCE problem, it is important to note that this problem is not perfectly suited for graph summarization. Indeed, having access to all quasi-cliques can create redundancy in the summary and impact the performances. To deal with this issue, we adapt the concept of visibility, introduced in [9] for cliques, to detect and avoid redundant quasi-cliques. The visibility of a maximal quasi-clique Q with respect to a set of maximal quasi-cliques S is a value between 0 and 1 which describes how well Q

is represented by the quasi-cliques of $\mathcal{S}$. If the visibility is 1 then Q is in $\mathcal{S}$, on the contrary if the visibility is 0 then Q does not overlap with any quasi-clique of $\mathcal{S}$.

Definition 3 (Visibility). *Given a graph G, $\gamma \in [0,1]$, $Q \in \mathcal{M}_\gamma(G)$, the set of all maximal γ quasi-cliques of G, and $\mathcal{S} \subseteq \mathcal{M}_\gamma(G)$, the $\mathcal{S}$-**visibility** of Q: $\mathcal{V}_\mathcal{S}(Q)$, is defined as:*

$$\mathcal{V}_\mathcal{S}(Q) = \max_{Q' \in \mathcal{S}} \frac{|Q \cap Q'|}{|Q|}.$$

This leads to a variant of the QCE problem where instead of finding all the quasi-cliques one only have to ensure that all maximal quasi-cliques have a sufficient visibility. We propose to use a redundancy-aware quasi-clique mining algorithm to summarize graphs. This will give direct access to the dense subgraphs in the summary while avoiding redundant quasi-cliques. This will improve not only compression rates but also allows to speed up the computation.

To implement this approach, we rely on the following observation. Given a graph G and two quasi-cliques Q, Q'. If Q and Q' are very similar ($\mathcal{V}_{\{Q\}}(Q')$ is close to 1) then most of the edges in $G[Q']$ are already covered in $G[Q]$. On the contrary if Q and Q' are very different then $G[Q]$ and $G[Q']$ do not share many edges. Hence, to avoid redundancy, after finding the quasi-clique Q one can remove from G all the edges inside $G[Q]$, thus the quasi-cliques of G similar to Q become sparse and are skipped by the pruning techniques of the Quick algorithm. Removing such edges usually does not impact other different quasi-cliques since these quasi-cliques do not share many edges with Q. This method depicted in Algorithm 2 outputs a set of quasi-cliques $\mathcal{S}$ which covers all edges contained in quasi-cliques.

Algorithm 2. Redundancy_aware(G, γ, m)

---

run Quick(G, γ, m)
    each time a quasi-clique X is found:
        remove from G all edges of $G[X]$

---

4 Performances

We compare the performances of three different quasi-clique mining algorithms: the Quick algorithm presented in Sect. 1, Algorithm 2 presented in Sect. 3.2 and finally a greedy quasi-clique mining algorithm, which serves as a basis of comparison. We implemented the three algorithms in Python 3 using a Windows 10 machine with a 2.5 GHz Intel(R) Core (TM) i5 processor and 8 GB RAM. We tested each algorithm with and without our pre-processing technique, namely Algorithm 1, on several types of graphs. We present, here, our results on real graphs from social networks such as Facebook (with 4039 vertices and 88,234 edges), Twitter (with 81,306 vertices and 1,768,149 edges) and Google+ (with 107,614 vertices and 13,673,453 edges), from the Stanford large network dataset collection [6]. We measure not only the runtime of the algorithms but also how well they are suited to graph summarization. For these experiments, we vary the number of nodes in the considered graphs from 0 to 1750 by growing subgraphs from the considered graphs. The obtained results are the average on several experiments.

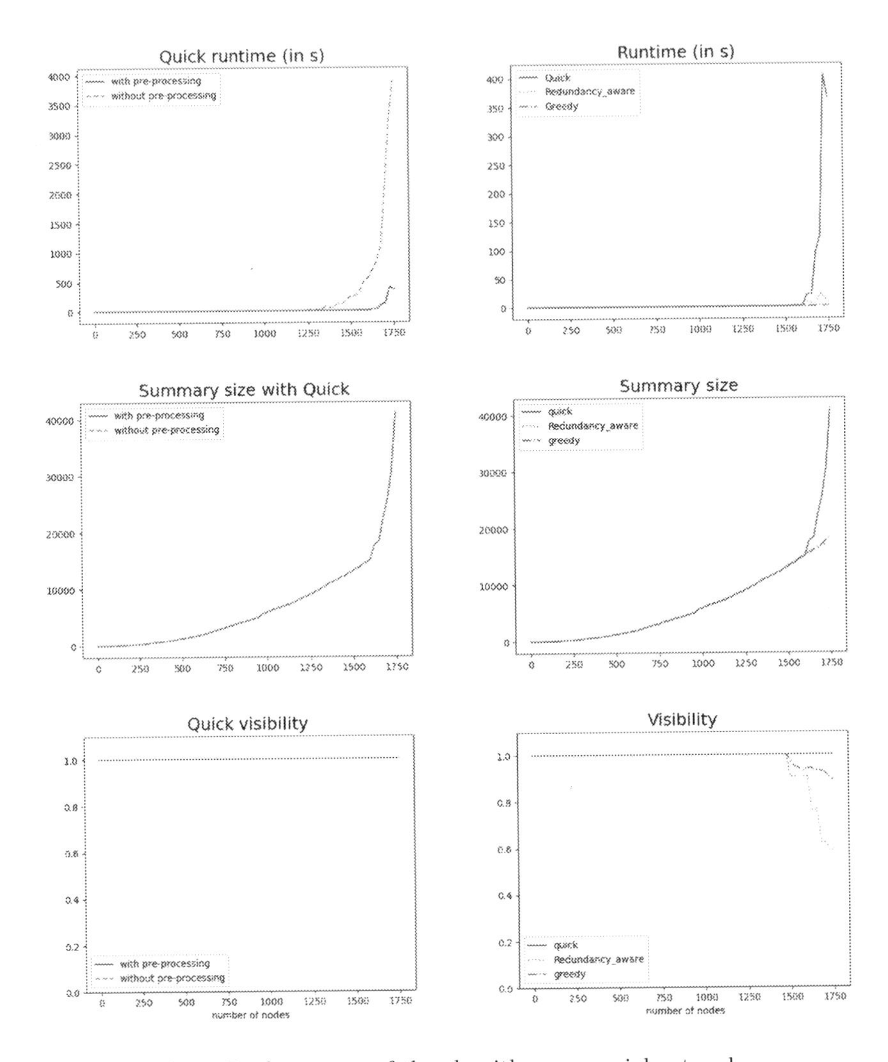

Fig. 2. Performances of the algorithms on social networks.

Figure 2 presents our results. The left side of the figure presents the impact of our filtering technique on the Quick algorithm, on three metrics: runtime, size of the summary and its visibility computed as in Definition 3. The results prove the utility of our filtering technique with a significant decrease of the execution time, up to a factor 10 with 1750 nodes, while not impacting the compression rates nor the visibility. The right side of Fig. 2 presents a comparison of the three algorithms, with our filtering technique, on the same metrics. The results show that even if Algorithm 2 seems efficient regarding execution time and compression rates, the visibility obtained with this algorithm is worst than the others dropping around 0.6 while the others stay greater than 0.9. Finally, the greedy algorithm seems to perform surprisingly well on these graphs, being highly efficient in time and compression rate while keeping a good visibility.

5 Conclusion

In this paper, we focused on graph summarization using dense subgraphs and more precisely quasi-cliques. This approach requiring a quasi-clique mining algorithm, we show how to improve the already existing quasi-clique enumeration algorithms introducing a very effective pre-processing technique. Also, we adapt and introduce a quasi-clique mining algorithm designed to avoid redundancy while improving performances both in runtime and compression rate.

References

1. Ahmad, M., Beg, M.A., Khan, I., Zaman, A., Khan, M.A.: SsAG: Summarization and sparsification of attributed graphs (2021)
2. Baril, A., Dondi, R., Hosseinzadeh, M.M.: Hardness and tractability of the γ-complete subgraph problem. Inf. Process. Lett. **169**, 106105 (2021)
3. Boldi, P., Vigna, S.: The webgraph framework i: Compression techniques. In: Proceedings of WWW2004. pp. 595–602 (2004)
4. Lagraa, S., Seba, H., Khennoufa, R., MBaya, A., Kheddouci, H.: a distance measure for large graphs based on prime graphs. Pattern Recogn. **47**(9), 2993–3005 (2014)
5. Lee, K., Jo, H., Ko, J., Lim, S., Shin, K.: SSumM Sparse summarization of massive graphs. In: Proceedings of the 26th ACM SIGKDD. pp. 144–154 (2020)
6. Leskovec, J., Krevl, A.: SNAP datasets: stanford large network dataset collection (2014). http://snap.stanford.edu/data
7. Liu, G., Wong, L.: Effective pruning techniques for mining quasi-cliques. In: Daelemans, W., Goethals, B., Morik, K. (eds.) ECML PKDD 2008. LNCS (LNAI), vol. 5212, pp. 33–49. Springer, Heidelberg (2008). https://doi.org/10.1007/978-3-540-87481-2_3
8. Navlakha, S., Rastogi, R., Shrivastava, N.: Graph summarization with bounded error. In: Proceedings of the 2008 ACM SIGMOD Conference. pp. 419–432 (2008)
9. Wang, J., Cheng, J., Fu, A.W.C.: Redundancy-aware maximal cliques. In: Proceedings of the 19th ACM SIGKDD Conference. pp. 122–130 (2013)
10. Wang, L., Lu, Y., Jiang, B., Gao, K.T., Zhou, T.H.: Dense subgraphs summarization: an efficient way to summarize large scale graphs by super nodes. In: 16th International Conference on Intelligent Computing Methodologies, Italy. pp. 520–530 (2020)

Author Index

Printed in the United States
by Baker & Taylor Publisher Services